# PRACTICAL GENETICS

# PRACTICAL GENETICS

edited by

## P. M. SHEPPARD

M.A. D.Phil. F.R.S.
Professor of Genetics
University of Liverpool

BLACKWELL SCIENTIFIC PUBLICATIONS

OXFORD LONDON EDINBURGH MELBOURNE

© 1973 Blackwell Scientific Publications
Osney Mead, Oxford
3 Nottingham Street, London W1
9 Forrest Road, Edinburgh
P.O. Box 9, North Balwyn, Victoria, Australia

ISBN 0 632 08750 1

First published 1973

Distributed in the U.S.A. by
Halsted Press
A division of John Wiley & Sons Inc.
605 Third Avenue, New York, N.Y. 10016

Printed in Great Britain by
Adlard & Son Ltd
Bartholomew Press, Dorking
and bound by
The Kemp Hall Bindery
Oxney Mead, Oxford

# CONTENTS

# Chapter 3. Quantitative Genetics
## M.J.Lawrence and J.L.Jinks · page 86

# Chapter 4. Cytogenetics
## S.Walker · page 130

## Chapter 5. Genetical Experiments with Fungi
### J.Croft and J.L.Jinks · page 173

## Chapter 6. Bacterial and Bacteriophage Genetics
### R.C.Clowes · page 225

## Chapter 7. Population and Ecological Genetics
### P.M.Sheppard · page 296

# CONTRIBUTORS

J.ANTONOVICS Department of Botany, Duke University, Durham, North Carolina, U.S.A.

B.W.BARNES Department of Genetics, University of Birmingham, Birmingham, England

R.C.CLOWES Division of Biology, University of Texas at Dallas, P.O. Box 30365 Dallas, Texas, 75230, U.S.A.

J.CROFT Department of Genetics, University of Birmingham, Birmingham, England

J.L.JINKS Department of Genetics, University of Birmingham, Birmingham, England

M.J.KEARSEY Department of Genetics, University of Birmingham, Birmingham, England

M.J.LAWRENCE Department of Genetics, University of Birmingham, Birmingham, England

P.M.SHEPPARD Department of Genetics, University of Liverpool, Liverpool, England

S.WALKER Department of Genetics, University of Liverpool, Liverpool, England

# PREFACE

One of the most difficult aspects of teaching genetics is to design a stimulating and informative practical class in the laboratory. Nevertheless, such a class is essential if the majority of students are to really understand the principles of the subject. All too often it is found that those who one thought had a firm grasp of the subject had given that impression as the result of hazy understanding coupled with rote learning.

One of the best ways of curing this state of affairs is to run a practical class in which the student has to interpret the data he has obtained for himself; that is to say, an open-ended practical. Such a class is immensely difficult to design and dovetail with the lecture course, even when the part of genetics being studied is one's own speciality. It is well nigh impossible if it is not, since one then does not have at one's fingertips the necessary techniques. Furthermore, to acquire these requires much reading in the research literature as well as a great deal of trial and error.

There seems, therefore, a real need for information on how to design and execute a wide range of laboratory experiments in genetics. Because no two lecture courses can be the same, since different emphasis and different standards are required in each, it is unprofitable to try to design a particular set of practical classes. This book, therefore, is intended to illustrate various experiments in nearly all branches of genetics so that the teacher can pick out and if necessary modify those experiments and techniques which fit his purpose best. Consequently, a number of people with considerable experience in laboratory work have written papers dealing with some aspect of their own specialities. This has led to some differences between the chapters, both in style and presentation. However, as editor, I have decided to modify the manuscripts as little as possible other than to put in a few cross references to other chapters, and have left the references appropriate to each chapter at the end of that chapter. Thus each is complete and in it will be found experiments which can be used at all levels of teaching, from the school to the university postgraduate course. For simplicity many of these experiments are described in relation to their purpose and the conclusion which can be drawn from them is given. However, when they are used for class purposes, it should always be borne in mind that a greater depth of understanding is likely to be gained by the student if the actual experiment is presented in such a way that he has to draw his own conclusions before there is a general discussion of the class results. Furthermore, an even better result may be obtained if the student, where possible, has to design his own experiment first, without the advantages (and disadvantages) of the methods suggested in this book. Later he can compare his approach with the one presented here.

Finally, if the specialist finds that his own favourite experiments and topics are not mentioned, he should remember that more excellent experiments have had to be omitted than could be described. Those that have been included have been picked not merely because they are informative to the student, but also because they can help a teacher inexperienced in a particular branch of genetics to become experimentally competent in it.

*Department of Genetics*                                          P.M.SHEPPARD
*University of Liverpool*
July 1973

# CHAPTER 1

# AN INTRODUCTION TO

# *DROSOPHILA MELANOGASTER*

# GENETICS

B. W. BARNES AND M. J. KEARSEY

## I. INTRODUCTION

The fruit fly, *Drosophila melanogaster*, was first used for genetic studies by T.H.Morgan around 1910. It is an extremely versatile organism whose special features make it a very convenient tool for genetic investigations. These features together with the genius of the early 'Drosophilists' contributed enormously to our knowledge of heredity.

Genetic investigations are concerned with the relationships between parents and offspring in successive generations. Organisms most suited for such studies have a short generation time and produce large families. Certain other considerations are of importance in the choice of organism. It should be relatively easy to handle and culture under controlled conditions, inexpensive to maintain and economical with regard to space. *Drosophila*, as we shall see, has all these features.

*Drosophila melanogaster* has a chromosome number of four. The complement consists of three pairs of autosomes and a pair of sex chromosomes. This low chromosome number greatly facilitated the construction of linkage maps, and at the present time about 1000 loci have been described and assigned to positions on the chromosomes. There is an $XY$ sex chromosome system, the normal female and male being $XX$ and $XY$ respectively. Sex is determined by the proportion of $X$ chromosomes to autosomes. An important and experimentally useful feature of meiosis is the absence of crossing over in the male. Stocks also exist bearing various chromosome aberrations. Initially these provided additional evidence for the chromosome theory of heredity and are now invaluable for carrying out chromosome manipulations.

*Drosophila* is then a useful and indeed a very popular organism for

I

teaching the basic laws of heredity. However, good results will be obtained only if good techniques of handling, rearing and scoring are practised. The purpose of this chapter is not to give a comprehensive account of the various genetical experiments that can be carried out using *Drosophila*, but to describe the important features of technique by reference to a few experiments and to illustrate some of the common pitfalls encountered by the inexperienced.

## II. LIFE CYCLE

The optimum temperature for most practical work with *Drosophila melano-gsater* is 25°C. At this temperature the life cycle is such that the experimenter need work only a five-day week, and this has an obvious appeal, particularly when *Drosophila* is used for teaching.

However, *Drosophila melanogaster* can tolerate a wide range of temperatures from near freezing up to 30°C. At low temperatures its physiology is slowed down and it becomes almost quiescent, while temperatures above 30°C can be lethal and will result in sterility, particularly if the period of exposure is prolonged. For this latter reason it is important to protect cultures from strong sunlight.

There are four principal stages of development, egg, larva, pupa and imago. The number of eggs laid by a single inseminated female varies widely with the stock. Some wild material may lay as many as 1000 eggs during their lifetime, whilst most mutant or highly inbred stocks will lay considerably fewer. The development times given below (table 1.1) will vary also, to a small degree, between stocks and on different occasions since they are influenced both by genotype and environment.

The eggs are about 0·5 mm long, white, and have a pair of filaments which act as anchors to prevent the egg sinking below the surface upon which it is laid. Fertilized eggs normally hatch 22 to 24 hours after laying. Occasionally eggs develop in the uterus and hence hatch much earlier, but this is likely to occur only if the female is denied a suitable laying site and is forced to store fertilized eggs. It should, perhaps, be emphasized that virgin females do lay eggs, albeit unfertilized and incapable of further development, and thus egg production, by itself, is no indication of non-virginity.

The larval period extends over four days and involves two moults occurring approximately 25 and 48 hours after hatching. These moults divide the larval period into three stages or 'instars'. Apart from the moults the larvae feed continually and reach a size of about 4·5 mm by the end of the third instar. Their burrows in the medium are the first sign that a culture is successful.

Towards the end of the third instar, the larvae crawl out of the food medium to find a dry site to pupate, normally the side of the container. Pupation occurs within the last larval skin, which initially is light in colour but gradually darkens. This phase of the life cycle lasts about four days and towards the end, the morphology of the adult fly can be seen through the pupal case.

The adult emerges through the anterior end of the pupal case: it is very pale in colour and the wings are unexpanded. After a further hour the wings are fully expanded and the body steadily darkens until after 8 to 10 hours the mature body colour is seen.

The faeces are stored during the pupal phase and not released until two or three hours after emergence. In immature flies this faecal spot can be clearly seen in the centre of the ventral side of the abdomen. This combination of very pale body colour and the faecal spot allows the separation of young flies from their older contemporaries in a culture.

Females will not mate before about 12 hours after emergence and do not lay eggs until they are two days old.

The whole life cycle is summarized in table 1.1 with approximate times. The convenience of the timing of this life cycle may readily be seen from this table. If matings are set up on a Wednesday or Thursday and then the flies transferred to lay in fresh cultures on a Friday afternoon (day o), their

Table 1.1 Life cycle at 25°C

| Day | Stage |
| --- | --- |
| 0 | Eggs laid |
| 1 | Eggs hatch |
| 2 | 1st moult |
| 3 | 2nd moult |
| 5 | Pupation begins |
| 10 | Adults emerge |

progeny will start emerging ten days later on a Monday morning. This allows Monday through Wednesday for collecting, scoring and mating the flies. On the Friday even the youngest females (those emerging on Wednesday) will have started to lay, and the whole cycle starts again.

Frequently a slower turnover is required, especially of stock material, to allow one to circumvent holiday periods, etc. This is most conveniently achieved by maintaining the material at 18°C, at which temperature the generation time is extended to 18 days.

## III. CULTURE METHODS

In nature *Drosophila melanogaster* is found in abundance in soft fruit growing areas, the adults and larvae feeding on well ripened fermenting fruit. An important component of the adult diet seems to be live yeast. In the laboratory *Drosophila* may be cultured on a medium that will support the growth of yeast, and a very simple medium, used by the early workers, consisted of ripe banana seeded with a live yeast suspension. At the present time many types of media are available. However, we will concentrate on one that has been used very successfully over a number of years in this laboratory.

*Drosophila* may be cultured in either 75 × 25 mm flat bottomed glass specimen tubes or half-pint (285 ml) milk bottles. The quantities of the components for the medium, given below, are sufficient for ten bottles or 50 tubes: medium grade oatmeal 72 g, agar-agar (Kobe) powder 6 g, Nipagin (10 per cent solution in absolute alcohol) 6 ml, black treacle 35 g.

The medium is made up as follows, the whole process being completed easily within an hour. The oatmeal is soaked in 120 ml of water for approximately 15 min. Dissolve the agar by boiling in 400 ml of water, and add the Nipagin cautiously as its boiling point is less than that of water. The purpose of the Nipagin is to reduce the growth of fungi and bacteria. Nipagin, a trade name for methyl-p-hydroxybenzoate, may be obtained from Nipa Laboratories Ltd, Treforest Industrial Estate, Pontypridd, U.K. The oatmeal–water mixture is then added to the agar and brought to the boil for 5 min. The mixture must be stirred continuously during this period to prevent burning. Finally, the treacle, in 38 ml of water, is added to the other ingredients and boiled for approximately 10 min.

The medium is then poured into bottles or tubes, which have previously been heat sterilized at 160°C for 1 hour. The sterilization helps to keep down fungal contamination, and in addition kills off any larvae or pupae that may be present in the bottles or on the bungs. The stoppers used for the cultures are made from cotton-wool covered with butter-muslin. Bottles should be papered before use; that is, a 100 mm square of sterilized filter paper is folded and pushed into the medium. This helps to keep the cultures dry, and also provides an additional area for pupation. Finally, bottles and tubes must be seeded with a few drops of a creamy suspension of baker's yeast in water. The yeast must be allowed to dry before the flies are introduced into the cultures.

## IV. APPARATUS

Fortunately the apparatus required for handling *Drosophila* for breeding work is very simple and, with the exception of a binocular microscope, can be 'home-made'. The lack of sufficient microscopes in schools does not present an insuperable difficulty, as many of the mutant characters used can be seen with the naked eye or with the aid of a hand lens.

If we are to use flies for genetic studies then they must be anaesthetized so that they can be examined and sorted. This can be achieved most easily

**Plate 1.1**. Top. Adult female *Drosophila melanogaster*. The position of the sterno-pleural chaetae on the thorax is shown by the arrow. Bottom. Adult male. (Photograph by P.N.CUMING, Department of Genetics, University of Birmingham, and C.H.WILLS F.R.M.S., Wednesbury College of Photography.)

[*Facing p.* 5]

using ether. A very efficient and simple etherizer may be made as follows. A cork is bored so that the neck of a 75 mm diameter plastic funnel is held securely in position in the mouth of a half-pint (285 ml) milk bottle. The bottom of the neck of the funnel is closed with a small piece of wire gauze. The gauze is easily secured in position by heating. A small quantity of ether is poured on to a pad of cotton-wool in the base of the bottle. It is important to ensure that when the etherizer is inverted no liquid ether escapes from the cotton-wool pad.

The culture containing the flies to be anaesthetized is tapped on a rubber or plastic foam pad. This will keep the flies at the bottom of the culture while the bung is removed. The culture is then inverted into the mouth of the funnel, and the flies are transferred into the etherizer by tapping. Care must be taken not to over-etherize the flies, particularly if they are to be used for mating. Normally, a period of 1 min in the etherizer is quite sufficient. Flies which have been seriously over-etherized have their wings elevated and the legs extended. When the etherizer is not in use the mouth of the funnel should be corked.

The flies are examined and sorted on a white tile, and to prevent damage all manipulations are carried out with a fine paint brush. If the flies begin to wake up before the examination is completed, they may be re-etherized on the tile by covering them with a Petri dish lid which has attached to the centre some absorbent paper on to which a few drops of ether have been placed. This method has the advantage that the flies which have been sorted do not have to be disturbed. In order to minimize the risk of contamination flies which have been examined and are no longer required should be disposed of immediately into a morgue, a small dish or jar containing oil.

A much more convenient and rapid method of handling flies, particularly when large scale breeding experiments are carried out, can be achieved using carbon dioxide. The apparatus required is again very simple. The carbon dioxide is distributed from a cylinder (12–13 kg), through a reduction valve in 10–15 mm polythene tubing. Two outlets are required by each student. Carbon dioxide is introduced into the cultures, without removing the bungs, through a hypodermic needle (serum size). The anaesthetized flies are then tipped on to a diffusion tile. This consists of a fritted glass filter plate (120 mm diameter) cemented on to an aluminium Petri dish cover. The carbon dioxide, introduced through an adapter fixed into the Petri dish cover, diffuses up through the filter plate. The flow of carbon dioxide can be regulated by a screw-clip on the rubber-tubing.

This method has several advantages over ether. First, flies can be anaesthetized in the culture containers. Second, they can be kept anaesthetized for considerably longer periods, up to 20 min. Finally, the flies recover from the effects of the carbon dioxide very rapidly, and will mate almost immediately.

## V. SEXING

It is relatively easy to distinguish the sexes of adult flies with the naked eye
even when they have only recently emerged from the pupa. However, the
beginner should make recourse to the microscope or hand lens. The essential
features of the external morphology of males and females are shown in
plate 1.1, facing p. 5, and reference should be made to this figure in the
following discussion.

The abdomen of the female is normally pointed posteriorly, while that
of the male is more rounded. The width of the dark banding on the dorsal
side of the abdomen is fairly uniform on all segments in the female. In the
male, on the other hand, the band increases in width posteriorly and the
terminal segments are almost completely black. In immature flies these
characters are less useful since the final body shape and colour are not
developed. Males may then be identified by the sex combs on the anterior
legs (this requires a microscope) and by the genital region which is clearly
visible as a dark disk.

For all experimental matings, apart from the routine maintenance of true
breeding stocks, it is essential that virgin females are used. Even with stock
matings, virgin female and male collection is advisable, in that their removal
from stock cultures at an early stage decreases the chances of fungal and mite
contamination. Also it encourages one to inspect flies and check that the
stock is in fact breeding true.

Females will mate repeatedly, but once a female has been inseminated
the sperm is stored and may be used to fertilize eggs some time later. Thus
although it is possible to mate the same female to different males the resulting
progeny may well contain representatives of all previous matings.

The simplest method of collecting virgins is to clear the culture completely
early on the first day of adult emergence. If the culture is cleared subse-
quently every 8–12 hours all the resulting female offspring will be virgin. The
main problem occurs at night when there is usually a period in excess of 12
hours between the evening and morning collection. Virgins can be identified
amongst the progeny by the method described in section II, i.e. by their pale
colour and the faecal spot. Alternatively, the culture can be kept at a lower
temperature (18°C) overnight, and this will delay mating long enough for
all resulting females to be considered virgin.

When using this repeated clearing technique to collect virgins it is essential
that all adults are removed each time. Flies often get temporarily trapped
in the culture bottle and their presence may become apparent only some
minutes later. It is always worth checking that the culture is clear.

Flies should be collected in dry culture tubes, and never be allowed to
drop on to the medium whilst still anaesthetized. Preferably these tubes
should be left permanently on their sides, so that even after the flies have

recovered from the anaesthetic, the chance of their accidentally being stuck to the medium is minimized.

The period of time during which virgins and males are collected will of course depend on the number required. The fortnightly cycle discussed earlier, leaves Monday through Wednesday available for this purpose.

## VI. MAINTAINING STOCK CULTURES

If *Drosophila* is used for teaching purposes it will be necessary to maintain stock cultures of the wild type and the mutants. These stocks, normally maintained in bottle cultures, must be sub-cultured frequently, every two weeks at 25°C or every three weeks at 18°C. The number of parents required for each culture will depend on the viability of the particular line, usually five to ten pair matings are required. In order to maintain a regular cycle the matings are set up in tubes in the middle of the week and transferred to fresh bottles on a Friday. The progeny will begin to emerge in ten days at 25°C or in 18 days at 18°C. The phenotypes of the parental flies should be checked each generation to reduce the possibility of contamination. Two cultures of each line should be set up as a precaution, and the previous generation kept until the sub-cultures are established (this can normally be determined within two days). All bottles must be labelled with the genotype of the line and the date of transfer.

The two main difficulties encountered in maintaining stocks are contamination of the cultures by fungi and by mites. The first of these is a problem only in lines with extremely low viability. The second, contamination by mites, can become a serious problem for all lines. However, this can usually be avoided by frequent transfer of the stocks, and by the immediate cleaning and sterilization of discarded culture bottles. If cultures do become contaminated they can be cleaned up by successive generations of sub-culture using as parents only the first few flies to emerge. During this period the contaminated cultures should stand in oil to prevent the spread of the infection. However, this procedure is to be recommended only if the infected stock cannot be replaced from another source. Whenever new stocks are brought in from an external supplier always check for mites, and maintain the cultures over oil until it is certain that they are uncontaminated. Foundation stocks may be obtained from commercial suppliers and certain University Genetics Departments.

## VII. BREEDING EXPERIMENTS

The experiments described below are designed to illustrate the basic laws of heredity. In the text no discussion of elementary genetic principles has been given as this may be found in any elementary genetics text book, such as

Sinnott, Dunn & Dobzhansky (1958) or Srb, Owen & Edgar (1965). As we saw earlier there are very many mutant characters available in *Drosophila*, and a complete survey of these, together with a phenotypic description, may be found in Lindsley & Grell (1968). However, it is important when choosing mutants for class work that they have good viability and are easily classified. The mutants suggested in table 1.2 fulfil these considerations. Each mutant character is abbreviated by a symbol. Recessive mutants are denoted by lower case letters, while the symbols for dominant or semi-dominant mutants begin with a capital letter. The wild type allele at a particular locus will be denoted by a +, for example *bw* + for the brown locus.

However, even using these recommended mutants ambiguous results may be obtained if insufficient attention is paid to rearing and scoring the progenies. In an attempt to avoid some of the difficulties usually encountered by beginners the first experiment will be described in detail.

**Table 1.2** Description of mutant characters

| Symbol | Name | Location | Description |
|---|---|---|---|
| *B* | Bar | 1 – 57·0 | Eye restricted to a narrow vertical bar. |
| *bw* | brown | 2 – 104·5 | Eye colour brown. |
| *cn* | cinnabar | 2 – 57·5 | Eye colour bright red, like *st*. |
| *Cy* | Curly | 2 – 6·1 | Wings curled upward. Homozygous lethal. |
| *D* | Dichaete | 3 – 40·7 | Wings extended at 45° from body axis and elevated. Alulae missing. Homozygous lethal. |
| *dp* | dumpy | 2 – 13·0 | Wings truncated and reduced. |
| *f* | forked | 1 – 56·7 | Hairs and bristles gnarled and bent. |
| *m* | miniature | 1 – 36·1 | Wings small. |
| *Pm* | Plum | 2 – 104·5 | Eye colour brown. Allele of *bw*. Homozygous lethal. |
| *Sb* | Stubble | 3 – 58·2 | Bristles reduced in length and thicker than wild type. Homozygous lethal. |
| *sc*[8] | scute | 1 – 0·0 | Loss or reduction of certain bristles. |
| *se* | sepia | 3 – 26·0 | Eye colour brownish at eclosion darkening to sepia. |
| *st* | scarlet | 3 – 44·0 | Eye colour bright scarlet. |
| *vg* | vestigial | 2 – 67·0 | Wings vestigial. |
| *w* | white | 1 – 1·5 | Eye colour white. |
| *w*[a] | white–apricot | 1 – 1·5 | Allele of *w*. Male eye colour yellowish pink, female more yellow. |

## EXPERIMENT 1. A MONOHYBRID CROSS

This experiment illustrates the inheritance of a character controlled by two alleles at a single autosomal locus. Any of the following mutants is suitable; *dp*, *vg*, *cn*, *bw*, *se*, *st*. The experiment will be illustrated using *dp*.

True breeding stock cultures of wild type and dumpy wing are bottled on the Friday ten days prior to the week the experiment is set up. When the progeny start emerging virgin females and males are collected from each line. Relatively vigorous stock cultures will yield approximately 50 flies of each sex in the first $2\frac{1}{2}$ days of emergence. The following matings are set up in tubes by each student.

$$5 ♀♀ \frac{dp\,+}{dp\,+} \times 5 ♂♂ \frac{dp\,+}{dp\,+}$$

$$5 ♀♀ \frac{dp}{dp} \times 5 ♂♂ \frac{dp}{dp}$$

$\left.\right\}$ To maintain the true breeding stocks

$$5 ♀♀ \frac{dp\,+}{dp\,+} \times 5 ♂♂ \frac{dp}{dp}$$

$$5 ♀♀ \frac{dp}{dp} \times 5 ♂♂ \frac{dp\,+}{dp\,+}$$

$\left.\right\}$ To produce the reciprocal $F_1$s

These are usually referred to as five pair matings and, unless otherwise stated, will be used throughout. This number of parents per mating usually ensures a good yield of progeny. It is important that a consistent system of labelling is used. The essential information is the genotype of the parents, conventionally the female parent is given first, and the date on which the mating is set up.

The parents are transferred to bottle cultures or fresh tubes without anaesthetization on the Friday, and finally discarded on the following Tuesday. There are two reasons for removing the parents from the cultures at this time. First, to avoid the parents and progeny being confused when scored. Second, any further progeny produced will emerge too late to be included in the experiment, and may result in excessively crowded cultures. Viability disturbances are increased with high densities and the cultures often become very humid so that removing the progeny becomes difficult.

When the progeny begin to emerge virgin females and males are collected from each line. On Wednesday the reciprocal $F_1$ progeny are scored. An accurate classification of the phenotypes will be made easier for beginners if individuals from the two stock lines are available for comparison.

The female and male progeny from both reciprocal $F_1$s are seen to be wild type. Several conclusions can be drawn from this result. The character dumpy wing could be inherited in two distinct ways, either cytoplasmically or chromosomally. If it is cytoplasmically determined, then all the $F_1$ progeny will resemble the mother. In this experiment this is seen not to be the case. We can conclude, therefore, that dumpy wing is chromosomally determined.

Furthermore, as there are no reciprocal differences in the $F_1$ males, the heterogametic sex, we can conclude that the character is not sex linked. Finally, the $F_1$ progeny are all wild type and hence wild type wing is dominant to dumpy wing.

Now that we have established that dumpy wing is not sex linked, it is no longer necessary to distinguish the reciprocal $F_1$s in setting up the $F_2$ and backcross generations. The following matings are set up in tubes; as the $F_1$ individuals are genetically uniform, virgin females need be used only for the backcross:

$$F_1 \times F_1 \qquad \text{to produce the } F_2$$

$$F_1 \times \frac{dp}{dp} \qquad \text{to produce a backcross.}$$

These crosses are transferred to bottles, and the parents discarded, as described previously. If it is necessary for the students to examine the whole experiment in one practical, due to time-tabling difficulties, then parental and $F_1$ cultures should be set up at this stage by the teacher. In this way all generations will be available on the same occasion.

In this experiment no further matings are set up and it is therefore unnecessary to collect virgins and males from the $F_2$ and backcross generations. It is important when dealing with segregating generations to ensure that the flies scored are a random sample of those which have emerged over several days. This is because different genotypes may vary in their development rate and hence flies collected over a restricted emergence time would contain a very high proportion of one genotype.

The following procedure is recommended when scoring a segregating culture. The anaesthetized flies are spread in a thin line on the tile. The flies are then separated into two lines on the basis of sex. Next, each sex is divided, one line for dumpy wing and one for wild type. This method, whereby each character is classified in turn, is easier and quicker than a complete phenotypic classification of each individual.

A set of results for this experiment are shown in table 1·3.

The observed results for the $F_2$, summing over sexes, are 80 wild type to 20 dumpy wing. The expected proportions of wild type to dumpy progeny are $\frac{3}{4}$ to $\frac{1}{4}$. We must decide now whether our observed results agree with expected. This is a basic problem which occurs in all experiments.

Let us consider briefly the possible reasons why our observed results may differ from expected. Basically, there are two causes. First, the hypothesis on which the expected numbers are calculated, i.e. $\frac{3}{4}$ wild type to $\frac{1}{4}$ dumpy wing, is correct, and all deviations result solely from chance or sampling variation. This variation occurs because the progeny scored represent only a small

**Table 1.3**

| Generation | Phenotypes | | | |
|---|---|---|---|---|
| | ♀♀ | | ♂♂ | |
| | Wild | Dumpy | Wild | Dumpy |
| $F_2$ | 43 | 12 | 37 | 8 |
| Backcross | 25 | 29 | 20 | 26 |

sample drawn from a large population of gametes. Although the two alleles $dp^+$ and $dp$ are equally frequent in this population, the equality will hold only very infrequently in small samples drawn from the population. This will result, therefore, in a departure from a 3 to 1 ratio. Second, the hypothesis is incorrect. The progeny are not segregating in the proportion $\frac{3}{4}$ wild type to $\frac{1}{4}$ dumpy wing. For example, dumpy wing flies may have a very low viability. In order to discriminate between these alternatives it is necessary to carry out a statistical test of significance. A detailed discussion of the appropriate test is obviously outside the scope of this book. In this experiment, where we are concerned with frequency data, the appropriate test is the $\chi^2$ test. This form of $\chi^2_{(N)}$ is defined as

$$\frac{(\text{observed–expected numbers})^2}{\text{expected numbers}}$$

summed over all classes, where $N$ is the number of independent comparisons that can be made amongst the classes, and is equal to one less than the number of classes (see also p. 15). As the deviations of observed from expected increase so also will the $\chi^2$. The method of calculation is illustrated below.

| | Wild type | Dumpy wing | Total |
|---|---|---|---|
| Observed numbers, $a$ | 80 | 20 | 100 |
| Expected proportion | $\frac{3}{4}$ | $\frac{1}{4}$ | 1 |
| Expected numbers, $m$ | 75 | 25 | 100 |
| Deviation, $(a-m)$ | 5 | -5 | 0 |

$$\chi^2_{(N)} = \sum \frac{(a-m)^2}{m}$$

$$\chi^2_{(1)} = \frac{5^2}{75} + \frac{-5^2}{25} = 1\cdot33$$

This value of $\chi^2$ gives us a measure of the deviations of observed from expected. The probability that as bad a deviation, or indeed a worse deviation, results solely from chance variation can be obtained from tables of $\chi^2$ (e.g. Lindley & Miller 1970). In this example there are two classes, wild type and dumpy wing, and therefore the $\chi^2$ takes one degree of freedom. The probability for a value of $\chi^2$ of $1 \cdot 33$ with one degree of freedom lies between $0 \cdot 3$ and $0 \cdot 2$. Let us consider what this probability means. If we suppose that the hypothesis of a 3 to 1 ratio is correct, then on repeating the experiment a large number of times with the same sample size, in this case 100 flies, disagreement as bad or worse than that observed would result with a probability of $0 \cdot 3$ to $0 \cdot 2$ or in 20 to 30 per cent of the trials, due to chance alone. In other words, the observed discrepancy is quite a likely event, and we have no reason to believe that it is due to anything but chance. If, on the other hand, the probability had been $0 \cdot 01$, that is only 1 in 100 of such repeat experiments would give as bad or worse a discrepancy due to chance alone, then we would say that such a discrepancy is so unlikely that our hypothesis is wrong. In practice, a probability of $0 \cdot 05$ or less (the 5 per cent significance level) is considered sufficient to indicate an unlikely event and results are considered to differ significantly from expected.

We should note at this point that a test of significance does not prove or disprove an hypothesis. All we can conclude is that the data do (or do not) depart significantly from the hypothesis. This does not exclude the possibility that the data may agree also with other hypotheses giving similar expectations. There is the problem, for example, of distinguishing a 1 to 1 segregation from a 9 to 7 segregation. However, in a particular case the simplest hypothesis is tested. Further discussion of tests of significance may be found in Mather (1951, 1964) and Snedecor & Cochran (1967).

Turning now to the backcross data, the number of dumpy and normal winged flies is 45 and 55, while equal numbers are expected. If the observed and expected results are compared as before, a $\chi^2$ of $1 \cdot 0$ for one degree of freedom is obtained. This has a probability of $0 \cdot 3$. The data do not, then, depart significantly from expected.

In these experiments 100 individuals were scored from the $F_2$ and backcross generations. In a class experiment each student's data can be analysed separately or all the results may be pooled. The second approach is valid only if the individual scores are consistent. This may be tested by a heterogeneity $\chi^2$, see Mather (1964) or Snedecor & Cochran (1967).


## EXPERIMENT 2. INDEPENDENT SEGREGATION OF TWO AUTOSOMAL GENES

This can be demonstrated using mutants on the second and third chromosomes. Suitable combinations are *dp* or *vg* (chromosome II) with *se* or *st* (chromosome III).

For the purpose of illustration the mutants dumpy wing (*dp*) and scarlet eye (*st*) will be used. The following crosses are set up, using five pair matings as before.

$$\frac{dp}{dp}; \frac{st\,+}{st\,+} \times \frac{dp}{dp}; \frac{st\,+}{st\,+} \quad \begin{array}{l}\text{dumpy wing,}\\ \text{normal eye}\end{array} \left.\begin{array}{l}\\ \\ \\ \\ \\ \end{array}\right\} \text{to maintain stocks;}$$

$$\frac{dp\,+}{dp\,+}; \frac{st}{st} \times \frac{dp\,+}{dp\,+}; \frac{st}{st} \quad \begin{array}{l}\text{normal wing,}\\ \text{scarlet eye}\end{array}$$

$$\frac{dp}{dp}; \frac{st\,+}{st\,+} \times \frac{dp\,+}{dp\,+}; \frac{st}{st} \left.\begin{array}{l}\\ \\ \\ \\ \end{array}\right\} \text{to produce the reciprocal } F_1 \text{s.}$$

$$\frac{dp\,+}{dp\,+}; \frac{st}{st} \times \frac{dp}{dp}; \frac{st\,+}{st\,+}$$

The following results should be noted from the reciprocal $F_1$ progeny:
1. There are no differences between the reciprocals and therefore neither character is sex linked.
2. All the progeny are wild type, and hence both dumpy wing and scarlet eye are recessive to wild type.

The $F_1$ females and males are mated to produce the $F_2$, and as the $F_1$ is genetically uniform, non-virgin females may be used.

The method described in experiment I should be used for scoring the $F_2$ progeny. The flies are sorted first into two rows on the basis of sex, and then each row is further divided into dumpy versus normal wing. Finally, these four rows are subdivided according to eye colour.

A typical set of results for the sexes combined is shown in table 1.4 and the analysis is:

$$\chi^2_{(N)} = \sum \frac{(a-m)^2}{m}$$

$$\chi^2_{(3)} = \frac{(-1 \cdot 75)^2}{60 \cdot 75} + \frac{(5 \cdot 75)^2}{20 \cdot 25} + \frac{(-1 \cdot 25)^2}{20 \cdot 25} + \frac{(-2 \cdot 75)^2}{6 \cdot 75} = 2 \cdot 88$$

$$P = 0 \cdot 3 \text{ to } 0 \cdot 5.$$

This probability tells us that the data do not depart significantly from the expected 9 : 3 : 3 : 1 ratio. The deviations can be accounted for solely by chance variation.

Let us consider now some results (table 1.5) produced by a similar experiment using vestigial wing and scarlet eye.

**Table 1.4**

| | Phenotypes | | | | |
|---|---|---|---|---|---|
| | Normal wing, normal eye | Normal wing, scarlet eye | Dumpy wing, normal eye | Dumpy wing, scarlet eye | Total |
| Number observed, $a$ | 59 | 26 | 19 | 4 | 108 |
| Expected proportion | $\frac{9}{16}$ | $\frac{3}{16}$ | $\frac{3}{16}$ | $\frac{1}{16}$ | 1 |
| Expected number, $m$ | 60·75 | 20·25 | 20·25 | 6·75 | 108 |
| Deviation, $(a - m)$ | −1·75 | 5·75 | −1·25 | −2·75 | 0 |

**Table 1.5**

| | Phenotypes | | | | |
|---|---|---|---|---|---|
| | Normal wing, normal eye | Normal wing, scarlet eye | Vestigial wing, normal eye | Vestigial wing, scarlet eye | Total |
| Observed numbers, $a$ | 80 | 24 | 13 | 3 | 120 |
| Expected proportion | $\frac{9}{16}$ | $\frac{3}{16}$ | $\frac{3}{16}$ | $\frac{1}{16}$ | 1 |
| Expected numbers, $m$ | 67·5 | 22·5 | 22·5 | 7·5 | 120 |
| $(a - m)$ | 12·5 | 1·5 | −9·5 | −4·5 | 0 |

These give:

$$\chi^2_{(3)} = 9·13 \qquad P = 0·05 \text{ to } 0·02$$

In this cross there is a significant departure from a $9:3:3:1$ ratio. In general, we may consider three causes for the disturbed segregation; reduced viability or poor manifestation at one or both loci, or linkage. The $\chi^2_{(3)}$ may be partitioned into three independent $\chi^2_{(1)}$ testing for these effects. In the present example, however, the linkage test is irrelevant as the mutants were chosen from different linkage groups. Both reduced viability and poor manifestation will disturb the corresponding single factor ratios, i.e. they will depart from a $3:1$ ratio. This is tested as follows:

| | Normal eye | Scarlet eye |
|---|---|---|
| Observed | 80 + 13 = 93 | 24 + 3 = 27 |
| Expected | 90 | 30 |

$$\chi^2_{(1)} = 0·4 \qquad P = 0·7 \text{ to } 0·5$$

Thus the scarlet locus is segregating as expected.

|          | Normal wing | Vestigial wing |
|----------|-------------|----------------|
| Observed | $80 + 24 = 104$ | $13 + 3 = 16$ |
| Expected | 90 | 30 |

$$\chi^2_{(1)} = 8 \cdot 71 \qquad\qquad P = 0 \cdot 01 \text{ to } 0 \cdot 001$$

This indicates a highly significant deficiency of vestigial winged progeny. We can conclude that the significant departure of the $F_2$ from a $9:3:3:1$ ratio results from a disturbed segregation at the vestigial locus. This disturbance is unlikely to result from poor manifestation for this particular character. However, reduced viability is occasionally a problem with vestigial, especially under crowded culture conditions. In class experiments it may result also from the incomplete removal of the vestigial progeny from the cultures. In cases where there is no previous information concerning the viability and manifestation of the characters used, then discrimination between these two causes of the disturbance can be achieved only by further breeding experiments.

## EXPERIMENT 3.   THE SEGREGATION OF TWO LINKED AUTOSOMAL GENES

The mutants cinnabar eye colour and vestigial wing on chromosome II are suitable for demonstrating linkage. The $F_1$ is produced by the following cross.

$$\frac{cn^+ \; vg^+}{cn^+ \; vg^+} \times \frac{cn \; vg}{cn \; vg}$$

Virgin $F_1$ females and males are collected and backcrossed to the homozygous cinnabar vestigial stock. A backcross is used here to demonstrate linkage, in preference to an $F_2$, as it is more efficient for both the detection and estimation of linkage for characters in which dominance is complete. A set of results given in table 1.6 are analysed below.

First, let us consider the data from the cross involving the $F_1$ female. The $\chi^2$ with three degrees of freedom testing the agreement with a $1:1:1:1$ segregation has a value of $116 \cdot 32$. This is highly significant. The departure of observed from expected may result from one, or a combination, of the following; reduced viability or poor manifestation at one or both loci, or linkage. This $\chi^2_{(3)}$ may be partitioned into three separate $\chi^2$ each for one degree of freedom testing for these effects.

The two $\chi^2$ testing the single factor ratios are obtained in the usual way (see previous experiment). The linkage $\chi^2$ is obtained as the difference between the $\chi^3_{(3)}$ and the sum of the two $\chi^2_{(1)}$ testing the single factor ratios.

**Table 1.6**

|  | Phenotypes | | | |
|---|---|---|---|---|
|  | Cinnabar eye, vestigial wing | Cinnabar eye, normal wing | Normal eye, vestigial wing | Normal eye, normal wing |
| Expected proportion (no linkage) | $\frac{1}{4}$ | $\frac{1}{4}$ | $\frac{1}{4}$ | $\frac{1}{4}$ |
| $F_1 \times \frac{cn\ vg}{cn\ vg}$ | 84 | 14 | 10 | 92 |
| $\frac{cn\ vg}{cn\ vg} \times F_1$ | 78 | 0 | 0 | 95 |

**Table 1.7**

| Item | d.f. | $\chi^2$ | $P(\%)$ |
|---|---|---|---|
| $vg^+ : vg$ | 1 | 0·72 | 50 – 30 |
| $cn^+ : cn$ | 1 | 0·08 | 80 – 70 |
| Linkage | 1 | 115·52 | $\ll$ 0·1 |
| Total | 3 | 116·32 | $\ll$ 0·1 |

From this analysis we can conclude that linkage alone causes the departure from the expected segregation. In the original cross the two dominant genes were introduced by the same parent, i.e. in the coupling phase. This may be seen also from the backcross data as there is an excess of the two classes normal wing with normal eye and vestigial wing with cinnabar eye. The classes normal wing with cinnabar eye and vestigial wing with normal eye result from recombination during gamete formation in the $F_1$ females. We can contrast this with the results obtained from the backcross involving the $F_1$ as male parent; here the two parental classes alone were recovered. This is brought about by the absence of crossing over in the males.

Now we have established that the loci vestigial and cinnabar are linked, we may go on to estimate the intensity of the linkage. This is measured in terms of the frequency of crossing over between the two loci. The greater the distance between the loci the higher will be the frequency of crossing

over, and loci that are very close together will recombine only very infrequently. Since a single crossover event results in 50 per cent recombinant
gametes, the frequency of crossing over, in a backcross, is estimated by the
proportion of recombinant progeny (recombination fraction, $p$). The expected numbers for each class are shown in table 1.8.

### Table 1.8

| | Phenotypes | | | | |
|---|---|---|---|---|---|
| | Cinnabar eye vestigal wing, | Cinnabar eye normal wing, | Normal eye vestigial wing, | Normal eye normal wing, | Total |
| Observed | $a_1$ | $a_2$ | $a_3$ | $a_4$ | $n$ |
| Expected | $\frac{1}{2}n(1-p)$ | $\frac{1}{2}np$ | $\frac{1}{2}np$ | $\frac{1}{2}n(1-p)$ | $n$ |

The recombination fraction, $p$, is then estimated as

$$p = \frac{\text{recombinants}}{\text{total}} = \frac{a_2 + a_3}{n}$$

In the present example $a_2 = 14$, $a_3 = 10$ and $n = 200$. These give an
estimated value for $p$ of $24/200 = 0.12$ or 12 per cent. The standard error
of this estimate may be calculated as

$$s(p) = \sqrt{\frac{p(1-p)}{n}} = 0.02298 \text{ or } 2.298 \text{ per cent.}$$

The estimated value of $p$ does not differ significantly from the accepted
value of 9.5 per cent.

In this example we have been concerned with the detection and estimation of linkage in the absence of both viability disturbances and partial
manifestation. A general treatment of this problem which takes into account
disturbances of the single factor ratios is given by Mather (1951) and Bailey
(1961). The conditions under which the simple treatment given here is valid
are summarized below.

1. If there is no disturbance at either locus, then the detection and estimation of linkage is as above.
2. If one locus alone shows disturbed viability, then again the procedure is
as above.
3. If one locus alone shows poor manifestation, then the test described is
valid, but the method of estimating $p$ is invalid.
4. If both loci are disturbed then neither the test nor the method of
estimation given above is valid.

## EXPERIMENT 4.  SEX LINKAGE

There are many genes suitable for demonstrating the inheritance of characters carried on the $X$ chromosome. Perhaps the easiest to classify is white eye, and this will be used for the purpose of illustration.

The following crosses are set up to produce the reciprocal $F_1$s. The symbol $\rightarrow$ is used here to represent the $Y$ chromosome.

$$\frac{w}{w} \times \frac{w+}{\rightarrow}$$

$$\frac{w+}{w+} \times \frac{w}{\rightarrow}$$

The phenotypes of the female and male progeny must be recorded separately for each reciprocal cross. The progeny from each reciprocal are sib-mated to produce the $F_2$.

The expected results are shown in table 1.9

Table 1.9

|  | $\frac{w}{w} \times \frac{w+}{\rightarrow}$ | | $\frac{w+}{w+} \times \frac{w}{\rightarrow}$ | |
|---|---|---|---|---|
|  | ♀♀ | ♂♂ | ♀♀ | ♂♂ |
| $F_1$ | Wild | White | Wild | Wild |
| $F_2$ | $\frac{1}{2}$ wild<br>$\frac{1}{2}$ white | $\frac{1}{2}$ wild<br>$\frac{1}{2}$ white | Wild<br>Wild | $\frac{1}{2}$ wild<br>$\frac{1}{2}$ white |

## EXPERIMENT 5.   MAPPING THREE SEX-LINKED GENES

Females from the triple recessive stock $w\,m\,f$ are crossed to wild type males. This will produce female progeny heterozygous at the three loci, and $w\,m\,f$ males. These progeny are sib-mated to produce what is essentially a back-cross of the heterozygous females to the triple recessive males. In the progeny of this cross eight phenotypes will occur, the two parental types, four single recombinants and two double recombinants. At this stage of the experiment we have no information concerning the correct order of the three loci, and therefore they are shown in table 1.10 in an arbitrary sequence.

### Table 1.10

| Phenotype | | Observed |
|---|---|---|
| white, miniature, forked | $w\,m\,f$ | 71 |
| wild type | $w^+\,m^+\,f^+$ | 89 |
| miniature, forked | $w^+\,m\,f$ | 45 |
| white | $w\,m^+\,f^+$ | 40 |
| white, miniature | $w\,m\,f^+$ | 27 |
| forked | $w^+\,m^+\,f$ | 18 |
| white, forked | $w\,m^+\,f$ | 12 |
| miniature | $w^+\,m\,f^+$ | 8 |
| | Total | 310 |

Before proceeding to estimate the recombination fractions between each pair of loci it is necessary to test the single factor ratios. The results of these tests are as follows:

| | $\chi^2_{(1)}$ | $P(\%)$ |
|---|---|---|
| $w : w^+$ | 0·323 | 70 − 50 |
| $m : m^+$ | 0·206 | 70 − 50 |
| $f : f^+$ | 1·045 | 50 − 30 |

As the single factor ratios are not disturbed the recombination fractions can be estimated as the proportion, recombinants/total.

Thus for white and miniature the recombination fraction is:

$$\frac{45 + 40 + 12 + 8}{310} = 0\cdot3387 \text{ or } 33\cdot87 \text{ per cent.}$$

Similarly, for
$w$ and $f$:
$$\frac{45 + 40 + 27 + 18}{310} = 0\cdot4194 \text{ or } 41\cdot94 \text{ per cent.}$$

$m$ and $f$:
$$\frac{27 + 18 + 12 + 8}{310} = 0\cdot2097 \text{ or } 20\cdot97 \text{ per cent.}$$

These results show that $w$ and $f$ are the furthest apart of the three loci, and hence the correct order is $w, m, f$. A linear map of the three loci is;

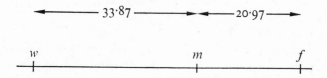

The estimated recombination fraction between $w$ and $f$ of $41 \cdot 94$ per cent is less than the sum of the distances between $w$ and $m$, and, $m$ and $f$ because of double crossover events. For this reason recombination fractions are additive only over small regions of the chromosome. In this experiment the frequency of detectable double crossovers was $(12 + 8)/310 = 0 \cdot 065$ or $6 \cdot 5$ per cent.

The recombination fractions estimated in this experiment do not differ from the accepted values of $34 \cdot 6$ per cent for $w$ and $m$, and $20 \cdot 6$ per cent for $m$ and $f$.


## EXPERIMENT 6. THE PRODUCTION OF SEX-LINKED LETHAL MUTATIONS BY X-RADIATION

This experiment is designed to investigate the relationship between the dose of X-rays and the induction of recessive lethal mutations on the X-chromosome. Two techniques have been devised for such assays by H.J.Muller. These are referred to as the ClB and Muller-5 tests. The latter is now most widely used as it has several advantages.

The $X$ chromosome of the Muller-5 stock carries the markers $sc^8$, $w^a$ and $B$, together with two extensive inversion sequences. The effect of these inversions, in females heterozygous for M5 and normal $X$ chromosomes, is to prevent the formation of gametes containing recombinant $X$ chromosomes.

If male flies are exposed to X-rays a proportion $(u)$ of their gametes will contain $X$ chromosomes bearing induced recessive lethal mutations. The purpose of this experiment is to estimate this proportion for particular X-ray doses. When virgin M5 females are crossed to such irradiated males, then of the $N$ female progeny produced, $uN$ are expected to contain recessive lethal $X$ chromosomes. The females heterozygous for such a recessive mutation may be identified by crossing all the $F_1$ females, singly, to their full sibs.

$$\frac{\text{M5}}{X} \times \frac{\text{M5}}{\rightarrow}$$

We would expect such crosses to yield the following $F_2$ progeny;

$$\frac{\text{M5},}{\text{M5}} \quad \frac{\text{M5},}{X} \quad \frac{\text{M5},}{\rightarrow} \quad \frac{X}{\rightarrow}$$

However, if the female parent is carrying a recessive lethal mutation on the $X$ chromosome, then none of the male progeny will be wild type. The progeny of all $N$ females are raised separately and $u$ is estimated as the proportion of such cultures which fail to yield wild type male progeny.

The experiment is carried out as follows. Five-day-old wild type males are exposed to the following X-ray doses, o, 1000, 2000, 3000 and 4000 rad. During irradiation the males are stored in small gelatine capsules. The males from each dose are then mass mated to virgin M5 females. The female progeny are crossed to their male sibs, and hence need not be virgin, and then separated ONE female per culture. In the next generation cultures are scored for the presence or absence of wild type males. If one wild type male is present in a culture, then we may conclude that the respective mother was not carrying a lethal $X$ chromosome. However, a problem arises when cultures produce very few males, all of which are M5. If no lethal $X$ chromosome is present, we expect half the male progeny to be M5 and half to be wild type. The probability that all $k$ males, in such a culture, are M5 is $(\frac{1}{2})^k$. When seven or more males are scored this probability is less than 1 per cent, and hence we can conclude that the absence of wild type males results from a recessive lethal mutation on the $X$ chromosome. This conclusion is not valid when less than seven males are scored, and the culture should be ignored.

In order to obtain reliable estimates of the mutation rate, the number of $F_1$ females tested must be large. This is particularly important at low doses and 100 females tested should be regarded as a minimum. The relationship between the frequency of lethal mutations and this range of X-ray doses is found to be linear (C.E.Purdum 1963).


# EXPERIMENT 7. DETERMINING A LINKAGE GROUP

The technique described allows a new mutant to be assigned to a particular chromosome. In the crossing programme the stock

$$\frac{Cy \quad Pm+}{Cy+ Pm} ; \frac{D \quad Sb+}{D+ Sb}$$

is required. This stock is particularly useful in many crossing programmes involving chromosome manipulations since both the second and third chromosomes are marked with dominant genes which are lethal when homozygous, and have extensive inversion sequences. This combination of homozygous lethal mutations and inversions maintains the stock as a pure breeding line.

Females from the stock homozygous for the unknown gene are crossed to the marked tester stock and the $F_1$ progeny are examined for this character. Three situations are possible. (1) The character is expressed in both sexes.

B

In this case we can conclude only that it is dominant. (2) Neither sex expresses the character, and therefore it is recessive and autosomal. (3) It is expressed in the male progeny only and hence is recessive and located on the X-chromosome.

Let us first consider the case where the character is dominant. Any one of the four types of $F_1$ male progeny, i.e. *CyD*, *CySb*, *PmD* or *PmSb* are crossed to homozygous wild type females. If the unknown gene is sex linked then in the next generation it will be expressed only in the female progeny. However, if it is autosomal, the following outcomes are possible. If it is carried on chromosome II, then it will segregate independently of the third chromosome markers, and conversely a gene carried on chromosome III will segregate independently of *Cy* and *Pm*. Finally, a gene carried on the fourth chromosome will segregate independently of both second and third chromosome markers.

In the second case, where the gene is recessive and autosomal, one of the four types of $F_1$ males is crossed to females homozygous for the mutant gene. The interpretation of the data is as above.

## EXPERIMENT 8.   POLYGENIC INHERITANCE

In the previous experiments we have been concerned with variation controlled by alleles at a single locus. This type of variation is discontinuous as individuals can be placed unambiguously into a distinct phenotypic class. A further important type of variation is determined by alleles at many loci (polygenic inheritance). In these cases the segregation of individual genes cannot be followed in successive generations. The analysis relevant to such variation will be discussed in detail in chapter 3 (Quantitative Genetics).

In *Drosophila*, as in most other organisms, most of the naturally occurring variation is of this type, for example, body weight, body length, wing length, fecundity, sternopleural and abdominal chaeta number, competitive ability and behavioural traits. However, the most suitable character for class experiments, for technical reasons, is the number of sternopleural chaetae (bristles). This character requires a binocular microscope with a magnification of 40 times. The sternopleural chaetae occur on the thorax, and their position is indicated by an arrow in plate 1·1. The female and male illustrated have ten and nine bristles respectively on their right sides. Conventionally, both right and left sides are scored and summed to give a single value for each individual.

In natural populations there is little apparent variation for this character. However, directional selection experiments have revealed a considerable amount of genetic variation in such populations. It has been possible to obtain lines with average values ranging from seven up to 50 bristles.

Experiments have shown genes affecting bristle number to be located on the *X*, II and III chromosomes.

In chapter 3 (Quantitative Genetics) class experiments for studying the inheritance of this character in generations derived from two inbred lines are discussed in detail.

## VIII. REFERENCES

BAILEY N.T.J. (1961) *Introduction to the Mathematical Theory of Genetic Linkage.* Oxford University Press.

LINDLEY D.V. & MILLER J.C.P. (1970) *Cambridge Elementary Statistical Tables.* Cambridge University Press.

LINDSLEY D.L. & GRELL E.H. (1968) *Genetic Variations of* Drosophila melanogaster. Carnegie Institution of Washington.

MATHER K. (1951) *The Measurement of Linkage in Heredity.* Methuen, London.

MATHER K. (1964) *Statistical Analysis in Biology.* University Paperbacks, Metheun, London.

PURDUM C.E. (1963) *Genetic Effects of Radiations.* Newnes, London.

SINNOTT E.W., DUNN L.C. & DOBZHANSKY TH., (1958) *Principles of Genetics,* 5th Edn. McGraw Hill.

SNEDECOR W.G. & COCHRAN W.G., (1967) *Statistical Methods,* 6th Edn. Iowa State University Press.

SRB A.M., OWEN R.D. & EDGAR R.S. (1965) *General Genetics.* W.H.Freeman & Co, San Francisco and London.

# CHAPTER 2

# TEACHING BASIC GENETICS IN HIGHER ORGANISMS

J. ANTONOVICS

## I. INTRODUCTION

The science of genetics was founded on the basis of the laws of inheritance proposed by Mendel. These laws, although they are in some ways an over-simplification, are still fundamental to genetics. Nevertheless they can be difficult to understand since they can be described in highly abstract terms, unfamiliar to many biologists. There is therefore a strong argument for teaching elementary genetics through a practical medium: the laws of genetics have their basis in real events and operate in all higher organisms.

There are three aims in teaching basic practical genetics. Firstly, the principles of inheritance in higher organisms must be explained. Too often a student can quote the 'magical' ratios of 3 : 1 or 9 : 3 : 3 : 1, but he has no understanding of how they are generated. Secondly, these principles must be demonstrated to illustrate their ubiquitous nature. Thirdly, basic genetics should be related to other aspects of the subject, such as biochemical, population or biometrical genetics, and so to wider areas of biology.

Basic genetics nowadays involves more than just the Mendelian Laws, for, as genetics has progressed so the scope of elementary genetics has increased. An understanding of basic genetics would not be complete without some mention of multiple alleles, linkage, gene interaction, and phenomena relating to gene expression, such as pleiotropy, penetrance and expressivity. Moreover, any comprehensive practical course in genetics cannot ignore the statistical problems associated with either sampling errors or viability disturbance.

Fundamental genetics can be taught in many organisms, but to avoid the presentation of a miscellaneous list of higher organisms showing the basic genetic phenomena (for the number of examples is almost infinite!), no attempt will be made to be exhaustive.

This chapter has been divided into two sections. The first section outlines types of investigation and includes several examples. The second section contains course suggestions for teaching basic genetics at various levels. Clearly the two sections are complementary, and should be considered together. This chapter should also be considered in relation to the other parts of this book so that a fully comprehensive, interesting course can be produced geared to the needs of the student.

## II. TYPES OF INVESTIGATION

### 1. CROSSES

Experiments in genetics are limited to a large extent by the generation time of the organisms involved. Genetics courses usually last a relatively short time, and even a year's course is only adequate for one generation in most higher organisms. Nevertheless, several organisms with a relatively short generation time are suitable, and although they may not appear to be as ideal as *Drosophila*, they each have certain advantages. They all have the advantage of showing that genetics isn't something that just happens in *Drosophila*! One of the simplest ways of teaching elementary genetics is to have the students perform crosses using not living organisms but coloured beads or counters, where these represent either genes or chromosomes (see also p. 296).

### (a) Beads

Beads, counters or children's building units form very useful vehicles for teaching basic genetics. One and two gene models are possible, and if interlocking beads are available, more sophisticated models involving linkage can be developed.

The sequence of operations for a simple one gene case would be as follows:

1.  Sort out the beads into two colour groups of about 20 each, one group representing genes of one homozygous parent, the male, the other group genes of the other homozygous parent, the female.
2.  Pair up the beads within each colour group to indicate the diploid state.
3.  Separate the beads to indicate the process of meiosis, and place the single beads (representing gametes) into the appropriate bag (one labelled male, the other female).
4.  Pick out one bead (gamete) from each bag at random and pair the two together to represent fertilization. Record the genotype of this offspring (combination of beads) and then return each bead to its appropriate bag. Repeat till a sufficiently large family has been produced. The resulting list indicates the genotypes of the $F_1$.

5.  Now pair up the beads in the $F_1$ combination and let one half of the pairs represent males (or male cells) and the other half females (or female cells).

6.  Now repeat operation 3. Separate the beads to indicate meiosis and place the single beads from the male into one bag, those from the female into another bag.

7.  Now repeat operation 4. That is to say, pick out one bead from each bag at random and pair the two together (fertilization). Record the genotype and return the beads to their correct bags. Repeat the process till an array of genotypes representing the $F_2$ is obtained.

8.  Sort out the different classes of genotype to indicate the frequency with which each occurs in the $F_2$.

These classes will not necessarily occur in the precise expected frequency, since the selection of beads at stage 7 (fertilization) was at random. The $F_2$ phenotypes are assessed on the assumption of a certain type of gene expression (e.g. dominance of one colour class).

The sequence for the two gene case is similar, and is best done using interlocking beads.

1.  First define which colours represent which alleles, e.g. gene 1: allele $A$, allele $a$; gene 2: allele $B$, allele $b$, where $A$, $a$, $B$, $b$ represent beads of different colours (or beads of two colours with $A$ distinguished from $a$ and $B$ from $b$ by a distinguishing mark such as a razor notch or ink spot).

2.  Match up the beads to represent the appropriate parental type, e.g. $AB/AB$ and $ab/ab$.

3.  Now separate the beads to indicate meiosis. Note that there is separation of homologous chromosomes, i.e. the resulting gametes must be of the types $AB$ and $ab$, not $AA$ or $BB$, etc.

Now interlock $A$ with $B$ and $a$ with $b$ to indicate that they are part of one gamete, and place those from one parent into one bag (labelled male), those from another into another bag (labelled female).

4.  Pick out one set (gamete) from each bag at random and pair the two sets together (fertilization). Record the genotype and return the sets to their appropriate bag. Repeat until enough $F_1$ genotypes have been obtained.

5.  Let half of these represent males, half females.

6.  The next step is meiosis and gamete formation, and during this there is recombination. This is achieved by pairing up sets of interlocked beads and exchanging partners between a required proportion (depending on recombination rate) of them. Thus if there is 50 per cent (maximum) recombination in both male and female, half the paired sets are selected from each bag and one pair of beads is exchanged between sets. Any other proportion may be chosen to represent different levels of linkage. Now separate the sets and place the sets from the males into one bag and the sets from the females into another bag.

7. These sets are then sampled in the usual way from each and joined in pairs at random to give the $F_2$.

8. The different types are recorded to get the frequency with which each genotype occurs in the $F_2$.

9. The $F_2$ phenotypes can now be assessed based on certain assumptions of gene interaction and expression.

Again the results do not necessarily fit the theoretical expectation because of random effects. The results of individual students can also be pooled into class data, which can then be examined by everyone.

This 'quick' way of performing 'crosses' emphasizes the role of chance in producing genetic ratios: the student can actually perform random union of gametes, and not just imagine it happening on a chequerboard diagram. As the student is dealing with 'genes' he will also realize, particularly in the two genes case, that the genotypic ratio is always the same, and that the phenotypic ratios that are produced depend entirely on the nature of expression and interaction of the genes. He can postulate different interactions (e.g. that gene $A$ required gene $B$ for its expression) which will lead him into developmental and biochemical considerations. Or he can assign quantitative values to the different genes, and so lead into quantitative genetics. Since beads are in common use for demonstrating principles of population genetics, Mendelian and population genetics can be related (p. 296). The main disadvantage of this method of teaching is that it is still highly abstract, but when combined with more conventional crossing it does provide a background that is less theoretical than mere paper work.

### Sources of material

Poppet beads from T. Gerrard & Co Ltd, East Preston, Littlehampton, Sussex; available in two colours (red and yellow).
Lego single brick building units, from almost any toy shop; available in red, yellow, blue and white.

### Reference

Nuffield Biology Teachers' Guide V (1967) *The Perpetuation of Life*, pp. 26–27. Longmans/Penguin.

### (b) Mice

Of the commonly used genetic organisms, mice are the most closely related to man and therefore of intrinsic interest. They are particularly valuable for teaching genetics to children who can keep them as pets at the same time. An additional advantage of mice is that their size and individuality makes them ideal for studying the nature of variable gene expression.

Rearing, handling and breeding mice are relatively easy and adequate methods for keeping, feeding and handling mice, together with accounts of how to detect mating and how to preserve the offspring by deep-freezing them are given in Nuffield Biology Teachers Guide V, pp. 31–35.

Useful mutants for teaching the genetics of mice are given in the following list.

*Monohybrid ratios with dominance*

Coat colour: agouti (white tip to coat hairs), *A*, dominant to
   non-agouti, *a*;
   black, *B*, dominant to brown, *b*;
   intense, *D*, dominant to Maltese dilution, *d*.
Eye and coat colour: black-eyed intense, *P*, dominant to pink-eyed dilution, *p*.
Ear size: normal long ears, *Se*, dominant to short ears, *se*.
Hair type: wavy hair, *Ca* (caracul) or *Re* (rex), dominant to normal straight
   hair, *ca* or *re*.

*Monohybrid ratio with no dominance*

Coat colour: agouti, *A*, crossed with tan belly, $a^t$, gives a heterozygote that is agouti and has a tan belly.

*Dihybrid ratio with dominance*

Ear size and eye-colour: short eared, black-eyed mice (*se P*) crossed with long eared pink-eyed mice (*Se p*) will produce a 9 : 3 : 3 : 1 ratio in the $F_2$.

*Linkage*

Danforth's short tail (*Sd*) wellhaarig (*we*, wavy hair) and non-agouti (*a*) are on chromosome V in the following order: *Sd*—40 per cent—*we*—14 per cent—*a*

*Pleiotropy* (one gene affecting several characters)

Eye and coat colour: the gene, albinism, *c*, produces a pink eye and white coat, while its allele extreme chinchilla, $c^e$ produces a black eye and a pale coffee coloured coat. Crossing $c^e c^e \times c\ c$ gives in the $F_2$ black-eyed coffee coloured, black-eyed white and pink-eyed white mice in the ratio 1 : 2 : 1. With regard to eye colour, pink eye is recessive to black eye, while with regard to coat colour white coat is dominant to coffee. This is an excellent illustration that dominance is a property of the character and not the gene.
Coat colour and behaviour: the varitint waddler, *Va*, produces both a coat

streaked with black, grey, and white, and mice which tend to waddle and shake.

Ear size and kidney defect: the short ear mutant, *se*, causes both a short ear and hydronephrosis of the kidney. The kidney appears enlarged, watery and balloon-like when the mouse is dissected. However, only a proportion of short ear mice show this enlarged kidney because of variable penetrance (only a certain proportion of individuals show the character, even though all have the gene), sex limited penetrance (one sex shows the character more than another) and variable expressivity (character expressed to different degrees in different individuals). The hydronephrosis characteristic of short eared mice, *se*, illustrates all three phenomena. The balloon-like kidney is not seen in all mice, it is more common in males than females, and sometimes it is obvious while in other mice hardly noticeable. The short ear character on all these mice confirms the presence of the same gene.

## Modifying genes

Number of digits: the effect of modifying genes is best seen in the mutant polydactyly, *py*, where the feet have from 5–8 toes and different numbers of bones to the extra toes. Stocks showing high and low incidence of this character are available and illustrate modifying genes either enhancing or repressing a character.

Coat colour: an example of a specific modifier is yellow coat, $A^y$, which reduces the size of the white areas in mice carrying genes for spotting (e.g. dominant pied, *W*).

## Variable dominance

The gene Danforth's short tail, *Sd*, results in a short tail length, and the heterozygote varies in its tail length depending on its genetic background: stocks where this character is recessive, dominant or shows no dominance are available. A very illuminating class study would be possible if one group in the class were given mice where the character was dominant, and another group where the character was recessive. The value of this particular example is that it demonstrates variable expression, modifiers, and can lead into a discussion of the evolution of dominance and other types of gene expression.

## Sex linkage

The gene Tabby, *Ta*, produces a striped appearance in the coat (cf. tabby cats) when present in the heterozygous state (*Ta ta*) in the female. *Ta Ta* females and *Ta Y* males do not show the tabby coat: instead the coat is greasy and absent from the tail or behind the ears. The greasy phenotype is controlled by a sex linked gene, the striped tabby mice being the female carriers of the gene for greasy coat.

*Multiple linkage and three point test crosses*

The multiple mutant stocks supplied by Harris Biological Supplies (see Sources of Materials) can be used to set up test crosses involving several genes.

*Sources of materials*

Stocks are available from the following sources:

Department of Education and Science Laboratories, Ivy Farm, Knockholt, Sevenoaks, Kent. They supply: *AA; aa; BB; bb; DD; dd; PP; pp; SeSe; sese; CaCa; caca; ReRe; rere; $a^t a^t$.*

Harris Biological Supplies, Oldmixon, Weston-super-Mare, Somerset. They supply *AA; aa; $a^t a^t$; $Aa^t$; AAy; bbaa; ddaa; ccaa; $c^e c^e aa$; $cc^e aa$;* (*aa* indicates that colour types are non-agouti); *pp; PP; dd pp bb aa* (dilute, pink eye dilution, brown, non-agouti, together=silver champagne); *sese; seSe; Tata*(X), *TaY; pypy unun wewe $a^t a^t$ papa* (multiple mutant); *wa-2wa-2 vtvt $c^{ch} c^{ch}$* (multiple mutant).

Dr M.E.Wallace, Department of Genetics, Cambridge University, will supply *Sd* lines showing variable dominance on request, although supplies cannot be guaranteed.

T. Gerrard & Co Ltd, Worthing Road, East Preston, Littlehampton, Sussex, supply a range of hair colour mutants.

*References*

FALCONER D.S. (1963) The use of mice in teaching genetics. In Darlington, C.D. & Bradshaw A.D. (Eds) *Teaching Genetics in School and University*, pp. 44–49. Oliver and Boyd, Edinburgh. (The use of mice, particularly in developmental genetics.)

NUFFIELD BIOLOGY TEACHERS' GUIDE V (1967) *The Perpetuation of Life*, pp. 30–35. Longmans/Penguin, London. (Best account of basic mouse technology.)

WALLACE M.E. (1965) Using mice for teaching genetics, I. *School Science Review, No.* 160, pp. 646–658. Using mice for teaching genetics, II. *School Science Review, No.* 161, pp. 39–52. (Excellent range of examples for teaching elementary and advanced concepts in genetics. Copies available from Association for Science Education, 52 Bateman Street, Cambridge, or Harris Biological Supplies, Oldmixon, Weston-super-Mare.)

WALLACE M.E., GIBSON J.B. & KELLY P.J. (1968) Teaching genetics: the practical problems of breeding investigations. *Journal of Biological Education*, 2, 273–303.

### (c) Arabidopsis

Once hailed as the *Drosophila* of the plant world, this plant has not been accepted as readily as was originally anticipated, because its short generation time of one month can only be achieved under ideal conditions of light and temperature, and because it is difficult to cross. The ideal conditions for growing *Arabidopsis* are 20°–25°C and daylight supplemented with anywhere between 3·5–5·5 klx to give continuous illumination. A life cycle of about two months can however be produced if the plants are grown at a reasonably warm temperature and with spring or autumn day length. Any fine sieved soil is suitable, but preferably it should be light and sandy. The soil should be sterilized by heating or steaming. Seed trays or 75 mm pots are suitable. Seeds may be collected about 14 days after pollination: these are usually still unripe and will germinate in a few days. The seeds ripen at about three weeks and when first collected may be dormant, and remain dormant for about two months. This dormancy can be broken by soaking the seeds in water for about an hour and then keeping them at between 0°C and 5°C for four days. Germination then occurs in 2–3 days. Seeds older than two months germinate without pretreatment in about three days at 20–25°C. The seeds should not be allowed to dry out after sowing and are best sown directly on to the soil.

The seedlings must be watered gently. They can be watered using a very

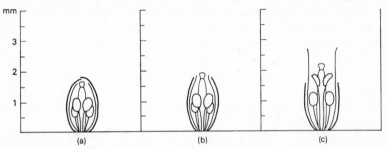

Fig. 2.1. Cross-section through a flower of *Arabidopsis* at different stages of development. Stages (a) and (b) are the preferred ones for emasculation. (From Muller, 1961.)

fine spray or the pots can be irrigated from below by standing them in shallow trays of water; in this way the water is absorbed into the soil from below, and the seed and seedlings are not disturbed. Another way is to cover the pots or seed trays with Saran Wrap. This is a thin cellophane-like material, which prevents water loss yet permits the passage of carbon dioxide and oxygen. If the soil in which the seed is sown is moist, then pots covered with Saran Wrap need watering only very occasionally.

*Arabidopsis* is normally completely self-pollinated, but the flower is slightly protogynous and emasculation can be performed by removing the anthers with a very fine pair of forceps. The best stages for removal of anthers are shown in fig. 2.1.

Because the flowers are tiny, it is helpful to use a binocular microscope (mounted on an arm and adjustable to take up a wide range of heights and orientations), a watchmaker's lens (×5–10), or lens-glasses. The flowers can be held still with the finger and thumb or a pair of forceps which have the tips wrapped in cotton wool or soft synthetic sponge. Success must not be expected to come easily and a few plants should be set aside for practising on. If the female parent carries a recessive gene, then this can be used as a marker: seed produced by selfing in the $F_1$ will show the recessive character and can be discarded. If the stigmas are smothered with pollen then it is not necessary to bag the flowers to prevent accidental crossing, as the chances of this happening are remote.

The recent discovery of a male sterile mutant, *ms*, with reduced anthers and only 0·3 per cent seed setting on self pollination, by A.J.Müller, should greatly increase the value of *Arabidopsis* in teaching and research. The male sterile mutant can be used as female parent without emasculation.

The following mutants have a high viability and easily recognizable characteristics.

*Recessive*

angulosa ($a_1$): leaves narrow, angular, serrated, pale green; stems slender.
apetala (*ap*): most flowers without petals, but variable number of petals on
    later flowers; only two sepals, large and persistent; pods erect.
axillaris (*ar*): two flowers growing from base of each pod; otherwise normal.
clavata–1 (*clv₁*): pods large, lumpy and club-like, rosette flat; flower buds
    visible at early stage of growth.
cordata (*cor*): smallish plant with heart shaped leaves, curled over; leaves
    pale green; rosette deep.
glabra (*g*): glabrous; vigorous growth; few hairs occasionally on leaf margins
    or stems.
longipetiola (*lp*): petioles long; rosette leaves elliptical; stem leaves linear.
maculosa–1 ($m_1$): rosette leaves lightly mottled, yellow green, paler at base.
nigra (*nig*): rosette leaves dark green and glossy; cotyledons small and dark;
    anthocyanin visible at base of stems; stems several, slender; height 175 mm.
pallida–1 ($pd_1$): rosette leaves yellow green, paler at base; stems pale green
    with anthocyanin; pods small.
serrata–1 ($se_1$): leaves serrated; cotyledons long; flower bud opening early;
    petals few; pods small.
variegata–2 ($v_2$): plant variegated white green; expression varying; pods
    small.
The mutants listed above also illustrate, to varying degrees, the phenomena of pleiotropy and variable expressivity.

*No dominance*

filicaulis–1 ($fi_1$): stems many, thin, crinkled; not exceeding 50 mm in height;
    pods long, thin with few seeds. Heterozygote intermediate.

*Linkage*

late ($gi_1$): late flowering (two months); large rosette.

patula (*pa*): dwarf; leaves dark green and curled; plant spreading and straggly; pods small.

chlorophyll-free ($ch_1$): no chlorophyll–*b;* yellow green plants.

These three linked in the order *gi–pa–ch* and separated by 25 and 8 map units respectively.

*Nutritional mutants*

pyrimidine requiring (*py*): requires 2:5 dimethyl–4 amino pyrimidine.

thiazole requiring (*tz*): requires 4–methyl–5 thiazole.

thiamine requiring (*th*): requires vitamin $B_1$.

These three mutants control the biochemical pathway shown below

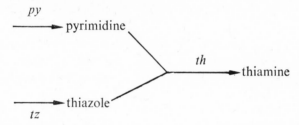

The techniques for growing the mutants in aseptic culture are given by Redei (1968).

*Sources of materials*

Professor A.D.McKelvie, Agricultural Botany Department, College of Agriculture, 581 King Street, Aberdeen, Scotland, supplies: $a_1$, *ap*, *ar*, $clv_1$, *cor*, *g*, *lp*, $m_1$, *nig*, $pd_1$, $se_1$, $v_2$, $fi_1$.

Professor G.P. Redei, Department of Genetics, University of Missouri, Columbia, Missouri 65201, U.S.A., supplies: $gi_1$, *pa*, $ch_1$, *py*, *tz*, *th*.

Dr A.J.Müller, Deutsche Akademie der Wissenschaften Zu Berlin, Institut für Kulturpflanzenforschung, Gatersleben, Berlin, Germany, supplies: male sterile mutant (*ms*).

*References*

McKELVIE A.D. (1962) A list of mutant genes in *Arabidopsis thaliana* (L.) Heynh. *Radiation Botany*, **1** 233–241. (A list of mutants.)

NUFFIELD BIOLOGY TEACHERS' GUIDE V (1967) *The Perpetuation of Life*, pp. 81–82. Longmans/Penguin, London. (Good account of basic techniques.)

REDEI G.P. (1968) *Arabidopsis* for the classroom. *Arabidopsis Information Service*, **5**, 5–7. (An account of a wide range of fairly advanced investigations, including nutritional mutants and aseptic culture.)

MÜLLER A.J. (1961) Zur charakterisierung der Blüten und Infloreszenzen von *Arabidopsis thaliana* (L.) Heynh. *Kulturpflanze*, **9**, 364–393.

MÜLLER A.J. (1968) Genic male sterility in *Arabidopsis*. *Arabidopsis Information Service*, **5**, pp. 54–55.

*Arabidopsis Information Service*, published by the Institut für Pflanzenbau und Pflanzenzüchtung, Universität Göttingen, Göttingen, Germany, is a useful source of information and further references.

### (d) Tribolium

The flour beetles (*Tribolium confusum and T. castaneum*) have for many years been used in the study of population dynamics, and only recently have their advantages as a genetic organism become appreciated. The techniques used for culturing them can be obtained from *Tribolium Information Bulletin*, I–3, pp. 3–13.

Useful mutants for teaching *Tribolium* genetics are given in the following list.

### *Tribolium castaneum*

#### *Monohybrid ratios with dominance*

Body colour: sooty, *s*, (black body gene) recessive to wild type (reddish brown);
  jet, *j*, (black body gene) recessive to wild type (reddish brown).
Eye colour: pearl eye, *p*, (whitish eye) recessive to wild type (black eye).
Head size: microcephalic, *mc*, (small narrow head) recessive to wild type (large head).
Antenna shape: antennapedia, *ap*, (antennae larger than normal and foot like; segments elongated) recessive to wild type (with short segmented antennae).

#### *Monohybrid ratio with no dominance*

Black *B*, body colour semi-dominant to reddish brown wild type body colour. Heterozygote has bronze colour.

#### *Dihybrid ratio with no linkage*

Pearl eye and sooty body, are on separate chromosomes and will give a dihybrid ratio.

#### *Linkage*

Jet, *j*, body colour and microcephalic, *mc*, are both on chromosome *V*. The cross over value is 25 per cent.

*Dominance and penetrance influenced by temperature*

The scar mutant, *sc*, has an engraved transverse groove on the lower body segment (metasternum) just anterior to base of leg (coxa), shows temperature dependent dominance and penetrance. The percentage of population of *sc/sc* genotypes showing scar phenotype increases between 33°C and 35°C (relative humidity 70 per cent) from 70–80 per cent to 90–100 per cent, and dominance (i.e., percentage of population of genotype *Sc/sc* showing *sc* phenotype) from 0–21 per cent to 12–77 per cent.

*Sex-linkage*

Red, *r* (light red eye colour) sex linked and recessive to wild type black eye colour.

*Recessive lethals*

Short antenna, *Sa*, and fused tarsi and antennae (*Fta*) are both homozygous lethals.

*Gene interaction*

The genes *Sa* and *Fta* are at different loci but when *Fta* and *Sa* occur in same genotype, that individual dies.

*Sex difference in recombination values*

Jet, *j*, and ruby, *rb*, eye colour show different recombination rates in backcrosses depending on whether the heterozygote is male (33 per cent) or female (21 per cent).

**Tribolium confusum**

*Monohybrid ratios with dominance*

Pearl, *p*, (white compound eye) recessive to wild type (black eye).
Ebony, *e₂*, (black body colour) recessive to wild type (reddish brown body colour).

*Linkage*

Pearl, *p*, eye and ebony, *e₂*, body colour are both on the same chromosome and have a recombination value of 2·5 per cent.

*Source of materials*

Stocks can be obtained from:
*Tribolium castaneum;* wild type, *s, j, p, mc, ap, B; Tribolium confusum; p,*

$e_2$ from T. Gerrard & Co Ltd, Worthing Road, East Preston, Littlehampton, Sussex.

*Tribolium castaneum, sc*, From Dr A.E.Bell, Population Genetics Research Institute, Purdue University, Lafayette, Indiana.

*Tribolium castaneum, r, Sa, Fta*, from Professor A.Sokoloff, Department of Genetics, University of California, Berkeley, California.

*Tribolium castaneum, rb*, from Dr A.A.Dewees, Population Genetics Research Institute, Purdue University, Lafayette, Indiana.

Food packs available from T. Gerrard & Co Ltd, Worthing Road, East Preston, Littlehampton, Sussex.

## References

*Nuffield Biology 'O' Level Five Year Course* catalogue available from T. GERRARD & Co LTD, Worthing Road, East Preston, Littlehampton, Sussex. (Good brief description of basic techniques and mutants.)

DEWEES A.A. (1967) Sex differences in recombination values for linkage group V of *T. castaneum*. *Tribolium Information Bulletin*, **10**, 89–90.

SOKOLOFF A. (1963) Two exercises demonstrating factor interaction in *Tribolium castaneum* Herbst. *Tribolium Information Bulletin*, **6**, 69–71.

BELL A.E., SHIDELER D.M. & EDDLEMAN H.L. (1964) Dominance and penetrance of the scar (*sc*) mutant in *Tribolium castaneum* as influenced by temperature. *Tribolium Information Bulletin*, **7**, 46–48.

SOKOLOFF A. (1966) *The Genetics of Tribolium and Related Species*. Academic Press, New York. (Advanced research text describing findings rather than techniques.)

HOSTE R. (1968) The use of *Tribolium* beetles for class practical work in genetics. *Journal of Biological Education*, **2**, 365–372.

HASKINS K.P. (1969) *Using Tribolium for Practical Genetics*. Harris Biological Supplies Ltd, Weston-super-Mare, England.

## (e) Other organisms

Very few other organisms are in common use for crossing experiments carried over several generations. Either the generation time is too long or there is a shortage of mutants. Where the generation time is too long, an investigation could be initiated in one year and carried on by a set of students in the following year, but in many ways this is unsatisfactory since the first generation of a cross is relatively uninformative. It is far better for the teacher to perform the initial cross himself and report the results when the students look at the $F_2$ generation.

The species described in the following section can be used for long-term investigations. Such investigations may be done as a class research project covering several years and aimed at investigating a complex of major gene characters such as are found in coat colours of mice, hamsters or guinea-pigs, or flower pigments and other characteristics of garden plants.

## 2. SEGREGATION DEMONSTRATED IN LIVING MATERIAL

The most potent effects of the laws of inheritance are seen in the second generation, and the demonstration of segregation in this generation is therefore one of the best ways of showing these laws in action. The parental lines and the $F_1$ can be indicated theoretically, kept from previous generations or demonstrated by using the appropriate phenotypes selected from the $F_2$.

When an $F_2$ generation is required, it is not normally necessary to start the whole sequence of crosses *de novo*. There are several methods whereby the $F_2$ can be obtained in a far shorter time.

(i)   If the dominance of one gene is incomplete such that the $F_1$ genotype can be recognized amongst the $F_2$ classes, this genotype can be maintained for production of another '$F_2$' in the following season.

(ii)   The $F_1$ can be maintained vegetatively and crossed to give an $F_2$ when required.

(iii)   The $F_2$ can be maintained as a collection of seeds which are sown and scored by the student as required. Where juvenile or seedling characters can be studied an $F_2$ can be obtained in a matter of a few weeks.

The latter two techniques are more or less restricted to plant material, which is therefore particularly suitable for this category of investigation.

(iv)   An alternative scheme, which can be combined with the above, is to use backcrosses, since even with dominance all the progeny of a backcross are of known constitution. The sequence of crosses is as follows: $AA \times aa \rightarrow Aa$; $Aa \times aa \rightarrow Aa$ and $aa$, which can then be intercrossed to give the same ratio again. This scheme can be used for two gene cases as well since

$$\frac{A\,B}{A\,B} \times \frac{a\,b}{a\,b} \rightarrow \frac{A\,B}{a\,b}; \quad \frac{A\,B}{a\,b} \times \frac{a\,b}{a\,b} \rightarrow \left[\frac{A\,B}{a\,b}\right], \frac{a\,B}{a\,b}, \frac{A\,b}{a\,b} \quad \& \quad \left[\frac{a\,b}{a\,b}\right].$$

The types in brackets can be picked out and used again. In this way, after the initial $F_1$ has been obtained, only 'one generation' is necessary for the production of a segregating backcross progeny.

### (a) Recognition of $F_1$ amongst the $F_2$ classes

### (i) Groundsel: *Senecio vulgaris* L.

A cross between groundsel without ray florets (normal form) and groundsel with long ray florets results in an $F_1$ with short ray florets which are intermediate in length. The long ray florets are longer than broad whereas the short ray florets are square shaped. The $F_2$ segregates in a 1 : 2 : 1 ratio for the three classes, and since groundsel is almost entirely self-pollinated the seeds from plants with short ray florets again produce the same ratio. The seeds are viable for several years. Groundsel will flower in about 6–8 weeks, in any soil, provided it is kept reasonably warm (i.e. greenhouse in winter). Groundsel readily sheds its seeds but this problem can be eliminated by digging up the appropriate genotypes and letting them ripen off in a vase, beaker or jam jar containing water. This can be placed indoors in a convenient draught-free position and seeds collected every morning.

### Sources of material

Department of Education and Science Laboratories, Ivy Farm, Knockholt, Sevenoaks, Kent.

### References

The only published works on this subject are some 50 years old, and include accounts of other characters in groundsel.

TROW A.H. (1912) On the inheritance of certain characters in the common groundsel—*Senecio vulgaris* Linn—and its segregates. *J. Genet.*, *Cambridge*, **2**, 239–276.

TROW A.H. (1916) On the number of nodes and their distribution along the main axis in *Senecio vulgaris* and its segregates and on albinism in *Senecio vulgaris* L. *J. Genet.*, *Cambridge*, **6**, 1–74.

A brief account of this work is given in Turrill W.B. (1958) *British Plant Life* 2nd Ed. Collins New Naturalist Series, p. 219.

### (ii) Seedling characters

The following seedling characters show intermediate expression in the heterozygote (see section (c)(ii) for further details and sources of materials).

Tomato: normal/lanceolate cotyledons
　　　　　white/green cotyledons
Radish: red/white hypocotyl
　　　　　round/long hypocotyl
Geranium (*Pelargonium zonale*): green/white seedling
Marrow Stem Kale: red/green hypocotyl
Soybeans: green/white leaves on seedlings.

### (b) $F_1$ maintained vegetatively

### (i) White clover (*Trifolium repens*)

White clover is of special interest, since its populations are polymorphic for several sets of major gene characters. Clover can be easily grown from cuttings (a few internodes long) pushed into ordinary garden soil or potting compost. It thrives best in cool bright conditions, and the leaf mark characters in particular are seen most clearly outdoors in spring, or in a cool greenhouse with artificial light. These conditions also reduce the chances of infection from aphids or red spider—an important consideration since greenhouse sprays and insecticides can easily damage clover leaves. Clover grows quickly and has to be replanted at least once a year, or cut back to prevent stolons trespassing into adjacent pots. Unwanted seed heads should also be removed before they are ripe, to prevent foreign seed germinating and 'contaminating' the existing plants.

Clover seed germinates slowly, and it is best *scarified* (weakening of seed coat) by immersing dry seed in *conc.* $H_2SO_4$ for 15 min, draining off the acid, and washing with plenty of water. Alternatively the seed can be rubbed with sandpaper. The seed is then sown directly or germinated on a moist filter pad and then sown as a small seedling.

The seedlings will flower if they are given summer (long) day length, but plants which have flowered once or adult plants collected in the field need short days and/or cold treatment (outdoors in winter) for about two months, followed by long days for flowering to occur. In natural conditions the plants normally flower between May and July. Clover is to all intents and purposes self-sterile so emasculation is unnecessary. Pollen is best transferred using the pointed tip of a cardboard triangle, which has been folded down the middle (see fig. 2.2). The keel is pulled away from the head with a pair of fine

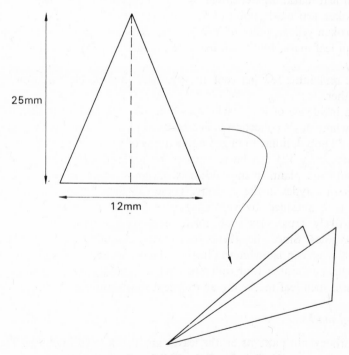

**Fig. 2.2.** Cardboard pollinator for clover plants.

forceps and the stigma is extruded for pollination. The plants or flowerheads can be covered in muslin to prevent stray pollination, or an insect-free green-house should be used. About four seeds per floret are produced after 4–6 weeks. Pollination is made easier if the number of florets per head is reduced to about 20 which are at an equivalent stage with their wing petals about to open.

The following sets of characters are of particular interest:

### V-shaped white leaf markings

The V-shaped leaf marks are determined by an allelic series, where absence of leaf mark is recessive to its presence. The different leaf marks show no dominance with respect to each other, so that where the marks are in different positions on the leaf, the heterozygotes can be distinguished (fig. 2.3) and can be propagated vegetatively.

The following leaf marks are known (their percentage frequency in wild populations in Britain is indicated):

$vv$: absence of leaf mark (23·4 per cent)

$V^l$: V mark in lower half of leaflet, size and position variable (72·2 per cent)

$V^h$: V mark extending into upper half of leaflet (0·4 per cent)

$V^f$: full leaf mark, area enclosed by V is all white (0·2 per cent)

$V^b$: broken leaf mark, point of V is absent (very rare)

$V^{by}$: broken yellow, point of V is yellow, arms are white (1·1 per cent)

$V^{ba}$: full leaf mark, but basal, narrower, longer and fainter than $V^f$ (0·4 per cent).

The remaining few per cent are types showing double marks of one sort or another.

The incidence of leaf mark types other than $v$, and $V^l$ is therefore rare and a whole day's collection is necessary to pick up some of the rarer types from wild populations. In mixed populations of $v$ and $V^l$ types, heterozygotes (phenotypically $V^l$) can be recognized by looking at the products of seed collected from plants in the wild: heterozygotes should show an appreciable number of $v$ types in the progeny. These can then be collected as a group which is maintained by vegetative propagation for future intercrossing. Alternatively crosses can be made between the types to generate the heterozygotes. Since the phenotypes closely resemble the genotypes other types of cross can be made to illustrate the consequence of say, crossing two genotypes containing four different alleles. Used purely for demonstration purposes, such leaf marks are an excellent illustration of an allelic series.

### Red leaf marks

Red anthocyanin patterns on the leaves are also known in clover. The most common type are the $R$-series of three alleles.

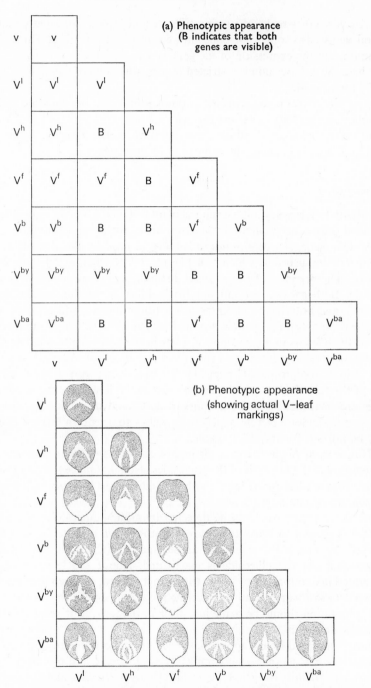

**Fig. 2.3.** V-shaped leaf marks in white clover (*Trifolium repens*).

$R^f$: red flecks or streaks of varying sizes, tending to be elongated parallel to lateral veins, also somewhat concentrated around the midrib. This is fairly common, but the expression of the gene is very variable.

$R^m$: here the anthocyanin is restricted to just either side of the midrib. This is rarely found wild.

$R^l$: here all the leaf is red. This too is rarely found wild but is often preserved in Botanical Gardens as var. *purpureum*.

All these alleles are dominant, and unlinked to the *V*-series. Again, cool bright conditions enhance the expression of these genes.

### Cyanogenesis

The ability to release cyanide when the plant is damaged is determined by two unlinked genes, and populations of clover are normally polymorphic for both genes. One gene, *Ac*, determines the ability to produce the glucosides, linamarin and lotaustralin, which will liberate HCN when acted upon by an enzyme. The presence of this enzyme, linamarase, is determined by another gene *Li*. Both genes are dominant (to inability to produce glucoside and enzyme), and both have to be present for full cyanide production.

The normal test for cyanide production is as follows. Crush and macerate two or three leaves in a few drops of water in the bottom of a small 50 mm × 15 mm glass tube (using a glass rod). Add two drops of toluene and insert a short slip of sodium picrate paper by jamming it between the cork and the side of the tube. The paper should be clear of the liquid at the bottom. The tubes are then incubated for 2 hours at 40°C, and if cyanide is released it will turn the sodium picrate papers from yellow to a reddish brown. This will only be positive for the *Ac Li* plants.

However, *Ac Li* plants can be distinguished from *Ac li* because spontaneous (non-enzymatic) hydrolysis of the glucosides also takes place but more slowly. If the tubes are left for at least another 24 hours, *Ac li* plants will also turn the sodium picrate paper brown.

The *ac Li* plants can only be detected by adding glucosides to see if the enzyme is present to break them down. Both *Ac li/li* and *Ac Li/-* plants can be used, because if the leaves are autoclaved (110°C for 25 min) then the enzyme but not the glucoside is inactivated. About 100 leaves should be macerated in about 25 ml of water, the extract filtered, and two drops (instead of two drops of water) used to macerate the cyanogenic leaves to test for *Li*. The preparation should be made just prior to the tests, as it keeps its activity for a few days only.

Once recognized by the appropriate tests, the genotypes can be propagated vegetatively. Crosses between completely acyanogenic and cyanogenic types will serve to identify the genotypes of the cyanogenic forms, and can serve as a backcross. Once these genotypes have been recognized, $F_2$ ratios can be generated by crossing genotypes heterozygous for either one (to give a 3 : 1 ratio) or both genes. If no attempt is made to distinguish the non-cyanogenic types a 9 : 7 ratio is obtained, but if *Ac li*, *ac Li* and *ac li* plants are distinguished a 9 : 3 : 3 : 1 ratio will be obtained. This is a good example of evidence of complementary gene action since both genes are necessary for the biochemical sequence to take place, and therefore for the character to be expressed.

The value of these various investigations in clover is that they provide a link between elementary genetics, population genetics and natural selection. The cyanogenesis character is also a clear demonstration of the way in which genes act in metabolism.

## *Source of materials*

Homozygotes and heterozygotes can be obtained from Dr W.E.Davies, Welsh Plant Breeding Station, Plas Gogerddan, Aberystwyth. However a supply of every possible type cannot be guaranteed. Many of the types can be obtained by collection from natural populations.

Sodium picrate paper is made by neutralizing a nearly saturated solution (0·05M) of picric acid with sodium bicarbonate $NaHCO_3$, filtering, and soaking filter paper in the solution. When dried the papers can be kept in a stoppered jar and used as required.

## *References*

DAVIES W.E. (1963) Leaf markings in *Trifolium repens*. In DARLINGTON C.D. & BRADSHAW A.D. (Eds) *Teaching Genetics in School and University*, pp. 94–98. Oliver & Boyd, Edinburgh.

PUSEY J.G. (1963) Cyanogenesis in *Trifolium repens*. In DARLINGTON C.D. & BRADSHAW A.D. (Eds) *Teaching Genetics in School and University*, pp. 99–104. Oliver & Boyd, Edinburgh.

## (c) $F_2$ maintained as seed

### (i) Peas

Mendel's experiments can be repeated or mimicked with different varieties of peas showing round/wrinkled or green/yellow seeds. Suitable varieties (most of which are available commercially) are shown in table 2.1 (pp. 44–47). This table also distinguishes tall and dwarf, and blunt pod and pointed pod varieties but these characters do not appear till later in the life cycle.

**Table 2.1** Basis of classification

| Pod type | Seed | Cotyledons | Pod shape | Stem | Group |
|---|---|---|---|---|---|
| Normal. *Pisum sativum* sub sp. *hortense* | Round | Yellow | Pointed | Under 450 mm. | I |
| | | | | 450–750 mm. | II |
| | | | | 750–1100 mm. | III |
| | | | | Above 1100 mm. | IV |
| | | | Blunt | Under 450 mm. | V |
| | | | | 450–750 mm. | VI |
| | | | | 750–1100 mm. | VII |
| | | | | Above 1100 mm. | VIII |
| | | Green | Pointed | Under 450 mm. | IX |
| | | | | 450–750 mm. | X |
| | | | | 750–1100 mm. | XI |
| | | | | Above 1100 mm. | XII |
| | | | Blunt | Under 450 mm. | XIII |
| | | | | 450–750 mm. | XIV |
| | | | | 750–1100 mm. | XV |
| | | | | Above 1100 mm. | XVI |
| | Wrinkled | Yellow | Pointed | Under 450 mm. | XVII |
| | | | | 450–750 mm. | XVIII |
| | | | | 750–1100 mm. | XIX |
| | | | | Above 1100 mm. | XX |
| | | | Blunt | Under 450 mm. | XXI |
| | | | | 450–750 mm. | XXII |
| | | | | 750–1100 mm. | XXIII |
| | | | | Above 1100 mm. | XXIV |
| | | Green | Pointed | Under 450 mm. | XXV |
| | | | | 450–750 mm. | XXVI |
| | | | | 750–1100 mm. | XXVII |
| | | | | Above 1100 mm. | XXVIII |
| | | | Blunt | Under 450 mm. | XXIX |
| | | | | 450–750 mm. | XXX |
| | | | | 750–1100 mm. | XXXI |
| | | | | Above 1100 mm. | XXXII |
| Edible Podded. *Pisum sativum* sub sp. *saccharatum* | Round | Yellow | Pointed | 450–750 mm. | XXXIII |
| | | | Blunt | Above 1100 mm. | XXXIV |
| | Wrinkled | Green | Pointed | Above 1100 mm. | XXXV |
| Purple. *Pisum sativum* sub sp. *arvense* | | | | Over 1100 mm. | XXXVI |

Against each name is the Roman number of the group in which it falls.

| | | | |
|---|---|---|---|
| Abundance | XVIII | Cropper | XVIII |
| Abundant | XXVII | Cropwell | XXVI |
| Achievement | XXVIII | | |
| Admiral Beatty | XXVIII | Daisy | XXV |
| Advance | XXVI | Daffodil | XVIII |
| Advance Guard | XXVII | Deg Grace | XXXIII |
| Alaska | XV | Delicatesse | XXVII |
| Alderman | XXVIII | Doux Provence | IX |
| Allotment Holder | XVII | Dreadnought | XXXI |
| American Wonder | XXX | Dr. Kitchen | XXVI |
| American Wonder | | Droitwich Victory | XXVI |
| Improved | XXX | Droitwich Wonder | XXVIII |
| Annonay | V | Duchess | XXVIII |
| Aristocrat | XXVII | Duke of Albany | XXVIII |
| Autocrat | XXXI | Duke of York | XXVIII |
| Aviator | XII | Duplex | XXVI |
| | | Dwarf Alderman | XXVI |
| Banqueter | XXVIII | Dwarf Champion | XXVI |
| Bantam, The | XVIII | Dwarf Defiance | XXVI |
| Beaconsfield | XXVII | Dwarf Exhibition | XXVI |
| Bedford Champion | XXVIII | Dwarf Prolific | XVIII |
| Benefactor | III | Dwarf Quite Content | XXVI |
| Best of All | XXVII | Dwarf Sugar | XXXIII |
| Better Times | XXVI | Dwarf Wonder | XVIII |
| Big Ben | XIV | | |
| Borderer | XXXI | Earl Grey | IV |
| Boston Unrivalled | XXVIII | Earl Haig | XVII |
| Bountiful | XI | Earle King | XXX |
| Breck, The | XXVII | Earliest of All | XV |
| British Lion | XII | Early Bird | VII |
| | | Early Custom | XXXI |
| Cambria | XXVI | Early Dawn | XI |
| Canners Perfection | XXXI | Early Dwarf | IX |
| Caractacus | VI | Early Giant | XX |
| Carrington Gem | XXVII | Early Giant Marrowfat | XXVIII |
| Carrington Perfection | XXXI | Early June | XXVI |
| Carrington Wonder | XXVI | Early Market | XXIV |
| Certain Satisfaction | XVII | Early Marrowfat | XVII |
| Challenger | XII | Early Morn | XX |
| Champion Marrowfat | XXVII | Early Onward | XXX |
| Chancelot | XXVII | Early Prolific | XXX |
| Charles Ist | XXVII | Early Superb | X |
| Cheltonian | XXVII | Early Wonder | XXVI |
| Chemin Long | VI | Eclipse | XV |
| Clipper | VII | Eight Weeks | I |
| Clucas, The | XV | Emerald | XXVIII |
| Comet | XXVI | Emigrant | XIV |
| Commonwealth | XXVI | English Wonder | XXX |
| Contentment | XXVII | Essex Star | IV |
| Continuity | XXXI | Essex Wonder | XXVIII |
| Coronation | XXVIII | Eureka | XXVI |

| | | | |
|---|---|---|---|
| Everbearing | XXVI | Kelvedon Maincrop | XXVI |
| Everest | XXII | Kelvedon Monarch | XXXI |
| Evergreen | XXVIII | Kelvedon Perfection | XXVI |
| Evergreen Gem | XXVIII | Kelvedon Spitfire | XXV |
| Evesham No. I | IX | Kelvedon Standby | XXVI |
| Exceptional | XXVI | Kelvedon Triumph | XXVI |
| Exhibition | XXVII | Kelvedon Wonder | XXVI |
| Exquisite | XXIV | King Edward | XXVI |
| Extra Early | III | King George | XXVIII |
| Favourite | XXVI | King Marrowfat | XX |
| Feltham Advance | XIV | King of Dwarfs | XXXI |
| Feltham First | IX | Lancashire Lad | XXVI |
| Feltham Forward | XV | Lancashire Lady | XXVI |
| Fenland Wonder | XXVI | Lancashire Pride | IX |
| Field Marshall | XXVI | Last of All | XXVII |
| Fillbasket | X | Late Duke | XX |
| Fin des Gourmets | IX | Late Queen | XXXI |
| First Early | VI | Lathom Wonder | XXVI |
| First of All | III | Laxtonian | XVIII |
| Foremost | XI | Laxton's Progress | XVII |
| Forerunner | XI | Laxton's Superb | X |
| French Canners | III | Leader | XXVI |
| Giant Laxtonian | XVIII | Liberty | XXVII |
| Giant Stride | XXVI | Lincoln, The | XVIII |
| Giant Sugar Pod | XXXIV | Little Giant | XXVI |
| Gilt Edge | XXVII | Little Hero | IX |
| Gladiator | X | Little Marvel | XXX |
| Gladstone | XXVII | Local Lady | IV |
| Gradus | XX | Loyalty | XVIII |
| Greenfeast | XVIII | Lord Chancellor | XXVII |
| Greenmantle | II | Magnum Bonum | XXVII |
| Gregory's Surprise | XXXI | Maid of Kent | III |
| Half-a-League | XXXI | Maincrop (Suttons) | XXVII |
| Harbinger Improved | XXII | Maincrop Marrowfat | XXVIII |
| Harrison's Glory | XIV | Manchester Man | XXVI |
| Harvestman | XXVIII | Manifold | XVIII |
| Highland Laddie | XXVI | Market Gem | VI |
| Holdfast | XXXII | Market Pride | X |
| Hundredfold | XVIII | Market Wonder | XXX |
| Hurst's 40 | IX | Marvellous | XXX |
| Ideal Dwarf | XXVI | Masterpiece | XVIII |
| Improved Harbinger | XXII | Matchless Marrowfat | XXVIII |
| Impudence | XVII | Meteor | IX |
| International | XX | Mid-Century | XVI |
| Invicta | XXVII | Miracle, The | XXVIII |
| Invincible Marrowfat | XVIII | Morse's Market | XXVI |
| Jap, The | IV | Munnings | XXVII |
| June Wonder | XX | My Own Marrowfat | X |
| Kelvedon Champion | XXV | New Surrey Star | XXVIII |
| Kelvedon Hurricane | XXVI | No. 457 | XXVI |
| | | Northumbria | XXVI |

| | | | |
|---|---|---|---|
| Olympic | XII | Stourbridge Marrow | XXVIII |
| Onward | XXXI | Stratagem | XXVI |
| Ormskirkian | XXVIII | Success | XVIII |
| Osmaston Pride | XXVI | Succession | XXVI |
| Osmaston Surprise | XXVIII | Successional | XI |
| | | Sugar Paramount | XXXIV |
| Perfected Freezer | XXXI | Sugar Tall | XXXV |
| Perfection Marrowfat | XXIV | Superlative | XXVII |
| Peter Pan | XVIII | Super Telegraph | XII |
| Petit Provençal | IX | Supremacy | III |
| Phenomenal | XXVI | Supreme | XXXI |
| Phenomenon | XXVI | | |
| Pilot | III | Tall Sugar | XXXV |
| Pioneer | XVIII | Taxpayer | XXVI |
| Poulwell | XVII | Telegraph | XII |
| P.P. | XXVII | The Bantam | XVIII |
| Premier | XXVI | The Breck | XXVII |
| Primo | III | The Clucas | XV |
| Prince Edward | XXVIII | The Jap | IV |
| Prince of Wales | XXIII | The Lincoln | XVIII |
| Priority | XVI | The Miracle | XXVIII |
| Prize | XXVIII | The Sherwood | XXVI |
| Profusion | XXIII | Thos. Laxton | XXIV |
| Progress No. 9 | XXV | Timperley Wonder | XVII |
| Provost | XXVIII | Tip Top | VII |
| Purple Podded | XXXVI | Tom Clucas | XI |
| | | Tomorrow | XI |
| Queen | XXVII | Tremendous | XX |
| Queen of The Marrowfats | XXVIII | Trophy | XXXI |
| | | | |
| Recorder | XXVII | Union Jack | XXVII |
| Referendum | XVIII | Unique | I |
| Rentpayer | XXVI | Universal | XXVI |
| Right Royal | XXVI | | |
| Rival | XVIII | Veitch's Perfection | XXXI |
| Rivenhall Wonder | XXVI | Victoria | VIII |
| Robust | XXVII | Victoria (Toogood) | XXXI |
| Roi des Conserves | XI | Victory Freezer | XXXI |
| Royal Favour | XXVII | Volunteer | XI |
| Royal Salute | XXVII | | |
| Royal Standard | XXVIII | What's Wanted | XI |
| | | William Hurst | XXVI |
| Senator | XXVII | William The First | III |
| Sensation | XVII | Witham Wonder | XXVI |
| Serpette d'Auvergne | III | Wireless | II |
| Shasta | XXXI | Wonder | XXVI |
| Sherwood, The | XXVI | Wonder Marrowfat | XVIII |
| Speed | VII | World's Record | XX |
| Springtide | VII | | |
| Standard | XXVIII | Yielder | XXVI |
| Steadfast | XXXI | Yorkshire Hero | XXXI |
| Stella | IV | | |
| St. George | XXVI | Zelka | XIV |

## (ii) Maize

Few plants have been more closely investigated from the genetic point of view than maize. Many genes affecting the 'seeds' (kernels) of maize are known and maize has the advantage that these kernels are all held together in a cob. These dried cobs can be kept for several years.

### Procedure for controlled pollination of maize

*Materials required*

An exercise book for recording pollination details
300–400 150 mm × 100 mm transparent paper bags
200 400 mm × 250 mm 1000 grade polythene bags (these can be washed and used again each year)
A sharp pocket knife
Stakes each bearing a maize family number
Paper clips
A chinagraph pencil
Support stakes and string

*Planting*

Seed is sown during the last week of April or the first week of May. A single maize seed is planted 350 mm apart from the next in rows which are 650 mm apart. Place a stake with the appropriate family number at the beginning of each family in the rows. Each family is limited to contain 20 plants.

*Covering of female flowers*

The female flowers appear about mid-July as a flat green sheath in the leaf axil (A1 fig. 2.4). Select only the female nearest the top of the plant (usually the least forward on the plant). Cut a slit on each side of the leaf base and, after ensuring that no stigmas or 'silks' have protruded, trim off the top half inch of the female flower. Immediately pull a transparent paper bag down over the female so that it is completely covered and the open end of the bag is securely tucked down in the slit between leaf base and stem (A2 fig. 2.4).

Check the plants at two or three day intervals to cover the topmost females as they appear and trim off any excessive sheath growth on those previously trimmed. Make sure that all are properly covered by the bags.

*Pollen collection*

The male flowers are at the top of the plant and are carried on a number of smaller stems. When pollen shedding begins on the required male, enclose as many of the flowers as can be easily contained in a transparent paper bag, fold over the excess opening, and secure with a paper clip (B fig. 2.4).

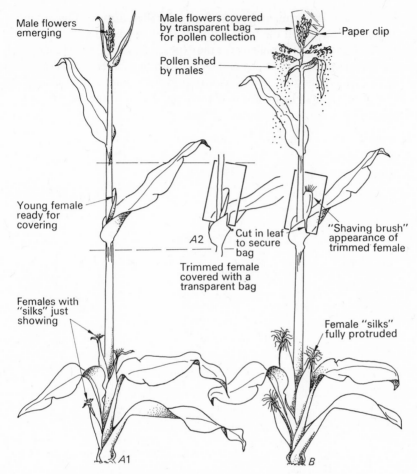

**Fig. 2.4.** Pollination techniques in maize (for explanation see text).

To collect a supply of pollen, cover the males (provided they are dry and shedding pollen freely) as described above during mid-morning and let the pollen accumulate in the bag until the afternoon.

*Pollination*

From the pollinating programme check the details of the pollination about to be made. Collect the pollen from the male flower by *carefully* bending the plant over (C fig. 2.5) and gently tap the bag so that a small heap of pollen is gathered. Keeping the bag in the same horizontal position, very carefully withdraw the inflorescence, leaving the pollen behind. Carry the bag of pollen (still horizontally) to the required female, tuck the top half of the plant behind

the free (usually left) arm, and when 'foreign' pollen disturbed by the move-
ment has cleared, rip off the top portion of the bag covering the female (C).
Quickly dust the collected pollen over the exposed female (C) and immediately
cover with a large polythene bag and secure with two paper clips (D fig. 2.5).

Write the pollination details on the polythene bag with a chinagraph
pencil (female × male) and also note the same details in the record book.

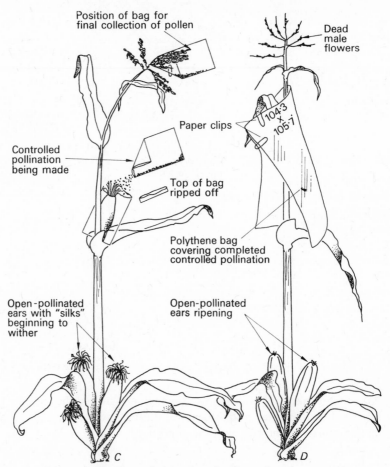

**Fig. 2.5.** Pollination techniques in maize (for explanation see text).

*Ripening and harvest*

Ripening time varies and is dependent on the weather conditions. All pollina-
tions should be completed by mid-August and covered ears left to ripen.
Stake and tie any plants that have been blown over by the wind. Harvesting
can begin in mid-October if the late summer and autumn have been warm and

sunny, but in bad conditions the ears may not be gathered until early November.

Detach the ear from the plant, strip off the sheath leaves and put back into its polythene bag for transport to the laboratory.

## Storage

In the laboratory, remove the ear from its bag and attach a label bearing the pollination details. Lay the ears out on a suitable rack or bench (near a radiator if possible) and leave to dry and harden. Check the ears from time to time in case any fungal growth starts (smaller patches of contamination can be eradicated by dipping the ear into 2 per cent lysol solution and putting out to dry again). The ears should be dried and hard after about six weeks and must be stored in dry, moth-proof containers (plastic bags sealed at the top or tins). Care should be taken to see that the ears are really dry before storage.

It should be noted that the maize plant tends to be rather brittle in all its parts so that all handling must be done with care.

## Source of materials

A very wide variety of corn cobs is available from Carolina Biological Supply Company, Burlington, N. Carolina 27215, U.S.A.
Purple: yellow, 3 : 1 and 1 : 1 (aleurone colour)
   Also available from: Faust Scientific Supply Co, Madison, Wisconsin 53713; and Genetic Products and Services, Goshen, Indiana 46526.
Smooth: shrunken, 3 : 1 and 1 : 1 (seed texture)
   Also available from: Faust Scientific Supply Co.
Starchy: sweet, 3 : 1 and 1 : 1 (endosperm texture)
White: purple, 3 : 1 (dominant colour inhibitor)
Starchy: waxy, 3 : 1 (endosperm character, stains with iodine)
Purple: yellow/starchy: sweet, 9 : 3 : 3 : 1
Purple: white/smooth: shrunken, 9 : 3 : 3 : 1, 1 : 1 : 1 : 1
   From Faust Scientific Supply Co; Genetic Products and Services; and Connecticut Valley Biological Supply Co Inc, Southampton, Massachusetts 01073.
Yellow: white/starchy: sweet, 9 : 3 : 3 : 1
Purple: red/yellow: white, 9 : 3 : 3 : 1
A range of cobs illustrating various types of gene interaction giving ratios of 12 : 3 : 1, 9 : 3 : 4, 9 : 7, 13 : 3.
Two point linkage: anthocyanin and shrunken endosperm on chromosome III
Three point linkage: purple, shrunken, waxy on chromosome IX
Four point linkage: as with three point linkage, but addition of yellow/green seedling factor.
Large numbers of these cobs are fairly expensive to buy, but they can be obtained by crossing and then easily preserved.

The following mutants are available on request from Dr H.L.K.Whitehouse, Botany School, Cambridge.

*Tu:* tunicate (bracts surrounding grains)

*P:* various alleles for pericarp colour (purple anthocyanin pigmentation of pericarp)

*A, C, R:* genes for aleurone pigmentation

*sh:* shrunken endosperm

*y:* white endosperm

*wx:* waxy endosperm and pollen

*su:* sugary endosperm and wrinkled seed

*Dt:* dotted (causes mutation of gene *a* to *A* for aleurone pigmentation)

*C* and *sh* are on the same chromosome with 7 per cent crossover between them.

A wide range of mutants including those mentioned above are available from:
Professor R.J.Lambert, Plant Breeding, College of Agriculture, University of Illinois, Urbana, Illinois 61801; and Mr Clarion Henderson, Illinois Foundation Seeds, Champaign, Illinois.

### $F_2$ maintained as seed (seedling characters)

A wide range of seedling characters are known which are suitable for demonstrating genetic segregation (table 2.2). Seeds of all the types mentioned germinate readily at room temperature or in the greenhouse and the seedlings are ready for inspection in 2–3 weeks. The seed can be sown in soil, on moist filter paper or on agar.

For simple one gene segregation all the mutants can be mixed in the appropriate ratio, and for some other seedling characters progeny segregating for a two-gene difference are available on the market.

It does not seem the place here to mention the crossing techniques for this wide range of plants, but if further information is wanted on this it can be had by writing to the sources of the material, the people mentioned in the references, or to horticultural and agricultural breeding stations (addresses available from local agricultural advisory services).

**Table 2.2(a)** Seedling characters

| Species | Character pair (* indicates dominant) | | Other comments | Source or trade name | References |
|---|---|---|---|---|---|
| Tomato (*Lycopersicum esculentum*) | (a) (i) | Red*—green hypocotyl | | L.K.Crowe | (1)(2)(3) (4)(5) |
| | (ii) | Purple*—green (*A, a*) hypocotyl | Chromosome II | P.G.L.† 5 | |
| | (iii) | Purple*—green (*Bls, bls*) | Chromosome III | P.G.L. 2, P.G.L. 6, P.G.L. 9 | |
| | (iv) | Purple*—green (*Ah, ah*) hypocotyl | Chromosome IX | P.G.L. 6 | |

Table 2.2(a).—*continued*

| Species | Character pair (* indicates dominant) | Other comments | Source or trade name | References |
|---|---|---|---|---|
| | (b) (i) Normal cut*—potato leaf | (leaflets entire in potato leaf type) | L.K.Crowe | |
| | (ii) Normal cut*—potato (*C,c*) leaf | Chromosome VI | P.G.L. 5 | |
| | (iii) Normal cut*—potato (*Sf, sf*) leaf | Chromosome III | P.G.L. 9 | |
| | (c) (i) White—green (*Xa*-2, *xa*-2) cotyledons (het: yellow green) | White seedlings die soon after germination Chromosome X | P.G.L. 4 | |
| | (ii) White—green (*Xa*-3, *xa*-3) cotyledons (het: yellow green) | Chromosome III | P.G.L. 8 | |
| | (d) Normal*—yellow (*Sy, sy*) cotyledons | *sy sy* has yellow leaf bases Chromosome III | P.G.L. 9 | |
| | (e) Hairy*—hairless (*Hl, hl*) stem and hypocotyl | Chromosome XI | P.G.L. 10 L.K.Crowe | |
| | (f) Normal—lanceolate cotyledons (het: intermediate) | Lanceolate lethal after germination | L.K.Crowe | |
| Radish (*Raphanus sativus*) | (a) Red—white hypocotyl (*F₁*: purple) | Gives 9 : 3 : 4 ratio in *F₂* not 1 : 2 : 1 as often stated | 'Scarlet globe' and 'Icicle' | (1) (6) |
| | (b) Round—long hypocotyl (*F₁*: intermediate) | | 'Scarlet globe' and 'Icicle' | (1) |
| Wallflower (*Cheiranthus cheiri*) | Red*—white hypocotyl | Correlated with red and yellow flowers | 'Vulcan', 'Cloth of Gold' | (1) (7) |
| Tobacco (*Nicotiana tabacum*) | Green*—white seedlings | | Harris Biological Supplies Ltd T.Gerrard & Co Ltd | |

c

**Table 2.2(a)**—*continued*

| Species | Character pair (* indicates dominant) | | Other comments | Source or trade name | References |
|---|---|---|---|---|---|
| Barley (*Hordeum vulgare*) | | Green*—white seedlings | | Harris Biological Supplies Ltd Department of Education and Science Laboratories | |
| Blackberry (*Rubus idaeus*) | | Glandular*— eglandular cotyledons | Tetraploid *GG gg* giving approximately 21 : 1, correlated with thorny/thornless stems | John Innes | (1) (8) |
| Stock (*Matthiola incana*) | | Dark*—light green cotyledons | Linkage with (i) pollen lethal gives 1 : 1 (ii) alleles for single and double flowers, maximum expression 9°C. | Hansen's varieties | (6) (12) |
| Snapdragon (*Antirrhinum majus*) | (a) | Straight*—twisted 50 mm stems | | L.K.Crowe | |
| | (b) | Green*—yellow first leaf | | L.K.Crowe | |
| | (c) | Round*—narrow first leaf | | L.K.Crowe | |
| Nasturtium (*Tropaeolum majus*) | | Green*—variegated first leaf | | L.K.Crowe | (6) |
| Wintercress (*Barbarea vulgaris*) | | Green*—variegated first leaf | | L.K.Crowe | (9) |
| *Pelargonium zonale* | | Green—white seedlings. Het: gold | White sub-lethal | 'Golden Crampel' | (7) |
| Marrow Stem Kale (*Brassica oleracea* var. *acephalia*) | | Red—green hypocotyl. Het: purple | Linked with incompatibility gene. Maximum expression 19°C | L.K.Crowe | (10) |

Table 2.2(a)—*continued*

| Species | Character pair (* indicates dominant) | Other comments | Source or trade name | Reference |
|---|---|---|---|---|
| Maize (*Zea mays*) | Lethal seedling traits: albinos, lemon whites, leuteous, Necrotic non-lethals: dwarfs, virescents, yellow greens, pale greens | (Temperature sensitive) | R.J.Lambert | (11) |

(1) CROWE L.K. (1963) Seedling characters. In Darlington C.D. & Bradshaw A.D. (Eds) *Teaching Genetics in School and University*, pp. 90–93. Oliver & Boyd.

(2) *Practical Plant Genetics*. Software accompanying tomato seed. 18 Harsfold Road, Rustington, Sussex.

(3) RICK C.M. & BUTLER L. (1956) Cytogenetics of the tomato. *Adv. Genet.*, **8**, 267–382. (List of genes affecting seedling characters p. 278.)

(4) CLAYBERG C.D., BUTLER L., RICK C.M. & YOUNG P.A. (1960) Second list of known genes in tomato. *J. Hered.*, **51**, 167–174.

(5) *Tomato Genetics Cooperative Reports*. Chairman of coordinating committee: RICK C.M., Department of Vegetable Crops, University of California, Davis, California.

(6) CRANE M.B. & LAWRENCE W.J.C. (1954). *The Genetics of Garden Plants*, 4th Ed. Macmillan, London.

(7) BATEMAN A.J. (1956) Cryptic self-incompatability in the wallflower: *Cheiranthus cheiri* L. *Heredity*, **10**, 257–261.

(8) CRANE M.B. & DARLINGTON C.D. (1932) Chromatid segregation in tetraploid *Rubus*. *Nature, Lond.*, **129**, 869.

(9) THOMPSON K.F. (1962) Breeding marrow-stem kale (*Brassica oleracea* var. *acephalia*). *Heredity*, **17**, 598.

(10) TILNEY-BASSETT R.A.E. (1963) Genetics and plastid physiology in *Pelargonium*. *Heredity*, **18**, 485–504.

(11) *Maize Genetics Cooperative Newsletter*.

(12) FISHER R.A. (1933) Selection in the production of the eversporting stocks. *Ann. Bot.*, **47**, 727–733.

Table 2.2(b) Ready made seedling crosses

| Tomato | | |
|---|---|---|
| P.G.L.† 2—(a) (iii) | $P_1, P_2, F_1, F_2, B_1$ | $F_2$ ratio 3 : 1 |
| P.G.L. 10—(e) | $P_1, P_2, F_1, F_2, B_1$ | $F_2$ ratio 3 : 1 |
| P.G.L. 4—(c) (i) | $P_1, P_2, F_1, F_2, B_1$ | $F_2$ ratio 1 : 2 : 1 |
| P.G.L. 5—(a) (ii) and (b) (ii) | $P_1, P_2, F_1, F_2, B_1$ | $F_2$ ratio 9 : 3 : 3 : 1 |
| P.G.L. 6—(a) (iii) and (a) (iv) | $P_1, P_2, F_1, F_2, B_1, B_2$ | $F_2$ ratio 9 : 7 |
| P.G.L. 8—(c) (ii) | $P_1, P_2, F_1, F_2, B_1$ | $F_2$ ratio 3 : 6 : 3 : 1 : 2 : 1 |
| P.G.L. 9—(a) (iii), (b) (iii) and (d) | $P_1, P_2, F_1, F_2, B_1$ | $F_2$ ratio loose linkage |

**Table 2.2(b)**—*continued*

Tobacco
Harris Biological Supplies Ltd ⎫
Faust Biological Supply Co  ⎬  $F_2$ green and albino, segregates 3 : 1 at seedling stage
Carolina Biological Supply Co ⎭

Maize
Connecticut Valley Biological Supply Co Inc ⎫
Genetic Products and Services, Indiana  ⎬  $F_2$ ratio 3 : 1 green : albino
Faust Biological Supply Co  ⎱  $F_2$ ratio 3 : 1 tall : dwarf (primary leaf rounded, wider
Carolina Biological Supply Co ⎰    and shorter)
Genetic Products and Services, Indiana $F_2$ 9 : 7 green : albino seedlings

Soybeans
Faust Biological Supply Co  ⎱  $F_2$ ratio for green/yellow cotyledons 3 : 1, leaf
Carolina Biological Supply Co ⎰  colour 1 : 2 : 1 green light-green yellow.
Connecticut Valley Biological Supply Co Inc
Genetic Products and Services

† P.G.L. stands for Plant Genetics Laboratory (see *Practical Plant Genetics*, p. 54).

## 3. SEGREGATION DEMONSTRATED IN PRESERVED MATERIAL

The use of preserved material to demonstrate segregation over several generations of crossing is familiar to most teachers of genetics. Material may be preserved from crosses done in previous years or obtained from a supply agency. The importance of trying to retain as much material as possible, either in preserved form or as photographs, cannot be over-emphasized. Elementary genetics in higher organisms other than *Drosophila* is either time consuming or a long term study, and preserving material for future years helps to demonstrate how widespread the laws of genetics are, and provides further examples for students to analyse.

This approach has a grave danger of being extremely dull, particularly if it is the only one that is available, but several precautions and some small effort can make it only a little less rewarding than getting the students themselves to perform the crosses.

Firstly, the material must be kept in a scrupulously clean and new-looking condition. Fragmented herbarium specimens and dusty maize cobs are not the most inspiring objects. If necessary the material should be remounted and cleaned each year.

Secondly, students should be encouraged to ask questions about the material, to dissect it and to examine it microscopically. Is the difference between round and wrinkled peas just in the seed coat? Do they taste different? What does the agouti factor of mice look like when a few hairs are put under the miscroscope? The ruptured anther of a $Wx/wx$ maize plant when stained in iodine in a watchglass, is an impressive sight when seen under a binocular microscope: most students will never have seen the 'release' of pollen 'laden with genes' demonstrated in such a dramatic fashion.

Thirdly, demonstrations are more attractive if the parental strains are available to show what they are like when alive. For most organisms this should provide no great difficulty and certainly adds interest to the 'musty demonstration sheet'.

### (a) Seed characters

Characters which show themselves in the seed are easily preserved simply by keeping the seed in reasonably dry conditions, and using them when required. Suitable characters have already been described in the Pea (p. 43) and in Maize (p. 48).

### (b) Pressed plants

Segregating progenies of plant material can be preserved on herbarium sheets as dried and pressed specimens. Distinct morphological differences are preferable but pigment differences are usually suitable. It is an advantage if the plants are small since a large sample can then be shown on only a few sheets: either seedlings (p. 56) or small adults such as *Arabidopsis* (p. 30) are suitable.

Plants are best pressed between sheets of blotting paper, either using weights (books or bricks) on top, or strapped between flat wire cake stands or wooden boards, with rope or two belts. The blotting paper is best changed after about two days, and if a less absorbent material like newspaper is used then changes should be made every two or three days for the first week. Two weeks is normally sufficient time for the specimens to dry.

### (c) Mounted insects

Insects, provided they are not too small, such as *Drosophila* can be mounted and generally retain their colours for many years.

They may be pinned (when still fairly fresh) directly on to cork mounting, or pinned or stuck on a stage (thin card or Bristol board) which is then pinned to the cork. Alternatively, the insects may be stuck directly on to white card. The most suitable adhesive is made by dissolving celluloid pieces in amyl acetate to produce a viscous glue (just does not drip off a rod). This is sufficient to hold the insect in place, but dries slowly so permitting manipulation of the insect. Model aeroplane cement dries quickly, but otherwise is suitable, as is colourless nail varnish. Care should be taken not to use too much adhesive. If the insect becomes deformed (e.g. wings emerge from elytra) on etherization, then it can be frozen prior to etherization.

### Suitable material

#### (i) Tribolium mutants (p. 34)

#### (ii) Ladybird (Ladybeetles) colour morphs

Ladybeetles show polymorphisms usually determined by multiple allelic series. *Adalia bipunctata* (two spot ladybird) has a wide range of types (see Creed 1966). The dominance order of this multiple allelic series is shown in fig. 2.6. The dark forms—mainly black with one, two, or three red spots per elytra are dominant to the red forms with one or two spots per elytra. One 'red' form is predominantly black with five red spots, but this form is extremely rare in Great Britain.

*Adalia decempunctata* (ten spot ladybird) has three main forms, determined by an allelic series with the dominance order as follows: *typica* dominant to 10-*pustulata*, dominant to *bimaculata*. *Typica* is the darkest form with only a small red area, 10-*pustulata* is the normal ten spot form, while *bimaculata* resembles the two spot ladybird. Ladybeetles are commonly found on lime trees (*Tilia*) and nettles (*Urtica*), the darker forms being commoner near town and industrial centres.

#### (iii) Lepidopteran polymorphisms

The butterflies and moths show a fairly wide range of polymorphisms, the best known being the melanic/non-melanic types of moths (e.g. *Biston betularia*). Melanism is usually dominant to non-melanism. Contrasting morphs can be collected and arranged to illustrate the inheritance of the character both as a demonstration of elementary genetics and as an introduction to population genetics.

Further information on genetics and evolution in Lepidoptera can be found in Ford (1953, 1967, 1971) and Robinson (1971).

Clearly the principles outlined here for a few insect groups are applicable to other insects and arthropods. If anyone has a particular interest in a polymorphic group then this can be turned to good account for demonstrating the genetics of that character.

### (d) Preserved skins

Preserved skins are a useful way of demonstrating segregation for factors determining coat colour in mice and other organisms such as rats or guineapigs. These skins can either be given to the student as such, or pinned out on to a board. The preservation of interesting features of crosses (e.g. variable expression of genes for coat colour patterns) is therefore possible, as well as the more standard type of demonstration.

**Fig. 2.6.** Some varieties of the two spot ladybird, *Adalia bipunctata:* 1. *sublunata*, 2. *quadrimaculata*, 3. *sexpustulata*, 4. *typical*, 5. *stephensi*, 6. *rubiginosa*, 7, *unifasciata* 8. *annulata*, 9. 12-*pustulata*. (From Creed 1966.)

Several techniques are available, and the following account is taken from Mahoney (1966).

(i)   Remove skin carefully with minimum amount of attached subcutaneous tissue.

(ii)   Fix in 5 per cent formalin for one week.

(iii)   Wash in running water for several hours.

(iv)   Stretch out and pin to board with hair face down.

(v)   While still wet, rub Lankroline FP4 (Lankro Chemicals, Eccles, Manchester) vigorously into skin. Lankroline does not affect human skin.

(vi)   Scrape away (with blunt instrument, e.g. table knife) any fat or subcutaneous tissue, apply more Lankroline and repeat till clean.

(vii)   Rinse skin in warm detergent (e.g. Teepol) solution.

(viii)   Pin out on board, fur uppermost, allow to dry and brush hair thoroughly when dry.

Skins prepared by this method are very soft.

### (e) Deep freezing and freeze drying

Small mannals such as mice can be easily preserved by putting them into a deep freeze ($-20°$ to $-30°C$) in suitable containers (refrigerator boxes) between layers of cellulose wadding or tissue paper. After preservation they are dry thawed using silica gel crystals. Silica gel crystals are dried either in an oven for 3 hours at 60°C or in a warm cupboard overnight. They are ready to use when deep blue.

Mice are dry thawed as follows. The day before the demonstration, half the mice in a box are transferred to another box; a tray (made, e.g. of blotting paper) of silica gel crystals is placed on the mice in each box. The lid is replaced tightly and the boxes kept overnight in a refrigerator or cold room. The next day the silica gel (now pink) is replaced with fresh crystals, and an hour before the class the box is put at room temperature. As the mice reach room temperature, they are spread out on to blotting paper and if necessary dried further with a dry cloth. The mice should appear similar to freshly killed material and can be gently handled by the class. As soon as they are finished with, they should be put into the deep freeze in the same way as before.

*Sources of materials*

Silica gel crystals can be obtained from most biological or chemical supply agencies.

*Reference*

Nuffield Biology Teachers' Guide, V. *The Perpetuation of Life*, pp. 8–10.

   Freeze drying is becoming increasingly used for drying birds, mammals and other soft-bodied animals which shrink if dried normally. The technique is a very useful one but quite elaborate apparatus is required and is best used if the apparatus is already available. (see Mahoney 1966).

## (f) Photography

An alternative to making permanent specimens is to take photographs (preferably coloured) of groups of progeny as they become available: these can then be displayed in place of demonstration material. The main advantage of photographs is that they are the quickest and probably in the long run the cheapest form of preservation.

## (g) Other methods and materials

### (i) Starchy and waxy pollen grains of maize

A useful demonstration in maize is that of genes segregating to produce two types of pollen grains or 'gametes'. The waxy mutant of maize (*wx*) alters the chemical nature of the starch produced by the plant so that instead of staining the normal blue black it stains a reddish colour with iodine. (Crystals dissolved in 70 per cent alcohol to give light brown solution.) This gene not only expresses itself in the adult plant but also in the pollen grain, where the carbohydrate metabolism is determined by the genotype of the pollen grain (not by the parental anther). Plants heterozygous for this gene (*Wx/wx*) therefore segregate into two types of pollen grain, half the pollen grains staining red and the other half staining blue with iodine. There are a few inviable pollen grains without cytoplasmic contents and these appear colourless.

A very illuminating practical exercise involves giving the students the genotype of the parent plant, explaining the nature of the waxy gene and then asking them to explain what they see when a squash of the anther is made in dilute iodine solution. One anther per student is adequate.

*Sources of materials*

Seeds or anthers of *Wx/wx* plants from:
Dr H.K.L.Whitehouse, Botany School, Cambridge University.
Carolina Biological Supply Company, Burlington, North Carolina 27215.

### (ii) Flower colour pigments

If flowers of several common garden species are dried rapidly (e.g. in a drying oven at 30°C), their pigment types are preserved so that they can be assessed either visually or by chromatography. One of the best examples is *Antirrhinum* the genetics of which has been studied for over 50 years.

*Antirrhinum*

The flower colours of *Antirrhinum* are analysed as follows:

(i)   Grind upper lips of two corollas in a test tube with 1 ml of 1 per cent HCl in ethyl alcohol. One of the pigments, aureusidin, is invariably present as a spot on the lower lip, and only the upper lip shows genetic variation for its presence or absence.

(ii)   Put four drops of the extract as a spot on some chromatography paper (e.g. Whatman No. 1), allowing one drop to dry before applying the next.

(iii)   Run chromatogram with one of the following solvents

(1)   butyl alcohol : acetic acid : water, in ratio 6 : 1 : 2

(2)   *m*-cresol : acetic acid : water, in ratio 50 : 2 : 48.

Any method can be used (either ascending or descending) as long as the chromatogram is run in an airtight container. A good method is to spot a large sheet of paper with many different extracts, clip the edges to form a cylinder and stand in the solvent at the bottom of a suitable airtight container such as a large jar or saucepan with tight fitting lid.

(iv)   From such chromatograms the pigments can be recognized by their colour. If the chromatogram is exposed to ammonia vapour the spots change to various colours and this helps their identification. In addition, the $R_F$ value (distance from starting point to centre of spot divided by the distance from starting point to solvent front) can be used to identify a spot. The larger the $R_F$ value the further the spot travels. Table 2.3 shows colours and $R_F$ values.

**Table 2.3**

| Pigment | Colour | | $R_F$ value | |
|---|---|---|---|---|
| | No treatment | NH$_3$ vapour | Butanol solvent | *m*-Cresol solvent |
| *Anthocyanins* | | | | |
| Cyanidin | Magenta | Blue | 0·32 | 0·28 |
| Pelargonidin | Pink | Mauve | 0·46 | 0·56 |
| *Flavonols* | | | | |
| Aureusidin | | | | |
| (two spots) | Green | Orange | 0·07 and 0·23 | 0·01 and 0·11 |
| Luteolin | Brown | Yellow | 0·48 | 0·19 |
| Apigenin | Dark yellow | Yellow | 0·64 | 0·36 |
| | | Green | 0·29 and 0·41 | 0·21 and 0·43 |

The pigments with an asterisk in table 2·3 also appear on the chromatograms but are of secondary interest. Apigenin with $R_F$ value in butanol of
0·64 is always found. Other apigenins are found always with pelargonidin.
Occasionally there are traces of pigments not mentioned here.

The following genes are responsible for pigment production.

*Y/-:* pigment produced, depending on other genes.

 *yy:* no pigment, irrespective of other genes.

*R/-:* cyanidin and pelargonidin formed.

 *rr:* anthocyanins absent.

*B/-:* (with R present) cyanidin only.

 *bb:* (with R present) pelargonidin only.

$I_A$-*:* aureusidin produced.

 *II:* no aureusidin.

The following scheme for pigment synthesis in *Antirrhinum* can be
suggested, although students can be invited to propose their own schemes.

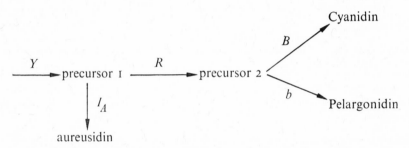

The varieties in table 2.4 (obtainable from Carter's Seeds Ltd, Raynes
Park, London) are suitable although clearly other varieties could be investigated.

**Table 2.4**

| Variety | Genes | Pigments |
|---|---|---|
| White | $r\,B\,I$ | None |
| Yellow | $r\,B\,I_A$ | Aureusidin |
| Nelrose (glowing pink) and Royal Cerise (rich carmine) | $R\,b\,I$ | Pelargonidin* |
| Scarlet Flame (fiery scarlet) and Guinea Gold (orange terracotta) | $R\,b\,I_A$ | Aureusidin and pelargonidin* |
| Mauve Beauty | $R\,B\,I$ | Cyanidin |
| Crimson | $R\,B\,I_A$ | Aureusidin and cyanidin |

* Where two colour varieties are grouped under one genotype there are differences in
intensity of expression of the colours determined by modifier genes. Royal cerise has more
pelargonidin than Nelrose, and Scarlet Flame has more pelargonidin but less aureusidin
than Guinea Gold.

Another single gene colour mutant type is 'delila'—'delila' varieties have a colourless or yellow corolla tube due to the absence of anthocyanidins in this region.

Crosses between varieties of *Antirrhinum* are easily made, since the anthers are large and easily removed from flowers just before they open. Crossing is prevented by bagging with fine muslin to exclude insects. Pollen can be transferred with a brush and one pollination produces numerous seeds.

Suggested crosses:

1  Single factor—$R\,b\,I \times R\,B\,I$, $R\,b\,I \times R\,b\,I_A$, $r\,B\,I \times R\,B\,I$, $r\,B\,I_A \times R\,B\,I_A$ and pure white × coloured, delila (colourless corolla tube) × coloured corolla tube

2  Two factors—$R\,b\,I \times R\,B\,I_A$, $R\,b\,I_A \times R\,B\,I$

3  Two factors with epistasis—$r\,B\,I \times R\,b\,I$, $r\,B\,I_A \times R\,b\,I_A$

4  Modifier complex—Nelrose × Royal Cerise, or Scarlet Flame × Guinea Gold.

## References

DAYTON T.O. (1956) The inheritance of flower colour pigments I. The genus *Antirrhinum*.
   *J. Genet.*, **54**, 249–260.
BRADSHAW A.D. (1963) Three teaching projects. In Darlington C.D. & Bradshaw A.D.
   (Eds) *Teaching Genetics in School and University*, pp 105–109. Oliver & Boyd.
WAGNER R.P. & MITCHELL H.K. (1964) *Genetics and Metabolism*, 2nd Ed, pp. 565–583.
   Wiley.

## Sources of materials

*Antirrhinum* colour varieties are obtainable from practically all nurserymen. Varieties quoted available also from Carter's Tested Seeds Ltd, Raynes Park, London.

## 2. *Impatiens balsamina*

Three genes are responsible for colour variation of the flowers of Balsam, but they affect the sepals and the petals in different ways. Another gene $W$ causes cream sepals in its recessive form ($w$), but this will not be dealt with in any detail.

The situation regarding the petal and sepal pigments is summarized in table 2.5. With regard to visible pigments, the following situation exists: in the petals when $L$ is present, malvidin is formed. The gene $H$ is responsible for the production of pelargonidin, but traces are also produced in $l\,h\,P^g$, and $l\,h\,P^r$ flowers. The gene $P$ determines the amount of pigment, this increasing with $p \rightarrow P^g \rightarrow P^r$. $P^r$ is dominant to $P^g$. In the sepals the situation is more complex. The gene $L$ again leads to the production of malvidin, but $L$ is epistatic to *H* such that *H* produces pelargonidin only in the genotype *ll*. In the absence of *L*, *Pg* plants produce peonidin, while *Pr* plants produce peonidin and cyanidin.

Table 2.5 Anthocyanin pigments in the sepals and petals of *Impatiens*

| | Sepals | | | | | Petals | | |
|---|---|---|---|---|---|---|---|---|
| Genes | Phenotype | PEL | CYA | PEO | MAL | Phenotype | PEL | MAL |
| *l h p* | White | — | — | — | — | White | — | — |
| *l h Pg* | Pale pink | — | — | + | — | Pale pink | + | — |
| *l h Pr* | Pink | — | + | + | — | Pink | + | — |
| *l H p* | White | — | — | — | — | Pink | + | — |
| *l H Pg* | Pink | + | — | + | — | Rose | + | — |
| *l H Pr* | Red | + | + | + | — | Red | + | — |
| *L h p* | White | — | — | — | — | Pale lavender | — | + |
| *L h Pg* | Lavender | — | — | — | + | Lavender | — | + |
| *L h Pr* | Purple | — | — | — | + | Purple | — | + |
| *L H p* | White | — | — | — | — | Rose lavender | + | + |
| *L H Pg* | Pink lavender | — | — | — | + | Pink lavender | + | + |
| *L H Pr* | Magenta | — | — | — | + | Magenta | + | + |

The above results are better understood if one looks at the chemistry of the anthocyanin compounds concerned.

Anthocyanins have the following basic structure:

The different types of anthocyanidins are distinguished according to the nature of the groups at *R′ R″ R‴* (table 2.6).

Table 2.6

| Pigment | *R′* | *R″* | *R‴* | Phenotype |
|---|---|---|---|---|
| Pelargonidin | H | OH | H | Red |
| Cyanidin | H | OH | OH | Blue |
| Delphinidin | OH | OH | OH | Purple |
| Peonidin | $OCH_3$ | OH | H | Rosy |
| Petunidin | $OCH_3$ | OH | OH | Purple—not present in *Impatiens* |
| Malvidin | $OCH_3$ | OH | $OCH_3$ | Mauve |

Because young buds and stems contain precursors of the anthocyanins cyanidin and delphinidin, these are probably the substances from which the other pigments are derived. We can therefore postulate the following schemes for pigment synthesis but students can be invited to formulate their own models. In actual fact the situation is probably more complex than the outline given here.

Petals

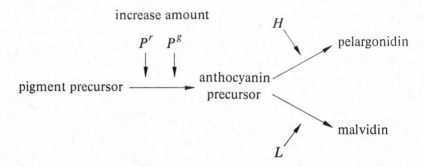

gene $L$—addition of groups at $R'$ $R'''$
gene $H$—addition of group at $R''$
If no $L$ present, pathway diverted to producing some pelargonidin even though $H$ not present.

Sepals

    (a) Pathway if $L$ present

        pigment precursor $\xrightarrow[P^r\,P^g]{}$ anthocyanin precursor $\xrightarrow[L]{}$ malvidin

    (b) Pathway if $L$ absent

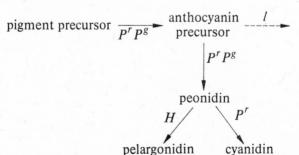

Anthocyanin extraction

1. Boil 4–5 freshly opened flowers (petals or sepals) with 1 per cent HCl in 95 per cent ethanol until no more visible extraction. Alternatively material preserved by freezing or rapid drying at 30°C can be used.
2. Concentrate extracts over steam bath.
3. Apply to chromatograms as short streaks on Whatman No. 1 filter paper using one or more of the following solvent mixtures:

$t$-butanol : acetic acid : water (3 : 1 : 1)
$m$-cresol : acetic acid : water (55 : 7 : 6)
acetic acid : hydrochloric acid : water (30 : 3 : 10)

Two per cent ethanolic $AlCl_3$ sprayed on to chromatograms accentuates the colour differences—cyanidin, delphinidin and petunidin turn deeper blue. The $R_F$ values of these pigments are given in table 2.7.

Table 2.7

| Pigment | Solvent | | |
| --- | --- | --- | --- |
| | $t$-butanol | $m$-cresol | HCl |
| Pelargonidin | 0·64 | 0·63 | 0·72 |
| Peonidin | 0·74 | 0·75 | 0·73 |
| Cyanidin | 0·41 | 0·33 | 0·53 |
| Malvidin | 0·36 | 0·85 | 0·67 |
| Petunidin | 0·30 | 0·48 | 0·50 |
| Delphinidin | 0·18 | 0·12 | 0·32 |

Since $R_F$ values do vary to some extent with temperature and conditions, pigments from species with standard anthocyanins can be used as controls:

Pelargonidin from *Pelargonium zonale* (but other pigments also present)
Pelargonidin from *Salvia splendens*
Cyanidin from *Centaurea cyanus* (blue)
Delphinidin from *Viola papilionacea* or *Delphinium ajacis* (blue)
Peonidin from *Paeonia lactiflora* (deep red)
Malvidin from *Parthenocissus quinquefolia* (fruits)
Petunidin with Malvidin from *Petunia hybrida* var *Alderman*.

*References*

ALSTON R.E. & HAGEN C.W. (1958) Chemical aspects of the inheritance of flower colour in *Impatiens balsamina* L. *Genetics*, **43**, 35–47.
CLEVENGER S. (1958) The flavonols of *Impatiens balsamina* L. *Arch. Biochem. Biophys.* **76**, 131–138.
CLEVENGER S. (1964) Flower pigments. *Scient. Am.*, June 1964, 84–92.

## (c) Other flower pigments

The techniques described here for extracting flower pigments can be used to analyse anthocyanin pigments in a wide variety of species, particularly if known pigments are available as controls.

### Sources of materials

*Impatiens* can be obtained from nearly all nurserymen, and colour charts, chemicals and chromatography paper from nearly all chemical or biological supply agencies.

# 4. HUMAN GENETICS

A practical approach to human genetics usually consists of a demonstration of single gene differences in human populations followed by case histories or pedigrees relating to the genes demonstrated in the class and to other features such as congenital abnormalities. Students may be asked to score their own families for certain characteristics. Most university students can only do this during the vacations. This type of practical may expose cases of illegitimacy or adoption. About 5–6 per cent of all children are illegitimate, about 1 per cent are legitimate but have been adopted, while an unknown number (estimated as high as 3 per cent) may be the children of men who are not their legal fathers. There is, therefore, a chance that three out of every 30 children will not be living with their real parents. This may not be known to some of the students concerned, and if the results of their genetic studies are inconsistent it could be the cause of considerable distress and suspicion. It is better, therefore, to avoid this approach to human genetics, although abnormal segregation is often due to abnormal expression or erroneous assessment.

### (a) Ability to taste PTC

Individuals differ in their ability to taste the substance PTC (phenylthiocarbamide or phenylthiourea), this character being inherited as a simple dominant trait but with minor complications.

The PTC can be tasted as a solution, but the most convenient method is to impregnate filter paper with it. The following methods are suitable.

(i) Dissolve 0·08 g of PTC in boiling water and make up to 1 litre. Use as such.

(ii) Dissolve 1 g of PTC in 100 ml of acetone (1 per cent solution) and soak filter paper in this solution. Remove and allow to dry. Cut into strips of 15 mm × 30 mm. Store in jar or envelope.

(iii) An appropriate method for advanced classes is to detect the lowest concentration that can be tasted by each individual and to plot these values on a histogram. $1 \cdot 3$ g of phenylthiourea in a litre of tap water is concentration No. 1. Take half of this and make it up to 1 litre by adding tap water. Repeat this procedure with concentration No. 2 and continue until you have 12 serial dilutions. Individuals are tested by dipping a clean glass rod into solution No. 12 and then tasting the liquid on the rod. If there is no distinctive taste, solution No. 11 is tried and so on until the phenylthiourea is detected. The value of this final solution is entered on the histogram. The result, if enough people are tested, is a bimodal distribution with an antimode at tube 4 or 5. This strength is then the distinction between taster and non-taster. The method demonstrates to the student that there is variation in threshold within each taster class and also that the two main classes show some overlap in threshold.

*Sources of materials*

PTC is available from most chemical and biological supply agencies.

### (b) Blood Groups

It is common knowledge that the blood of human beings can be classified into various groups, the most well known being the groups forming the ABO system. These blood groups reflect the nature of the 'proteins' on the red-blood cells, thus
A group contains 'protein' A
B group contains 'protein' B
AB group contains 'protein' A and B
O group contains 'protein' neither A nor B.

The liquid part of the blood, or plasma, contains antibodies which are the agents mainly responsible for combating foreign substances and organisms in the blood, whether they be infective bacteria, viruses or allergic substances from say pollen grains. Two such substances are 'anti-A' and 'anti-B' which cause clumping of the red-blood cells containing 'protein' A and B respectively. 'Anti-A' will, therefore, cause clumping (agglutination) of A and AB blood, but not B and O. A consequence of this is that blood of groups A and AB do not contain 'anti-A' otherwise the blood would clump and not flow freely. Nor does AB contain 'anti-B' since this too would cause clumping of AB. The ABO blood groups are determined by three alleles or forms of one gene, which have been termed $I^A$, $I^B$, and $i$. The situation is summarized in table 2.8.

The clumping of red-blood cells can be seen with the naked eye or under a binocular microscope, and the practical determination of blood groups,

**Table 2.8**

| Blood group | Substance on wall of red cell | Antibody in plasma | Genotypes | Blood can be transfused to |
|---|---|---|---|---|
| A | A | Anti-B | $I^A I^A, I^A i$ | A, AB |
| B | B | Anti-A | $I^B I^B, I^B i$ | B, AB |
| AB | A and B | Neither anti-A nor anti-B | $I^A I^B$ | AB |
| O | Neither A nor B | Anti-A and anti-B | $ii$ | All groups |

apart from certain precautions that have to be taken, is a straightforward process. Blood is obtained by piercing the thumb* with a sterile lancet in the following way.

(i) Sterilize a mounted needle by heating in a bunsen and cooling in sterile water. A far better alternative is to use *sterilized disposable lancets* for the purpose. Under no circumstances should a needle or lancet be passed from person to person since certain blood-borne diseases are very resistant to sterilization.

(ii) Shake the hand (which should have been washed in soapy water) vigorously downwards and apply tourniquet with a clean handkerchief round base of thumb so that blood is retained in the thumb (the end of the thumb becomes red). If the person is right handed, the left thumb should be used.

(iii) Wipe the thumb with cotton wool soaked in surgical spirit and allow to dry.

(iv) Make quick small jab through the skin; do not put needle on skin and press down.

(v) Squeeze the thumb and put one or two drops in two positions on a white tile, on separate microscope slides, or on blood testing cards (see below).

(vi) Wipe punctured area with surgical spirit and apply plaster. There should be no further bleeding but a plaster is a useful precaution against infection through contact with material used in the rest of the practical. Students should be told to remove the plaster afterwards as this hastens healing.

(vii) To one drop of blood add one drop of anti-A serum, and to the other anti-B, carefully noting which is which.

(viii) Mix with corner of slide, a glass rod, or sterile lancet (cleaning between mixing or using separate rods and lancets) and leave for about 10 min.

(ix) Observe clumping, using binocular if necessary. Classify blood groups accordingly (see table 2.9).

* Some people prefer to prick an ear lobe and then use sterile cotton wool not a plaster under (vi).—Ed.

**Table 2.9**

| Blood group | Clumping reaction with | |
| | Anti-A | Anti-B |
| --- | --- | --- |
| A | + | − |
| B | − | + |
| AB | + | + |
| O | − | − |

An alternative to using anti-A and anti-B serum is to use blood group testing Eldon cards. These cards have four spots on them impregnated with different anti-sera, so that the ABO and Rhesus (either Rhesus positive or Rhesus negative) blood groups can be identified if the blood is dropped on to the four spots.

During such a practical it is interesting to demonstrate the importance of blood groups in blood transfusion and disease, as well as (if Eldon cards are used) the importance of the Rhesus factor during child birth and modern methods of preventing Rh haemolytic disease of the newborn (see Clarke *et al.* 1966). The ABO system therefore, demonstrates a clear-cut multiple allelic difference, and shows one of the most interesting human polymorphisms with considerable practical relevance.

### *Sources of materials*

Blood group anti-sera can usually be obtained without difficulty by contacting the local hospitals. Blood group testing (Eldon) cards and disposable lancets can be obtained from most biological supply agencies.

### (c) Secretor testing

The ABO antigens as well as being found on the red-blood cells are in some people also present in the body fluids and secretions: they are most easily detected in the saliva. The presence or absence of such antigens in the secretions is genetically determined: about 78 per cent of the population of the United Kingdom are secretors (with genotype *SS* or *Ss*) while 22 per cent are non-secretors (with genotype *ss*). The *S* gene is inherited independently of the genes controlling the ABO system.

Only the same antigens are present in the blood as are found in the body fluids except that all the secretor types, including group O, produce an antigen called H which reacts with anti-H serum. This antigen is also found on all red-blood cells.

The presence of A, B or H substances in the saliva is determined by an Inhibition Test. The principle behind this is that the saliva is mixed with the appropriate anti-serum. If the antigen is present then it will inactivate the anti-serum which therefore won't cause clumping of the appropriate blood group type. A difficulty is that it is difficult to define the concentration of saliva that will precisely inactivate an often unknown concentration of anti-serum. The method adopted is therefore one of serial titration or dilution of the saliva.

A row of small (precipitin or analysis) tubes is required. To each tube except the first an equal volume of saline solution is added so that each tube is less than half full. This is best done using a pipette and teat with the appropriate volume marked off with a grease pencil. The empty first tube is filled with an equal volume of undiluted saliva. This is best collected in a test tube which is then placed in a beaker of boiling water for ten min. The saliva is then centrifuged for 4 min and the clear supernatant fluid used for the test. An equal volume of undiluted saliva is added to the second tube and the two are mixed by squeezing the liquid up and down the pipette. An equal volume is then removed from this tube and added to the third, mixed, and so on. When the last tube is reached half of the liquid (equal volume) is discarded. In this way are obtained equal volumes of saliva at successive dilutions of $1, \frac{1}{2}, \frac{1}{4}, \frac{1}{8}$, etc.

A drop of the appropriate anti-serum is now added to each. Which anti-serum to use is determined from the person's blood group, or alternatively anti-H can be used for all types. The tubes are left for 15 min and then a drop of blood of the appropriate blood group is added. After an hour the suspensions are examined microscopically. If no agglutination occurs in several of the tubes (where the saliva is more concentrated) then the person has secreted antigen, if agglutination occurs in all the tubes then the person is a non-secretor.

### Sources of materials

Blood, A and B anti-sera, and saline can all be obtained from local hospitals or blood-transfusion centres. For a class of 30 about 10 ml of each blood group type, a small vial of each anti-serum type and a litre of physiological saline should be sufficient. H anti-serum is obtainable from Baxter Laboratories Ltd, Caxton Way, Thetford, Norfolk. The anti-serum can also be made up by making an extract of the seeds of *Ulex europeaus* in physiological saline. This can be diluted appropriately and then can be frozen when not in use. Add O cells in preference to A or B when using anti-H.

### (d) Other human characteristics

The characteristics described below all have a very strong genetic component but the precise details of their inheritance are complex because either several alleles are involved, or their expression is easily modified by other genes.

Their use in pedigree studies is, therefore, to be regarded with caution, although they still provide an extremely useful demonstration of human genetic variability. More advanced students can try to construct pedigrees with these characters and try to interpret them bearing in mind the uncertainties about their inheritance.

(i)  Eye colour—blue eyes recessive to other colours, but expression very variable, inheritance complex.

(ii)  Hair colour—blonde hair recessive to dark hair, red hair recessive to darker colours, but expression very variable, inheritance complex.

(iii)  Freckles—dominant to non-freckling, expression enhanced by exposure to sunlight.

(iv)  Hair whorl direction—hair at back of head whorling in clockwise direction dominant to whorl in anti-clockwise direction.

(v)  Hair form—curly hair dominant to wavy hair dominant to straight hair, but expression variable.

(vi)  Presence of hair in middle segment of fingers (other than index finger) dominant to absence of hair—but probably several genes involved.

(vii)  Colour blindness can take quite a wide range of forms, but the commonest type (red-green colour blindness) is sex linked. Eight per cent of European men show this character but only 0·7 per cent of women. This can be tested by using the Ishihara Test Cards available from most biological supply agencies.

(viii)  Premature or pattern baldness which is the best known sex-limited (as opposed to sex-linked) trait in human beings. The character occurs to some extent in more than 40 per cent of the male population over the age of 34, but is extremely rare in females. In males the character is inherited as a dominant, hairiness being recessive. The character is transmitted by females, but not phenotypically expressed in them.

(ix)  Ear lobes. The ear lobes may be either free and pendulous, or adherent. The adherent type is recessive. Again the expression is variable.

(x)  Double-jointed thumbs. The loose ligaments in the thumb, permitting it to be bent sharply backwards, is a character dominant to the normal condition of tight ligaments.

(xi)  Ability to taste thiourea. Ability to taste this substance (tastes bitter) is dominant to inability to taste it, but not linked to ability to taste PTC.

(xii)  Taste of sodium benzoate. Paper impregnated with this chemical tastes either sweet, salty, bitter or not at all. Its inheritance is not understood. Note: Some older practical manuals recommend brucine as another substance which shows a tasting/non-tasting difference. This substance is not recommended because it is highly poisonous if swallowed.

(xiii)  Dimpled cheeks dominant to plain cheeks.

(xiv)  Widow's peak—a point of the hair line which extends down in the centre of the forehead—dominant to straight hairline.

(xv)   Tongue rolling—ability to roll the tongue so that the sides form a U is dominant to the inability to roll it.

An interesting addition to a practical on basic human genetics is to have students score themselves for a range of characters and then plot the results on an 'individuality chart' (fig. 2.7). This is done by shading in the section of the chart referring to the student's particular character, starting from the centre and working outwards. Two or more charts can be used to incorporate all the features listed above.

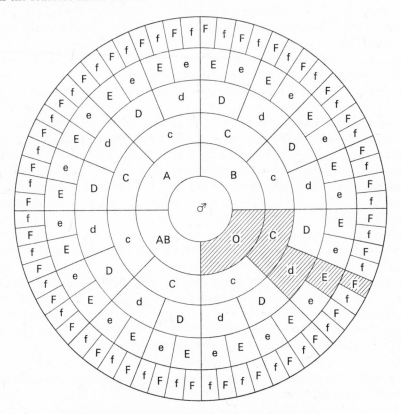

**Fig. 2.7.** Individuality chart (for explanation see text).

For example if
C/c refers to ability/inability to taste PTC,
D/d refers to ability/inability to roll tongue laterally,
E/e refers to dark/blue eyes,
F/f refers to free/attached ear-lobes,
then a male with blood group O, ability to taste PTC, inability to roll tongue, with dark eyes and free ear lobes would have a chart as in fig. 2.7. When all

individuality charts have been plotted, they can be compared to see if any two human beings are genetically identical (on the basis of the few clear-cut characters studied).

## 5. STATISTICS

For the teaching of elementary genetics to students with little mathematical knowledge it is fortunate that only one type of statistical test is absolutely necessary: this is the Chi-squared test (see also p. 11). There is very little that is really difficult about this test, provided that three things are accepted without explanation: the term 'chi-squared' the concept of 'degrees of freedom', and the conversion of the chi-square value into a probability value. The theory behind these is far too advanced to even be hinted at for most biologists without mathematical training.

From the teaching point of view, one of the best approaches is not to mention either the words 'chi-squared' or 'statistics' until the basis of the test has been made clear. One way of doing this is to get the student to do an almost trivial project involving different coloured beads. They should be asked to perform the following simple procedures: firstly, mix an equal number of beads of two colours (about 100) in a bag. Then pick at random samples of 20 beads (see p. 296) and note down the number of beads of each colour that occur in each sample. Ten samples should be adequate. The results of one of the students can then be chosen and written up on the blackboard.

A series of questions are then asked of the class. What do you notice about this data? You put in exactly equal quantities of the two colours, but what has happened? Is this what you expect? Would you not have expected equal quantity of each colour in each sample? How do you known that the differences are due to chance? How would you tell that the differences were due to chance *if you did not know* that you had put in equal quantities of the two colours? Let us for example look at the following set of data. (An extreme case is put up on the board where the ratios deviate very much from 1 : 1.) Would you not say that there was something wrong here with our presumption that the beads were thoroughly mixed in equal proportions? At this point the crucial question can be put, namely, where do you draw the line between a normal deviation by chance and a real deviation from the expected value.

The first step towards answering this question is to obtain some measure of how much the values deviate from the expected. This is done by calculating the following quantity for each colour class:

$$\frac{(\text{observed} - \text{expected number})^2}{\text{expected number}}.$$

The reason why this quantity is used can be explained by an appeal to common sense.

A value is obtained for each colour class in one sample and they are then added to give a chi-squared ($\chi^2$) for that sample. This chi-squared is then assigned a number called 'degree of freedom' which equals the number of expected classes minus one (i.e. for two-colour classes as in the bead example the degrees of freedom $= 2 - 1 = 1$ or say if it had been for the four classes in an $F_2$ segregation with dominance, the degrees of freedom $= 4 - 1 = 3$). Once chi-squared and its degrees of freedom have been estimated, we have a reasonable measure of degree of divergence from expectation. How then can this measure be used?

This measure can be used to calculate the probability with which we expect the observed deviation to arise by chance. This probability does not have to be calculated every time we want an estimate but can be read off from tables of chi-squared. The resultant probability gives the probability of obtaining the observed deviation or a greater one from the expected value by chance: in other words if the deviation is large, the chi-squared will be large, but the probability of getting such a deviation by chance will be small.

The smaller the probability the more we are likely to say that the original hypothesis about the expected values is wrong, and that the deviations have not come about by chance but indicate that we have made wrong predictions. In the particular example using beads this might mean either that equal numbers were not put in originally or that they were not mixed properly or that they were not selected at random. However, there is still the problem of where to draw the line: when does one say that the probability is so small as to make the truth of the hypothesis unlikely. The level usually chosen is 5 per cent (and most tables only give chi-squared in the range of probabilities immediately below and immediately above 5 per cent). This value is essentially arbitrary but provides a useful working norm generally accepted among scientists. When the probability value falls on or below 5 per cent then the deviation from expected is deemed 'significant'. This is a technical term rather than an emotive expression, and is used in statistics in this specific context.

So far chi-squared has only been applied to several classes in one sample: it can also be used for looking at the situation over many samples, to see if there is an 'overall deviation' from expected value or heterogeneity between samples. These more complicated extensions of the chi-squared test are not normally required in teaching very elementary genetics but can be found in any standard textbook.

Considerable space has been devoted to the problem of chi-squared. This has been done for two main reasons. Firstly, if elementary genetics is to be taught in a way that does not presuppose the answers (as has been recommended throughout this account) then it is important to bring home the idea of testing the validity of an hypothesis, i.e. testing for significant divergence from expected. Secondly, it is important that statistics is taught in a meaningful way and that its aims are understood and appreciated. There are two ways in which the meaning can be brought home. The subject must be taught in the context of an experiment (e.g. with beads, or actual data) and it must be fully explained in the sense that, where possible, the intuitive basis for a test must be pointed out and equally any less obvious aspect of the test should be explicitly pointed out so that the student knows what to take for granted, and what to learn to appreciate.

## III. COURSE DESIGN IN TEACHING BASIC GENETICS

## 1. INTRODUCTION

So far very little mention has been made of the content or planning of particular courses. Much of this will be determined by personal preference and availability of materials. Certain general principles should however be remembered. The teacher should be clear for what purpose each practical investigation is designed: is it to help the student understand the laws of genetics, is it to show that many organisms show a similar behaviour, or is it to relate elementary genetics to more advanced aspects of the subject? The investigation should be tailored to the purpose for which it has been chosen.

Investigations designed to explain the principles of gene behaviour should be kept as open-ended as possible. Even a simple investigation with a single pair of alleles should not be described as such, but as a cross between two organisms showing contrasting characters: whether the parents are from pure lines, whether one or many genes are involved and whether the genes are dominant or recessive is something that the student should be asked to discover.

Another essential is that the student should, after he has analysed his results, draw up a theoretical scheme to illustrate the mechanism involved. This scheme should indicate the genotype and phenotype of the parents and progeny, it should indicate the genotypes and ratios of the gametes, it should demonstrate random union of gametes, and finally a summary of ratios of the resultant genotypes and phenotypes. Clearly at a more advanced level many stages of the reasoning can be omitted, as long as the same general scheme of presenting the results is used.

Demonstrations may not appear at first sight to be amenable to such an open-ended project approach, but this is in fact not so. A demonstration should not be looked at with one fixed end point in view which is written down clearly for the student: he should be asked to observe, assess and comment. Fortunately this is readily possible in genetics, since the subject does have a logical theoretical basis.

The series of investigations outlined below are classified according to the general educational level of the students concerned. The rationale behind these course suggestions is also explained, so as to give some guidance when it comes to including investigations not specifically mentioned in this account.

## 2. SCHOOL LEVEL

### (a) Elementary

(i)   Breeding mice differing in coat colour at one locus.

(ii)   Demonstration of differences in various characters in human beings.

(iii)   Examination of human pedigrees for fully expressive single gene characters.

(iv)   Demonstration (e.g. groundsel ray/non-ray florets) of single gene segregation in plants.

This small series of investigations should serve to illustrate the main principles of heredity, and introduce them in the general context of biology. The investigation with mice could be a reasonably long-term one: in between the mice can be kept as pets. The first step is to show that say, black mice always produce black mice and brown mice always produce brown mice. Then a cross can be made between the stocks and carried on to the $F_2$ generation. This can be done throughout the school year not so much as a formal genetics experiment but almost 'just to see what happens'. The full explanation can then be given at the end of the year and then combined with the study of groundsel and human genetics in one or two simple practicals. Ideally the pupils should sow the groundsel $F_2$ themselves, but be provided with plants to show the parental and $F_1$ generations.

The human genetics can be part of a practical where chromosomes in general are examined and where there are photographs of human chromosomes, twins and other aspects of human genetics (e.g. sex determination).

### (b) Advanced

(i)   Breeding experiments with mice or *Arabidopsis*, differing in two pairs of unlinked alleles.

(ii)   Demonstration of 9 : 3 : 3 : 1 segregation in peas, tomato seedlings or some other material.

(iii)   Demonstration of complementary genes and an explanation in terms of development or sequence of biochemical reactions (e.g. maize cobs, cyanogenesis in clover).

(iv)   An introduction to linkage, preferably an extreme case, to contrast this with independent genes in the $F_2$. Mention of backcrosses as a better way of assessing linkage.

(v)   Allelic differences in crop plants, horticultural varieties, humans to show application of genetics.

This is a very similar scheme to the previous one but the two gene-case and its attendant complications of gene interactions (to be considered in biochemical terms rather than in abstract ideas of epistasis or complementation) and gene linkage have been introduced. Further extension is achieved by looking at the application of genetics to plant breeding: this can be done by showing the value of single gene changes, particularly with regard to dwarf varieties, flower colour, type of vegetable in the Brassicas, and disease resistance.

## 3. UNIVERSITY LEVEL

### (a) Elementary

(i)   Crossing mice, *Tribolium*, or *Arabidopsis* differing by two loci. An open-ended investigation where minimum information is given about the number of genes, their dominance or their linkage.

(ii)   The behaviour of genes as illustrated by experiments with coloured beads. The segregation of genes during gamete formation, demonstrated by $Wx/wx$ pollen. Demonstration of genetic ratios in a range of organisms (e.g. maize cobs, seedling characters, mice coat colours).

(iii)   Demonstration of relevance and economic applications of genetics. Human genetics, pedigrees, counselling. Single gene differences of economic importance (e.g. flower colour, disease resistance, morphological changes).

Universities are usually faced with a very diverse intake, particularly with regard to the extent of background that the students have had in the biological sciences, and in genetics in particular. The courses therefore have to be designed to cater 'for all tastes'. The best compromise in this situation is to have a course which starts at a very elementary level but goes fairly rapidly into an area which is fairly advanced and beyond the level of teaching at school. This rapid transition can be dangerous, since genetics is a progressive logical subject and if the initial arguments are not fully understood there is subsequent deterioration in understanding of the subject as the course progresses.

For this reason, the first investigation is the most critical part of the course. This first investigation should be in the form of a project involving a cross between strains differing by two loci. Different students should investigate different pairs of genes and even different organisms. This may seem rather an 'advanced' way to introduce the subject of genetics but it has two main advantages (common to all types of open-ended investigation): firstly, the student is allowed to proceed at his own speed without being rushed through specific procedures, and secondly the interpretation of the results demands a lot more thought and understanding simply because the answers are unknown.

The use of two genes in such a cross should not be too much of a complication if the student has done no genetics before, as long as he is told first of all to consider each character in turn. Each character considered separately should show in the $F_2$ a 3 : 1 segregation (assuming dominance is complete). This can be tested for using a chi-squared test. If there is deviation from a 3 : 1 ratio, then it suggests that one of the phenotypes is less viable than the other (see pp. 15, 75 for the full $\chi^2$ analysis).

Two approaches to the investigation are possible. It may either start with a cross between a wild type and a mutant strain carrying two mutant genes, or it may start with a cross between two different mutant strains. The former approach is the conventional one, and has the advantage that if both the mutant genes are recessive, a backcross to the double recessive is possible and can be used for estimating linkage. Where two (recessive) mutant strains are crossed a backcross is impossible unless a double recessive pure line is available. Linkage has to be detected by a significant deviation from independent assortment (see p. 15). The problem of estimating linkage from the $F_2$ can be brought to the attention of the students and they can be shown the advantage of backcrosses.

An investigation which starts from a cross between two mutants has the advantage that one need only maintain a relatively few pure lines to obtain a large number of paired combinations of characters. Thus with six pure strains 15 different crosses can be performed. This type of cross is also unusual in its effects. If both the mutants are recessive then the $F_1$ will appear wild type, so demonstrating clearly how several genes may be hidden in the recessive condition.

The aims and relevance of the other investigations mentioned above have been outlined in the previous sections. The experiment with coloured beads can be used as a model for the experiment with the actual organisms, and it can be used to illustrate the effects of gene interactions of various sorts or the different coloured beads may be assigned quantitative values and so serve as an introduction to genetics of quantitative characters (see chapter 3). The experiment with $Wx/wx$ pollen provides a striking illustration of 'different coloured beads' actually in nature.

The recommended demonstrations should serve to amplify the grounding the student gets from his main experiments and are therefore best considered towards the end of the course.

### (b) Advanced

Teaching basic genetics at an advanced level in University is almost a contradiction in terms, since in more advanced courses the basic aspects are developed into considerations of other branches of genetics. These various branches are considered in other parts of this book. Nevertheless teaching basic genetics at an advanced level has several aims.

### (i) Revision exercises

Very frequently genetics courses do not run sequentially, and there is much to be said in favour of a few revision exercises when the subject is reintroduced at a more advanced level. Students can fumble through more advanced genetics courses without being aware of the basic principles of the subject, these having been left behind several terms previously. Several important aspects of elementary genetics have not yet been mentioned and these are suitable for more advanced instruction and revision.

(i)   Three-point test crosses—this illustrates the process of genetic mapping, genetic interference and effect of double crossovers on recombination values. These are all important fundamental concepts. Suitable material is provided by mice, *Arabidopsis* and maize, although *Drosophila* is still probably the best material to use here (see chapter 1).

(ii)   A study of a few clover populations polymorphic for V-shaped white leaf markings and cyanogenesis. This exercise can bring basic genetics into relation with population (chapter 7) and biochemical genetics (chapters 5 and 6).

(iii)   Genetics of mice using genes of variable dominance, expressivity or penetrance. This exercise shows the importance of genetic background and modifier genes. Similar phenomena can be shown in human pedigrees. It also raises questions about polygenes/major genes, developmental genetics, evolution of gene expression and dominance, and the difficulties of genetic analysis of such characters.

(iv)   A study of the genetics and biochemistry of flower petal pigments. This can also serve to introduce the phenomenon of variable gene expression, and illustrate the value of genetic knowledge to plant breeding.

(v)   A study of temperature sensitive mutants in maize or *Tribolium*. Another important exercise in developmental and physiological genetics, shows the importance of genotype × environment interactions (p. 120).

Several points of importance should be noted about such exercises. Firstly, they should not repeat an investigation that has already been undertaken, but should introduce either a new problem or a new organism. Secondly they should be approached as open-ended investigations, preferably with different members of the class investigating different aspects of the same general phenomena. Thirdly the investigation should be angled towards the area in which the more advanced course will specialize. Clover is of value for a general advanced course, or for placing emphasis on evolution. The mouse investigation would be suitable for the beginnings of a course in human or developmental genetics, and the flower pigment one for a course on physiological genetics.

### (ii) Basic genetics research projects

The second value of basic genetics is that it can provide a useful source of ideas for minor research investigations. In many final year courses, and to an ever-increasing extent in lower years, the student is required to perform one or two minor research investigations and to write these up as a B.Sc. 'thesis' of one sort or another. Frequently basic genetics has by this time been left well behind, but there are many types of investigations related to basic genetics that could be suitable.

*Investigation of genetics of characters in short-lived plants*

Several plants species closely approach *Arabidopsis* with regard to the length of their life cycle, but few have mutants readily available although natural populations may show variation for some characters. These plants could be investigated with regard to the induction of mutations. A list of such short-lived plants (or tachyplants) is given by Postlethwait & Enochs (1967).

*Inheritance of varietal differences in horticultural plants*

The inheritance of flower colour, colour pattern and morphology in many horticultural varieties is imperfectly understood and an investigation of one or more aspects of such plants would provide an exacting project. Normally it would be long term but the student could be presented with preserved material (pressed, dried or deep frozen) covering several generations. Suitable ideas and techniques can be obtained by consulting McQuown (1963) which gives techniques for growing and crossing the following: *Dianthus, Campanula, Epiphyllum, Dahlia, Iris, Rhododendron, Rosa, Gladiolus, Narcissus, Tulipa, Hydrangea, Ilex, Lycopersicum, Primula, Pelargonium, Lathyrus, Lilium, Delphinium, Chrysanthemum, Fuchsia.* Crane M.B. & Lawrence W.J.C. (1954) *Genetics of Garden Plants*, 4th edn., published by Macmillan, London, is also useful.

## IV. CONCLUDING REMARKS

This chapter has attempted to show that it is possible to teach basic genetics using examples taken solely from the higher plants and animals, and without recourse to the classical organism, *Drosophila*. This has been done mainly in an attempt to bring the many diverse examples from higher organisms into a coherent plan and to indicate what type of investigation each organism can be used for, and where such a type of investigation can be put in a course. It is almost certainly not desirable to teach basic genetics without recourse to other organisms such as, of course, fungi and *Drosophila* which have distinct practical advantages; they are further illustrations of the range of organisms which are covered by the elementary laws of genetics, and it is important that students should handle organisms that are central to research and teaching. This chapter, although it has been presented in a comprehensive manner, should therefore not be considered in isolation.

The whole topic of basic genetics is a story of variations on a theme, that theme being the fundamental laws of genetics as proposed by Mendel. The understanding of this theme (and its more obvious variations such as linkage) requires that it be explained, demonstrated and extended into related fields, and this can best be done through the medium of practical work. The most difficult part of teaching genetics is that of explanation: it is here that practical work can be of paramount importance. A full explanation should result in a complete understanding, but this understanding can really only be achieved by active participation by the student. The importance of this participation has been recognized frequently in the form of theoretical problems presented to the student during the course. The practical work has generally been looked upon rather differently, even though its overall aims should be the same. This chapter has tried to emphasize the importance of maintaining the 'problem approach': every investigation should be regarded as a minor open-ended project, and practical schedules should contain the minimum of information about the expected results (and even in some cases about the types of cross the student should perform). The student should be allowed to work out for himself what must be going on in inheritance and if possible (as in a general school investigation on mice as pets) it is well worth while students having to make crosses without any prior knowledge of genetics, and then trying to get them to deduce what is happening from their results. Normally a theoretical and practical course are concurrent, so students have prior knowledge of the laws of inheritance, but as has been mentioned, their understanding is detached, and takes on a different complexion when it comes to solving a problem for themselves.

Teaching practical genetics therefore not only has techniques that need to be learnt but also certain attitudes and aims, all of which go towards the production of a satisfactory course. In this chapter some of these attitudes have been emphasized since teaching genetics at a basic level is normally not just a technological problem but also an educational one.

# V. ACKNOWLEDGEMENTS

I am extremely indebted to the following people who provided me with invaluable first hand information on many of the organisms mentioned in this chapter. Dr J.G.Pusey, Professor A.Sokoloff, Dr M.E.Wallace, Dr S.Matthews, Dr E.R.Creed, Dr B.Clarke, Dr R.A.E.Tilney-Bassett, Professor

J.R.S.Fincham, Dr L.K.Crowe, Dr A.D.McKelvie, Dr A.P.C.Seaton, Dr R.A.Beatty, Dr B.G.Cumming, Professor R.J.Lambert, Professor A.D.Bradshaw, Dr M.J.Lawrence, Professor G.P.Redei, Dr H.L.K.Whitehouse, Dr L.A.Darby, Dr J.D.Cash, and Mr J.K.Burras.

## VI. REFERENCES AND FURTHER READING

CLARKE C.A. *et al.* (1966) Prevention of Rh haemolytic disease: results of the clinical trial. A combined study from centres in England and Baltimore. *Brit. med. J.* **ii**, 907–914.

CREED E.R. (1966) Geographic variation in the two spot ladybird in England and Wales. *Heredity* **21**, 57–72.

DARLINGTON C.D. & BRADSHAW A.D. (1963) *Teaching Genetics in School and University*. Oliver & Boyd, Edinburgh.

FLOR H.H. (1956) The complementary genic systems in flax and flax rust. *Adv. Genet.* **8**, 29–54.

FORD E.B. (1953) The genetics of polymorphism in the Lepidoptera. *Adv. Gent.* **5**, 43–88.

FORD E.B. (1967) *Moths*. New Naturalist Series. Collins, London.

FORD E.B. (1971) *Ecological Genetics*. Chapman and Hall, London.

FOSTER M. (1965) Mammalian pigment genetics. *Adv. Genet.* **13**, 188–339.

GLUECKSOHN-WAELSCH S. (1951) Physiological genetics of the mouse. *Adv. Genet.* **4**, 2–52.

GRANT V. (1956) The genetic structure of races and species in *Gilia. Adv. Genet.* **8**, 55–87.

HEAD J.J. & DENNIS N.R. (1968) *Genetics for 'O' level*. Teachers' Guide. Oliver and Boyd, Edinburgh.

KEER W.E. & LAIDLAW H.H. (1956) General genetics of bees. *Adv. Genet.* **8**, 109–153.

KOMAI T. (1956) Genetics of ladybeetles. *Adv. Genet.* **8**, 155–188.

KRUG C.A. & CARVALHO A. (1951) The genetics of *Coffea. Adv. Genet.* **4**, 127–158.

LEVINE P. (1954) The genetics of the newer human blood factors. *Adv. Genet.* **6**, 183–234.

McQUOWN F.R. (1963) *Plant breeding for gardeners: a guide to practical hybridizing*. Collingridge, London.

MAHONEY R. (1966) *Laboratory Techniques in Zoology*. Butterworth, London.

NAGAO S. (1951) Genic analysis and linkage relationships of characters in rice. *Adv. Genet.* **4**, 181–212.

NUFFIELD BIOLOGY TEACHERS' GUIDE V. (1967) *The Perpetuation of Life*. Longmans/ Penguin, London.

OLDROYD H. (1963) *Collecting, Preserving and studying insects*. Hutchinson, London.

POSTLETHWAIT S.N. & ENOCHS N.J. (1967) Tachyplants—suited to instruction and research. *Plant Science Bulletin* **13**, 1–2.

RAE A.L. (1956) The genetics of the sheep. *Adv. Genet.* **8**, 189–266.

REMINGTON C.L. (1954) The genetics of *Colias* (Lepidoptera). *Adv. Genet.* **6**, 403–450.

RICHEY F.D. (1950) Corn breeding. *Adv. Genet.* **3**, 159–192.

RICK C.M. & BUTLER L. (1956) Cytogenetics of the tomato. *Adv. Genet.* **8**, 267–382.

ROBINSON, R. (1971) *Lepidoptera genetics*. Pergamon Press, Oxford.

SAWIN P.B. (1955) Recent genetics of the domestic rabbit. *Adv. Genet.* **7**, 183–226.

SEARS E.R. (1948) Cytology and genetics of the wheats and their relatives. *Adv. Genet.* **2**, 239–270.

SHRODE R R. & LUSH J.L. (1947) The genetics of cattle. *Adv. Genet.* **1**, 209–261.

TANAKA Y. (1953) Genetics of the silkworm, *Bombyx mori. Adv. Genet.* **5**, 239–317.

WHITE M.J.D. (1951) Cytogenetics of Orthopteroid insects. *Adv. Genet.* **4**, 267–330.
WHITING A.R. (1961) Genetics of *Habrobracon. Adv. Genet.* **10**, 295–348.
ZIEGLER I. (1961) Genetic aspects of ommochrome and pterin pigments. *Adv. Genet.* **10**, 349–403.

## INFORMATION SERVICES

Wheat Information Service
Laboratory of Genetics
Biological Institute
Kyoto University
Kyoto, Japan

Tribolium Information Bulletin
Alexander Sokoloff
Department of Genetics
University of California
Berkeley
California

Maize Genetics Co-operation News Letter
Department of Botany
Indiana University
Bloomington
Indiana

Mouse News Letter
Laboratory Animals Centre
M.R.C. Laboratories
Woodmansterne Road
Carshalton
Surrey

Report of the Tomato Genetics Co-operative
Department of Vegetable Crops
University of California
Davis
California

Arabidopsis Information Service
Institüt für Pflanzenbau und Pflanzenzüchtung
Universität Göttingen
Germany

D

# CHAPTER 3

# QUANTITATIVE GENETICS

## M. J. LAWRENCE AND J. L. JINKS

## I. INTRODUCTION

### 1. QUALITATIVE AND QUANTITATIVE CHARACTERS

The kinds of characters which Mendel studied in his classic experiments and which continue to be studied in elementary experiments in genetics are *qualitative* characters. Any individual in such experiments can usually be classified without ambiguity into one of two or more classes, *e.g.* tall or short in peas, red or white eyed in *Drosophila*. The distribution of such characters or variables in a population of pea plants, *Drosophila* or men is thus *discrete* or *discontinuous*. The inheritance of this kind of character has, of course, been analysed extensively in a wide variety of animal, plant and human populations and is well understood.

Not all characters, for which individuals in a population differ, are of this kind. Stature in human populations, for example, varies from around 1·5 m to 2·0 m, depending on sex, and geographical origin. But it is impossible to classify individuals into natural groups in respect of stature. This is because the distribution of this kind of variable is *continuous*, as is usually the case for *quantitative* or *metrical* variables such as stature, i.e. within the range of variation, all statures are possible.

The genetical analysis of these characters appears to be, at first sight, a formidable task for it is not possible even to classify the data unambiguously. Yet one is aware from observation on one's own and friends' families that stature tends to be inherited. Tall parents tend to have tall children and short parents, short children. This tendency for like to beget like, or in statistical terms, for the statures of parents and offspring to be positively correlated, is a sure sign of genetical determination. At the same time, one is also aware that stature is determined in part at least by the environment. Evidence from the Armed Services, from recruits in the two world wars, shows that the average stature of young men in the second war exceeded that of their fathers' generation in the first. This difference between generations in respect of stature is, of course, most probably due to the better nutritional standards obtaining in the 1920s and '30s. Stature then, appears to be subject both to heredity and the environment.

Now of course, all or nearly all characters of a living organism are affected by the environment. The magnitude of differences between individuals in respect of qualitative characters, however, are so great that the effects of the environment for the purpose of their genetical analysis can be effectively discounted. But since the environment is not infrequently a larger source of variance than the genotype in respect of quantitative characters, it is clear that we are not able to ignore the effects of the former in attempts to analyse the latter. In consequence statistical techniques must be employed to partition the total variability of a population, with respect to a particular quantitative character, into a part determined by the genotype and a part determined by the environment. Clearly, the genetical analysis of such characters is likely to be rather more laborious than that required for qualitative characters (see Mather and Jinks 1971 whose approach will be used throughout this chapter). Why, then, is the study of the inheritance of quantitative characters a matter of some importance?

## 2. THE IMPORTANCE OF QUANTITATIVE CHARACTERS

A first and practical answer to this question is that the majority of characters for which crop plants and domestic animals are raised, e.g. milk yield in dairy cattle, yield of tubers in potatoes, number of eggs in poultry are quantitative. Any attempt by plant or animal breeders to develop new varieties must therefore concern the selection of superior genotypes from populations consisting of an array of mostly average genotypes. An understanding of the inheritance of quantitative characters therefore appears to be a prerequisite for efficient breeding procedures.

A second and more theoretical answer to the same question concerns the nature of the differences between closely related species of animals and plants. Again, most of these are quantitative, e.g. flower size and shape; number of stridulation pegs on the hind legs in grasshoppers. This being the case, it is clear that the greatest impact of natural selection, at least so far as divergence between species is concerned, is on such quantitative characters. It can therefore be argued that quantitative differences between individuals in a population are likely to be selectively more important than gross, qualitative differences.

## 3. EXPERIMENTAL REQUIREMENTS IN QUANTITATIVE INHERITANCE

Since quantitative characters are influenced by the environment, it is clear that there is a need to control and to measure its effects in any investigation of the inheritance of such characters. This is most easily accomplished with small animals and plants. Indeed, these species have the further advantages that their life-cycles are usually short, permitting several generations to be studied in a term; that their size of family is large; and that material of known genetical composition is available. The last advantage, though not necessary, is desirable. Thus organisms such as *Drosophila melanogaster*, the mouse, *Arabidopsis thaliana* and several fungal species are particularly suitable for this purpose.

Many of the experiments described below concern *Drosophila melanogaster*. It is convenient to carry out experiments with this species at 25°C. At this temperature, a new generation will emerge within ten days. Allowing time for handling and for the collection of virgin females, experiments can be run on a 14-day cycle which allows easy planning within the confinement of the usual class timetables. Inbred lines, lines within which individuals are identically homozygous, are also readily available and therefore afford the most convenient starting material for experiments in quantitative inheritance. Lastly, techniques for raising large numbers of flies are well known (chapter 1).

## II. THE DEMONSTRATION OF QUANTITATIVE VARIATION AND ITS CAUSE

### EXPERIMENT 1

#### (a) Materials and methods

This experiment may be performed with any pair of inbred lines of *Drosophila* which differ with respect to an easily scored quantitative character. At Birmingham the inbred lines Oregon (O) and 6CL (C), are used which though wild-type, differ by approximately 11 sternopleural chaetae (bristles). Males, which are smaller than females, bear slightly fewer bristles. It is convenient therefore to score females only. Though bristle number is not difficult to score with an ×40 binocular microscope (see chapter 1, plate 1.1), students are advised to spend some time in practice before embarking upon an experiment. It is important to use virgin females for all matings, since illegitimate matings cannot easily be detected when parents are wild-type.

The results presented here were obtained by a class of seven students.*
Each student scored ten female flies from a pair of cultures of the crosses
listed in table 3.1. To ensure that each student has 28 cultures available it is
desirable to put up twice this number, expecially where the students themselves set up the cultures. The total number of 56 cultures for each student is
laid out on a single shelf of an incubator set to 25°C as a single, completely
randomized block.

**Table 3.1** The crossing programme for Experiment I

| Family | Cross | Number of cultures/student | Total number of cultures in class |
|---|---|---|---|
| $P_1$ | O × O | 2 | 14 |
| $P_2$ | C × C | 2 | 14 |
| $F_1$ | O × C | 2 | 14 ⎫ 28 |
|  | C × O | 2 | 14 ⎭ |
| $F_2$ | (O × C) × (O × C) | 2 | 14 ⎫ 28 |
|  | (C × O) × (C × O) | 2 | 14 ⎭ |
| $BX_1$ | O × (O × C) | 2 | 14 ⎫ |
|  | O × (C × O) | 2 | 14 ⎪ 56 |
|  | (O × C) × O | 2 | 14 ⎪ |
|  | (C × O) × O | 2 | 14 ⎭ |
| $BX_2$ | C × (O × C) | 2 | 14 ⎫ |
|  | C × (C × O) | 2 | 14 ⎪ 56 |
|  | (O × C) × C | 2 | 14 ⎪ |
|  | (C × O) × C | 2 | 14 ⎭ |
|  |  | 28 | 196 |

The order in which each student scores his or her cultures should be
random and scores should be entered tidily into a previously prepared book
(fig. 3.1). It is common practice to score the left-hand side of each fly first;
to enter the score for each side separately in the book; and then to sum
sides to obtain the total number of bristles for each individual.

When all cultures have been scored, each student calculates the mean
number of bristles and the sum of squares of bristle numbers about the mean
for each culture he or she has scored. It is useful also to group the scores into
bristle number classes for each family so that their distribution may be
examined graphically. The culture means, sums of squares and the group data
can then be pooled over the class so that these results are available for com-
munal analysis.

*We are grateful to the 1967–1968 M.Sc. and Honours B.Sc. students of the Department
of Genetics for these data.

Name _____ | Application

Date _____ | Page ____ of ____

| 2 | 4 | 6 | 8 | 10 | 12 | 14 | 16 | 18 | 20 | 22 | 24 | 26 | 28 | 30 | 32 | 34 | 36 | 38 | 40 | 42 | 44 | 46 | 48 | 50 |
|---|---|---|---|----|----|----|----|----|----|----|----|----|----|----|----|----|----|----|----|----|----|----|----|----|
| 1 | 1 | 1 | 1 | 1 | 1 | 12 | | 9 | | 13 | 10 | 14 | 11 | | 11 | 10 | | 11 | | 10 | | | | |
| | | | | | 2 | 13 | | 11 | | 11 | 11 | 12 | 12 | | 11 | 11 | | 9 | | 12 | | | | |
| | | | | | | 25 | | 20 | | 24 | 21 | 26 | 23 | | 22 | 21 | | 20 | | 22 | | | | |
| | | | | | | | | | | | | | | | | | | | | | | | | |
| 1 | 1 | 1 | 1 | 2 | 1 | 11 | | 11 | | 10 | 12 | 11 | 12 | | 12 | 13 | | 12 | | 10 | | | | |
| | | | | | 2 | 12 | | 11 | | 13 | 10 | 13 | 12 | | 11 | 12 | | 11 | | 13 | | | | |
| | | | | | | 23 | | 22 | | 23 | 22 | 24 | 24 | | 23 | 25 | | 23 | | 23 | | | | |
| | | | | | | | | | | | | | | | | | | | | | | | | |
| 1 | 1 | 1 | 2 | 1 | 1 | 19 | | 15 | | 18 | 14 | 17 | 16 | | 17 | 18 | | 18 | | 15 | | | | |
| | | | | | 2 | 15 | | 18 | | 17 | 17 | 17 | 16 | | 14 | 18 | | 17 | | 18 | | | | |
| | | | | | | 34 | | 33 | | 35 | 31 | 34 | 32 | | 31 | 36 | | 35 | | 33 | | | | |
| | | | | | | | | | | | | | | | | | | | | | | | | |
| 1 | 1 | 1 | 2 | 2 | 1 | 19 | | 16 | | 17 | 16 | 13 | 19 | | 17 | 15 | | 17 | | 16 | | | | |
| | | | | | 2 | 16 | | 15 | | 16 | 17 | 16 | 16 | | 19 | 17 | | 16 | | 16 | | | | |
| | | | | | | 35 | | 31 | | 33 | 33 | 29 | 35 | | 36 | 32 | | 33 | | 32 | | | | |
| | | | | | | | | | | | | | | | | | | | | | | | | |
| | | | | | | | | | | | | | | | | | | | | | | | | |

**Fig. 3.1.** One method of recording the data from Experiment 1. The form used to enter the scores here is a 'coding sheet' of the type used when the data is to be punched onto cards.

*Key*

Column

| | |
|---|---|
| 2 | experiment number (1) |
| 4 | block number (only 1 in this case) |
| 6 | scorer (1–7) |
| 8 | line (1 = Oregon; 2 = 6Cl, etc.) |
| 10 | replicate (bottle 1 or 2) |
| 12 | side of fly (1 = L.H.S. and 2 = R.H.S.) |
| 14–42 | the scores of the ten flies examined. |

## (b) Preliminary analysis

The results from the authors' class of seven students summarized in this way are shown in tables 3.2, 3.3, and 3.4. It is necessary to ensure that the students' scores are consistent, for if they are not this can result in some undesirable consequences in later analysis. It is essential, therefore, to carry out an analysis of variance of the differences between scorers in respect of culture means for each family. Taking, for example, the Oregon means, the Total sum of squares (with 13 degrees of freedom), is partitioned into a pair of items, one measuring differences between students (the Scorers sum of squares with six degrees of freedom) and the other measuring differences between replicate cultures (the Duplicates sum of squares with seven degrees

**Table 3.2** Culture mean bristle numbers of ten female flies in each culture

| Family | Cross | Rep. | Scorer | | | | | | | Cross means | Family means |
|---|---|---|---|---|---|---|---|---|---|---|---|
| | | | 1 | 2 | 3 | 4 | 5 | 6 | 7 | | |
| $P_1$ | O × O | 1 | 22.4 | 22.7 | 23.7 | 22.5 | 23.0 | 23.6 | 22.6 | | |
| | | 2 | 23.2 | 23.8 | 22.6 | 21.7 | 21.7 | 23.9 | 22.9 | 22.9 | |
| $P_2$ | C × C | 1 | 33.4 | 35.2 | 33.2 | 33.1 | 32.5 | 34.7 | 33.9 | | |
| | | 2 | 32.9 | 35.4 | 32.8 | 34.0 | 32.4 | 33.3 | 34.4 | 33.7 | |
| $F_1$ | O × C | 1 | 29.2 | 28.3 | 27.2 | 29.7 | 26.9 | 27.9 | 29.7 | | |
| | | 2 | 29.4 | 29.9 | 27.3 | 29.2 | 30.2 | 30.0 | 30.4 | 29.0 | |
| | C × O | 1 | 28.9 | 28.8 | 30.7 | 29.7 | 28.9 | 29.8 | 29.3 | | |
| | | 2 | 29.2 | 31.3 | 28.1 | 29.0 | 27.5 | 28.5 | 29.8 | 29.3 | 29.1 |
| $F_2$ | (O × C) | 1 | 26.7 | 28.1 | 27.9 | 27.0 | 29.2 | 28.0 | 27.0 | | |
| | × (O × C) | 2 | 28.1 | 28.9 | 27.6 | 26.0 | 28.6 | 28.0 | 27.2 | 27.7 | |
| | (C × O) | 1 | 31.5 | 32.3 | 29.6 | 28.4 | 27.5 | 31.2 | 29.3 | | |
| | × (C × O) | 2 | 27.5 | 31.2 | 28.4 | 29.2 | 29.8 | 31.0 | 32.0 | 29.9 | 28.8 |
| $BX_1$ | O × (O × C) | 1 | 24.7 | 26.7 | 26.4 | 26.0 | 24.9 | 26.3 | 25.2 | | |
| | | 2 | 25.7 | 25.0 | 23.4 | 26.4 | 25.3 | 24.5 | 25.4 | 25.4 | |
| | O × (C × O) | 1 | 26.9 | 27.9 | 26.3 | 26.5 | 27.3 | 26.9 | 27.3 | | |
| | | 2 | 27.2 | 25.7 | 25.7 | 29.0 | 27.3 | 26.9 | 24.7 | 26.8 | |
| | (O × C) × O | 1 | 26.3 | 27.5 | 25.0 | 27.6 | 25.3 | 28.0 | 26.3 | | |
| | | 2 | 24.9 | 26.3 | 24.3 | 25.0 | 26.7 | 26.4 | 27.4 | 26.2 | |
| | (C × O) × O | 1 | 26.9 | 27.3 | 24.6 | 24.2 | 25.7 | 27.6 | 26.1 | | |
| | | 2 | 25.5 | 26.4 | 25.6 | 27.0 | 26.3 | 27.4 | 25.4 | 26.1 | 26.2 |
| $BX_2$ | C × (O × C) | 1 | 31.5 | 29.6 | 31.9 | 29.5 | 32.2 | 30.3 | 31.4 | | |
| | | 2 | 29.3 | 30.5 | 29.6 | 29.0 | 29.1 | 31.8 | 30.3 | 30.4 | |
| | C × (C × O) | 1 | 31.7 | 31.2 | 33.8 | 32.8 | 32.3 | 30.9 | 32.4 | | |
| | | 2 | 31.4 | 33.7 | 32.1 | 33.1 | 31.6 | 32.6 | 32.2 | 32.3 | |
| | (O × C) × C | 1 | 32.0 | 31.0 | 30.1 | 31.4 | 32.6 | 32.8 | 29.8 | | |
| | | 2 | 31.2 | 30.4 | 30.3 | 31.0 | 32.7 | 31.6 | 31.8 | 31.3 | |
| | (C × O) × C | 1 | 31.8 | 30.2 | 33.1 | 29.0 | 31.5 | 31.9 | 31.2 | | |
| | | 2 | 30.9 | 31.0 | 32.2 | 30.1 | 29.2 | 31.1 | 30.5 | 31.0 | 31.3 |

of freedom). The analysis of the $F_1$, $F_2$ and $BX$ families is similar, except that in these there is, in addition, a Crosses sum of squares. These analyses of variance are shown in table 3.5.

It turns out that the Scorers item is significant in only two of the six analyses. Since, however, it is only just significant in both these cases, differences between these students in respect of their scoring ability is clearly not an important source of variance. It is most unlikely therefore, that one would be misled in the interpretation of the data by heterogeneous scores.

**Table 3.3** Within culture sums of squares of bristle number, and within culture variance

| Family | Cross | Rep. 1 | 2 | 3 | 4 | 5 | 6 | 7 | $\sigma^2_\omega$ | df |
|---|---|---|---|---|---|---|---|---|---|---|
| $P_1$ | O × O | 1  38.4 | 44.1 | 34.1 | 54.5 | 20.0 | 58.4 | 10.4 | | |
| | | 2  7.6 | 17.6 | 18.4 | 82.1 | 10.1 | 6.9 | 10.9 | 3.28 | 126 |
| $P_2$ | C × C | 1  26.4 | 27.6 | 41.6 | 54.9 | 16.5 | 16.1 | 20.9 | | |
| | | 2  38.9 | 34.4 | 43.6 | 90.0 | 16.4 | 8.1 | 16.4 | 3.59 | 126 |
| $F_1$ | O × C | 1  23.6 | 44.1 | 27.6 | 54.1 | 22.9 | 42.9 | 64.1 | | |
| | | 2  64.4 | 20.9 | 68.1 | 53.6 | 17.6 | 30.0 | 54.4 | | |
| | C × O | 1  24.9 | 85.6 | 32.1 | 34.1 | 18.9 | 55.6 | 40.1 | | |
| | | 2  55.6 | 24.1 | 62.9 | 108.0 | 22.5 | 46.5 | 51.6 | 4.96 | 252 |
| $F_2$ | (O × C) | 1  96.1 | 28.9 | 134.9 | 50.0 | 93.6 | 88.0 | 28.0 | | |
| | × (O × C) | 2  76.9 | 96.9 | 96.4 | 108.0 | 26.4 | 74.0 | 81.6 | | |
| | (C × O) | 1  24.5 | 84.1 | 52.4 | 142.4 | 50.5 | 25.6 | 102.1 | | |
| | × (C × O) | 2  68.5 | 79.6 | 88.4 | 207.6 | 127.6 | 100.0 | 162.0 | 9.50 | 252 |
| $BX_1$ | O × (O × C) | 1  62.1 | 60.1 | 116.4 | 110.0 | 84.9 | 42.1 | 85.6 | | |
| | | 2  84.1 | 32.0 | 120.4 | 14.4 | 62.1 | 68.5 | 64.4 | | |
| | O × (C × O) | 1  56.9 | 80.9 | 54.1 | 86.5 | 118.1 | 60.9 | 52.1 | | |
| | | 2  29.6 | 68.1 | 136.1 | 64.0 | 100.1 | 72.9 | 84.1 | | |
| | (O × C) × O | 1  56.1 | 34.5 | 56.0 | 76.4 | 104.1 | 78.0 | 64.1 | | |
| | | 2  64.9 | 42.1 | 60.1 | 70.0 | 26.1 | 34.4 | 170.4 | | |
| | (C × O) × O | 1  58.9 | 74.1 | 36.4 | 69.6 | 46.1 | 194.4 | 76.9 | | |
| | | 2  72.5 | 88.4 | 42.4 | 122.0 | 32.1 | 38.4 | 70.4 | 8.00 | 504 |
| $BX_2$ | C × (O × C) | 1  96.5 | 72.4 | 40.9 | 38.5 | 57.6 | 90.1 | 70.4 | | |
| | | 2  62.1 | 30.5 | 132.4 | 70.0 | 46.9 | 31.6 | 44.1 | | |
| | C × (C × O) | 1  24.1 | 69.6 | 27.6 | 53.6 | 74.1 | 54.9 | 78.4 | | |
| | | 2  72.4 | 80.1 | 22.9 | 90.9 | 42.4 | 54.4 | 47.6 | | |
| | (O × C) × C | 1  20.0 | 26.0 | 48.9 | 206.4 | 124.4 | 19.6 | 115.6 | | |
| | | 2  45.6 | 90.4 | 100.1 | 62.0 | 82.1 | 76.4 | 99.6 | | |
| | (C × O) × C | 1  51.6 | 47.6 | 116.9 | 260.0 | 56.5 | 100.9 | 71.6 | | |
| | | 2  16.9 | 62.0 | 39.6 | 90.9 | 37.6 | 74.9 | 54.5 | 7.69 | 504 |

Discussion on the interpretation of the significant Crosses items is deferred. Also, for the present, the parental, $F_1$ and $F_2$ families only are considered.

### (c) Genetic analysis

Turning then to the distribution of bristle numbers in these families (table 3.4 and fig. 3.2), it will be recalled that the parent lines, O and C, are inbred. Individuals within each line are therefore genetically uniform. Thus it follows that differences between individuals within each line must be due to small differences in the environment, i.e. the variance of the distribution of O and of C bristle numbers is solely environmental.

**Table 3·4** Distribution of bristle counts for each family. (The variances here are measures of dispersion of distributions as they stand. They are *not* to be taken as estimates of family variance, for they are inflated by differences between scorers, cultures, and where applicable, crosses)

| Family | 18 | 19 | 20 | 21 | 22 | 23 | 24 | 25 | 26 | 27 | 28 | 29 | 30 | 31 | 32 | 33 | 34 | 35 | 36 | 37 | 38 | 39 | 40 | 41 | 42 | Mean | Variance |
|---|---|---|---|---|---|---|---|---|---|---|---|---|---|---|---|---|---|---|---|---|---|---|---|---|---|---|---|
| Oregon | | 2 | 11 | 19 | 30 | 29 | 26 | 12 | 6 | 1 | 4 | | | | | | | | | | | | | | | 22·9 | 7·68 |
| 6CL | | | | | | | | | | | | 1 | 5 | 15 | 22 | 25 | 22 | 26 | 15 | 5 | 2 | 1 | 0 | 1 | | 33·7 | 12·01 |
| $F_1$ | | | | | | 4 | 6 | 6 | 23 | 32 | 47 | 30 | 44 | 40 | 32 | 10 | 1 | 2 | 1 | | | | | | | 29·1 | 15·90 |
| $F_2$ | | | | | 6 | 9 | 10 | 15 | 23 | 33 | 36 | 24 | 32 | 38 | 15 | 12 | 8 | 6 | 4 | 6 | 5 | | | | | 28·8 | 36·50 |
| $BX_1$ | 1 | 3 | 3 | 15 | 39 | 53 | 62 | 59 | 69 | 70 | 64 | 48 | 37 | 19 | 11 | 5 | 0 | 1 | 1 | | | | | | | 26·2 | 19·82 |
| $BX_2$ | | | | | 2 | 2 | 2 | 6 | 12 | 22 | 46 | 54 | 81 | 86 | 67 | 64 | 37 | 35 | 23 | 13 | 5 | 2 | 0 | 0 | 1 | 31·3 | 21·39 |

**Table 3.5** Analyses of variance of bristle counts for each family (analysis of culture totals)

| Source | df | SS | MS | F | P |
|---|---|---|---|---|---|
| *Oregon* | | | | | |
| Scorers | 6 | 376 | 62·6 | 1·58 | >0·20 |
| Duplicates | 7 | 278 | 39·8 | | |
| Total | 13 | 654 | 50·3 | | |
| | | | | | |
| *6CL* | | | | | |
| Scorers | 6 | 1043 | 173·9 | 7·00 | 0·05–0·02* |
| Duplicates | 7 | 174 | 24·9 | | |
| Total | 13 | 1217 | 93·6 | | |
| | | | | | |
| $F_1$ | | | | | |
| Crosses (C) | 1 | 63 | 63·0 | <1 | >0·20 |
| Scorers (S) | 6 | 776 | 129·3 | ≃1 | >0·20 |
| C × S | 6 | 538 | 89·7 | <1 | >0·20 |
| Duplicates | 14 | 1807 | 129·1 | | |
| Total | 27 | 3184 | 117·9 | | |
| | | | | | |
| $F_2$ | | | | | |
| Crosses (C) | 1 | 3344 | 3344·1 | 26·01 | <0·001*** |
| Scorers (S) | 6 | 1578 | 263·0 | 2·05 | 0·20–0·10 |
| C × S | 6 | 1066 | 177·6 | 1·38 | >0·20 |
| Duplicates | 14 | 1800 | 128·6 | | |
| Total | 27 | 7788 | 288·4 | | |
| | | | | | |
| $BX_1$ | | | | | |
| Crosses (C) | 3 | 1394 | 464·5 | 4·96 | 0·01–0·001** |
| Scorers (S) | 6 | 1350 | 225·0 | 2·40 | 0·05–0·01* |
| C × S | 18 | 1121 | 62·3 | <1 | >0·20 |
| Duplicates | 28 | 3183 | 113·7 | | |
| Total | 55 | 7048 | 128·1 | | |
| | | | | | |
| $BX_2$ | | | | | |
| Crosses (C) | 3 | 2519 | 839·6 | 7·68 | <0·001*** |
| Scorers (S) | 6 | 535 | 89·2 | <1 | >0·20 |
| C × S | 18 | 2353 | 130·7 | 1·36 | >0·20 |
| Duplicates | 28 | 2677 | 95·6 | | |
| Total | 55 | 8084 | 147·0 | | |

It can be supposed, for simplicity, that bristle number is determined by a single gene with two alleles, $B$ and $b$. Oregon flies are thus $bb$ and 6CL flies $BB$. Now consider the genotype of individuals of the $F_1$ cross between these lines. Clearly, all will be $Bb$, that is, uniformly heterozygous. Thus again, the variance of the $F_1$ distribution must be solely environmental.

On a single gene model, the prediction is that the $F_2$ family will consist

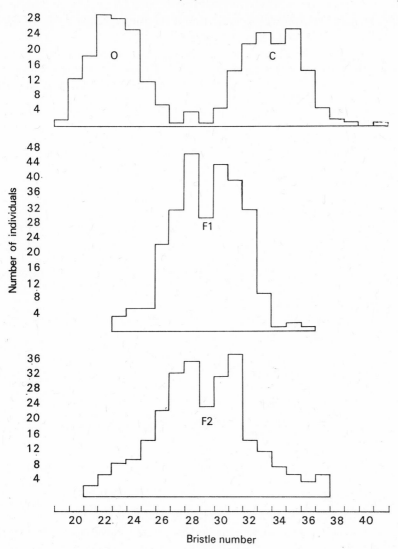

**Fig. 3.2.** The distribution of bristle numbers in the parental (O and C), F1 and F2 generations.

of three genotypes in the proportion $1/4$ *BB*, $1/2$ *Bb*, $1/4$ *bb*. Hence it can also be predicted that differences between $F_2$ individuals will be partly environmental, partly genetic. To put this another way, the variance of the $F_2$ generation will consist of two components, one genetical, the other environmental. This in turn implies that this variance will be larger than those of either the parental or $F_1$ generation, because it is inflated by the segregation of

$B$ and $b$. And indeed it will be noticed that the spread of the $F_2$ distribution is greater than that of the parents and $F_1$.

The $F_2$ generation allows us to estimate the proportion of the total phenotypic variance of this family that is due to genetic differences, known as the *heritability* of the character, is this case bristle number. Algebraically the argument appears thus.

| Family | | | Variance Expected | Observed* |
|--------|---|---|------------------|-----------|
| $P_1(O)$ | $\sigma^2_{P_1}$ | $=$ | $E$ | 3·28 ⎫ |
| $P_2(C)$ | $\sigma^2_{P_2}$ | $=$ | $E$ | 3·59 ⎬ 4·20 |
| $F_1$ | $\sigma^2_{F_1}$ | $=$ | $E$ | 4·96 ⎭ |
| $F_2$ | $\sigma^2_{F_2}$ | $=$ | $E + G_{F_2}$ | 9·50 |

*Taken from table 3.3

where $\sigma^2$ is the variance of the distribution, $E$ is the variance component due to the environment, and $G_{F_2}$ is the variance component of the $F_2$ generation due to genetical differences between individuals.

It is now a simple matter to estimate this genetical component, $G_{F_2}$. There are three independent estimates of $E$, provided by $\sigma^2_{P_1}$, $\sigma^2_{P_2}$ and $\sigma^2_{F_2}$ (where $\wedge$ denotes a sample estimate of). Given certain assumptions these are expected to be equal. In fact, they are not, so a weighted average is taken whereby one quarter of each of the parental variances is added to half the $F_1$ estimate, namely 12·87. By substitution one has

$$\hat{\sigma}^2_{F_2} = E + G_{F_2} = 4 \cdot 20 + G_{F_2} = 9 \cdot 50$$

and

$$G_{F_2} = 9 \cdot 50 - 4 \cdot 20 = 5 \cdot 30$$

The genetical proportion of the $F_2$ variance, the heritability, is thus

$$\hat{h}^2_{F_2} = 5 \cdot 30 / 9 \cdot 50 = 0 \cdot 558 \text{ or approximately 56 per cent.}$$

This method of estimating the genetical proportion of the variance of a segregating family is known as the *broad* heritability, because $G_{F_2}$ includes *all* genetical effects. Later on situations will be encountered in which a slightly different approach is possible.

### (d) Sex-linked inheritance

So far, the $F_2$ family variance has been partitioned into environmental and genetical components. The present data is however capable of yielding further information about the genetical control of bristle number. In arguing that the $F_1$ family variance is expected to be caused by environmental differences only, it has been assumed that (1) determination of bristle number is solely by autosomal genes, i.e. and not also by sex-linked genes, and (2) no extra-nuclear determination is involved. If either of these assumptions fail, then a difference between reciprocal $F_1$ families can be expected, i.e. the expression of the character depends on the way the cross is made.

The analysis of variance of the $F_1$ culture means (table $3 \cdot 5$) shows quite clearly that differences between reciprocals are no greater than those due to sampling variation or chance effects. The assumption concerning the absence of extra-nuclear effects therefore appears to be valid. The present $F_1$ data however provides no test of the assumption in respect of the absence of sex-linked gene effects because females only have been scored. Under these circumstances, since reciprocal differences between *male* $F_1$ progeny are expected with sex-linkage, one is clearly unable to conclude that such effects are absent. If sex-linked effects play a part in determining bristle number, however, then a difference between the female progeny of $F_2$ reciprocal crosses can be expected.

The analysis of variance of the $F_2$ means (table $3 \cdot 5$) shows that there is in fact a highly significant difference between reciprocal $F_2$ females. The paternal parent of the $(O \times C) \times (O \times C)$ females, an $O \times C$ individual, carries an Oregon $X$-chromosome. All his female progeny therefore receive this chromosome, together with another $X$-chromosome from their maternal parent, which because of recombination, will be partly Oregon and partly 6CL in origin. Similarly, of the pair of $X$-chromosomes which $(C \times O) \times (C \times O)$ females carry, one will be an intact 6CL $X$-chromosome received from their paternal parent, a $C \times O$ male. Because of sex-linkage therefore, the $(O \times C) \times (O \times C)$ females have a lower bristle number than their $(C \times O) \times (C \times O)$ reciprocals. Had male, rather than female, flies been scored throughout one would expect to detect a significant difference between reciprocal $F_1$ progenies, but not between those of the $F_2$ generation. Though we are not concerned with the backcross progenies for the moment, it is perhaps worth noting that in these too, there is clear evidence of sex-linked inheritance.

In summary, then, this experiment allows: (1) the partition of genetical and environmental components of total or phenotypic variation, and (2) the detection of extra-nuclear and/or sex-linked gene effects.

## III. SELECTION

## EXPERIMENT 2

The purpose of this experiment is to provide a further demonstration that bristle number is a heritable character. The initial material or *base population* can be any segregating population of flies, so that the $F_2$ family of the previous experiment can be used. It is desirable to set up separate cultures of this generation for the experiment. A convenient procedure is to provide each student in the class with a pair of $F_2$ cultures, one being $(O \times C) \times (O \times C)$ [culture (1)], the other $(C \times O) \times (C \times O)$ [culture (2)]. Each student then scores ten male and ten female progeny from each of these cultures. To establish the high bristle number selection line, the female individual with the highest score from culture (1) is mated with the male individual with the highest score from culture (2). Similarly, the high female from culture (2) is mated with the high male from (1). In this way, the high selection line is set up as a pair of parallel cultures in which matings are between double first cousins (table 3.6). This mating system causes less inbreeding than the simpler sib-mating of high males and high females from the same culture, with a better, more prolonged response to selection.

Table 3.6 Double first cousin mating system by which selection lines are maintained

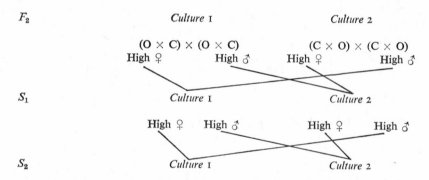

The low line is established in the same way except that, of course, the individuals selected have low bristle numbers.

This procedure is then repeated at each cycle of selection. Since only one from ten individuals (of each sex) is chosen to maintain the line, the intensity of selection is 10 per cent. It is advisable to duplicate the pair of parallel cultures in each selection line by mating the next most extreme individuals in case the first choice matings fail.

The results of a selection experiment carried out in this way by a class of six students are shown in fig. 3.3. In both high and low lines there is a steady response to selection. This response tells one, firstly, that bristle number is a heritable character, for, if the difference between the bristle numbers of the selected parents and their cultures means was purely environmental, no response to selection would be expected or obtained. Secondly, the steady nature of the response indicates that several genes must be involved in the determination of bristle number. Indeed, this conclusion may also be drawn from the distribution of $F_2$ family bristle number. Thus if bristle number is determined by a single gene it would be expected that three phenotypes in the $F_2$ family would be observed, corresponding to the three genotypes *BB, Bb bb*. Because of the 'blurring' effect of the environment, however, the $F_2$ distribution might be expected on this hypothesis to be a composite of three overlapping distributions. Nevertheless, each distribution could be expected to be discernible, since, as has been previously shown, the heritability of this character is 56 per cent. But the $F_2$ distribution of bristle numbers is approximately continuous. Thus it is clear that this simple hypothesis does not satisfactorily account for these observations. Hence more than one gene must be concerned in the determination of bristle number.

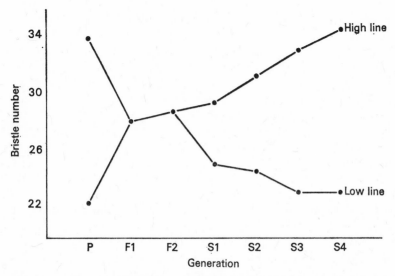

**Fig. 3.3.** Response to selection for high and for low bristle number. Each point represents the mean of 12 cultures, i.e. 240 flies.

Returning now to the results of the selection experiment, it is expected on the one gene hypothesis that a single generation of selection should cause complete fixation of the genes under selection, i.e. all high line individuals would be *BB*, all low line flies *bb*. Under these circumstances no further response could be expected or observed. Allowing for the effects of the environment on the expression of the bristle determining genes, response might not be complete after one generation only of selection. But again, with a heritability of 56 per cent, most of the potential response would occur in the first cycle of selection. We see however that response to selection in both high and low lines is reasonably steady. Hence, as before, more than one gene must be concerned in the determination of bristle number.

## IV. THE COMPONENTS OF FAMILY MEANS

### The model

In previous sections of this chapter a single gene model has been used to predict the genetical composition of the parental, $F_1$ and $F_2$ generations. This approach can now be extended to derive the expectations of the means of these and the backcross generations.


## EXPERIMENT 3

This experiment concerns the analysis of all the families shown in table 3.1, the data from which appear in table 3.2, i.e. the $BX$ families are now included. There are, of course, two types of backcross families; one concerning the cross $F_1 \times P_1$, the other $F_1 \times P_2$. But taking reciprocal $F_1$ families into account, each type of backcross family can be produced in four ways, depending on whether the female or male is taken from the inbred line, and depending on which reciprocal $F_1$ cross is used for the other parent.

In order to derive expectations for the mean bristle counts of all the families under consideration, one can imagine any bristle phenotype to be determined by two genetic components, whose magnitudes are $m$ and $d$. Firstly, there is the effect of the gene or genes specifically concerned in the determination of differences in bristle number. On the single gene model this is the effect of substituting $B$ for $b$ in homozygotes which is represented as $d$. Secondly, there is the average effect of all other genes determining the phenotype of the zygote, or the mean background effect represented by $m$. No genes operate in a vacuum and it has been seen that both sex and size affect bristle number. Thus it is clearly important to allow for this background effect in expectations of family means.

The effect of gene $B, b$ is defined in terms of a deviation (hence $d$) about the average background effect, $m$, in the following way*

| Family | Observed mean | Expected mean |
|--------|---------------|---------------|
| $O(P_1)$ | 22·9  $(\bar{P}_1)$ | $m-d$ |
| $C(P_2)$ | 33·7  $(\bar{P}_2)$ | $m+d$ |

Having two equations here, by equating observed and expected, one can then solve for $m$ and $d$.

Thus

$$\hat{m} = \tfrac{1}{2}(\bar{P}_1 + \bar{P}_2) = 28\cdot3 \text{ bristles}$$
$$\hat{d} = \tfrac{1}{2}(\bar{P}_2 - \bar{P}_1) = \phantom{0}5\cdot4 \text{ bristles.}$$

* Since $d$ by definition must always be positive the inbred line with the larger mean (in this case C) is always equated to $m+d$ and that with the smaller mean (in this case O) to $m-d$.

Turning now to the $F_1$ family, in which all individuals are $Bb$, one might expect $F_1=m$, the $d$'s as it were, cancelling out. But this assumes that the $F_1$ is exactly intermediate to the parent, inbred lines, i.e. that the gene manifests no dominance. Clearly a more general expectation is preferable so that allowing for dominance one can write

| Family | Observed mean | Expected mean |
|--------|---------------|---------------|
| $F_1$ | 29·1 $(\bar{F}_1)$ | $m+h$ |

where $h$ measures dominance and can take sign according to whether $B$ is dominant to $b$ ($h$ positive) or the other way round ($h$ negative).

Since previously, it has been found that $m=28\cdot3$, $h$ is by substitution found to be

$$\hat{h}=\bar{F}_1-\tfrac{1}{2}(\bar{P}_1+\bar{P}_2)=0\cdot8 \text{ bristles.}$$

Now the $F_2$ and $BX$ families comprise, on this single gene model, proportions of the three genotypes $BB$, $Bb$ and $bb$. Since expectations have been derived for each of these genotypes separately, it is a straightforward matter to derive the expectations for the $F_2$ and $BX$ family means (table 3.7).

**Table 3.7** The expected composition of generation means

| Family | Observed mean | Genetic composition | Expected mean |
|--------|---------------|---------------------|---------------|
| O | 22·9 | $bb$ | $m - [d]$ |
| C | 33·7 | $BB$ | $m + [d]$ |
| $F_1$ | 29·1 | $Bb$ | $m + [h]$ |
| $F_2$ | 28·8 | $\tfrac{1}{4}BB\ \tfrac{1}{2}Bb\ \tfrac{1}{4}bb$ | $m + \tfrac{1}{2}[h]$ |
| $BX_1$ | 26·2 | $\tfrac{1}{2}Bb\ \tfrac{1}{2}bb$ | $m - \tfrac{1}{2}[d] + \tfrac{1}{2}[h]$ |
| $BX_2$ | 31·3 | $\tfrac{1}{2}BB\ \tfrac{1}{2}BB$ | $m + \tfrac{1}{2}[d] + \tfrac{1}{2}[h]$ |

### (a) Predicting family means

The adequacy of our model can now be tested by *predicting* the $F_2$ and $BX$ family means, using for this purpose the estimates of $m$, $d$ and $h$ obtained from the non-segregating families.

For example, the predicted mean for $BX$ is obtained as

$$\overline{BX}_1=m-\tfrac{1}{2}d+\tfrac{1}{2}h=28\cdot3-\tfrac{1}{2}\times5\cdot4+\tfrac{1}{2}\times0\cdot8=26\cdot0 \text{ bristles.}$$

The remaining predictions are

| Family | Observed mean | Predicted mean | Discrepancy |
|--------|---------------|----------------|-------------|
| $F_2$ | 28·8 | 28·7 | +0·1 |
| $BX_1$ | 26·2 | 26·0 | +0·2 |
| $BX_2$ | 31·3 | 31·4 | −0·1 |

The agreement between observed and predicted means here is remarkably good. This appears to imply that a single gene model adequately accounts for the variation between generation means. Yet, as has been seen, the data when examined in other ways suggest that more than one gene is concerned in the determination of the character. Why then is agreement between the observed and predicted values of the generation means so good here?

It turns out that where a character is determind by several genes, $d$ measures their net effect. Thus the genotype of 6CL individuals might in fact be *BBCC* and that of Oregon *bbcc*. In this case $d$ is a measure of the joint effects of $B$ and $C$ in homozygotes. For instance, if $B$ and $C$ have equal effects, each would contribute 2·7 to the value of $d=5·4$. Alternatively, the genotype of 6CL flies might be thought of as *BBcc* and that of Oregon *bbCC*. If the effect of $B$ were the same as $c$ then $d$ would equal zero. Clearly in this case, since $d \neq 0$, the effect of $B$ would have to be greater than that of $C$. The point here, then, is that $d$ is a measure of the *net* effect of all the genes determining bristle number in the homozygote. Since $d=0$ does not necessarily imply the genetical identity of the inbred lines, it is important to remind oneself of this fact by writing the contribution of genes in homozygotes as [$d$] rather than $d$. In the same way, where several genes are concerned in the determination of the character, $h$ is a measure of their net dominance effects. This effect can therefore be written as [$h$].

### (b) Scaling tests

Now in the present case, agreement between the observed and predicted values of the family means is sufficiently good to persuade us that a simple genetical model accounts for nearly all the observed differences between generation means. But this will not always be the case. What is required therefore is a procedure which permits a more objective test of the adequacy of a model, a facility provided by the so-called *scaling tests* of Mather (1949).

Now the $F_2$ was predicted by using estimates of $m$, [$d$] and [$h$] obtained from $\bar{P}_1$, $\bar{P}_2$ and $\bar{F}_1$. Formally, this prediction is

$$\bar{F}_2 = \tfrac{1}{4}\bar{P}_1 + \tfrac{1}{2}\bar{F}_1 + \tfrac{1}{4}\bar{P}_2$$

or
$$\tfrac{1}{4}\bar{P}_1 + \tfrac{1}{4}\bar{P}_2 + \tfrac{1}{2}\bar{F}_1 - \bar{F}_2 = 0$$

Clearing fractions and calling this quantity, $C$, the prediction now is that

$$C = \bar{P}_1 + \bar{P}_2 + 2\bar{F}_1 - 4\bar{F}_2 = 0$$

Similarly, for the $BX$ families

$$\overline{BX}_1 = \tfrac{1}{2}\bar{F}_1 + \tfrac{1}{2}\bar{P}_1$$

and
$$\overline{BX}_2 = \tfrac{1}{2}\bar{F}_1 + \tfrac{1}{2}\bar{P}_2$$

or
$$A=\bar{P}_1+\bar{F}_1-2\overline{BX}_1=0$$

and
$$B=\bar{P}_2+\bar{F}_1-2\overline{BX}_2=0.$$

But these generation means, being estimates, are subject to sampling error; that is to say, they are sample estimates of unknown population parameters (hence the use of the symbol $\wedge$ to denote this fact). Thus the scaling test quantities $A$, $B$ and $C$ are also subject to sampling variation.

Assuming that the estimates of these generations means are uncorrelated and that the sampling variance of the generation means are

$$\hat{\sigma}^2_{\bar{P}_1},\ \hat{\sigma}^2_{\bar{P}_2},\ \hat{\sigma}^2_{\bar{F}_1},\ \hat{\sigma}^2_{\bar{F}_2},\ \hat{\sigma}^2_{\overline{BX}_1}\ \text{and}\ \hat{\sigma}^2_{\overline{BX}_2},$$

then the sampling variance of $A$, $B$ and $C$ are:

$$\hat{\sigma}^2_A=\hat{\sigma}^2_{\bar{P}_1}+\hat{\sigma}^2_{\bar{F}_1}+4\hat{\sigma}^2_{\overline{BX}_1}$$
$$\hat{\sigma}^2_B=\hat{\sigma}^2_{\bar{P}_2}+\hat{\sigma}^2_{\bar{F}_1}+4\hat{\sigma}^2_{\overline{BX}_2}$$
$$\hat{\sigma}^2_C=\hat{\sigma}^2_{\bar{P}_1}+\hat{\sigma}^2_{\bar{P}_2}+4\hat{\sigma}^2_{\bar{F}_1}+16\hat{\sigma}^2_{\bar{F}_2},$$

i.e. the sampling variances of the scaling test quantities are the weighted sums of the variance of generation means, the weights being the squares of the coefficient the mean carries in the scaling comparison.

The degrees of freedom of $\hat{\sigma}^2_A$, $\hat{\sigma}^2_B$ and $\hat{\sigma}^2_C$ are found as the sum (unweighted) of the degrees of freedom of the sampling variance of the generation mean.

The appropriate test of significance is therefore

$$z=\frac{A}{\hat{\sigma}_A}$$

where $z$ is the one-tailed normal deviate, or, where the number of degrees of freedom of $\sigma_A$ is small,

$$t_{(df)}=\frac{A}{\hat{\sigma}_A}$$

Let us now look at the data of the present experiment. From the values shown, i.e. table 3.2, it is found that

$$\hat{A}=22{\cdot}9+29{\cdot}1-52{\cdot}4=-0{\cdot}4$$
$$\hat{B}=33{\cdot}7+29{\cdot}1-62{\cdot}6=+0{\cdot}2$$
$$\hat{C}=22{\cdot}9+33{\cdot}7+58{\cdot}2-115{\cdot}2=-0{\cdot}4$$

The sampling variance of the generation means may be obtained from the analyses of variance of culture *totals* (table 3.5). To take, for example, the sampling variance of $\bar{P}_1$, $\hat{\sigma}^2_{\bar{P}_1}$. Note that the, Total $SS$ of the analysis of variance of the Oregon culture totals is 654 with 13 degrees of freedom. The best estimate of the variance of the distribution of Oregon culture *means* is therefore the total mean square of this analysis divided by $10 \times 10 = 100$, i.e. $50 \cdot 3/100 = 0 \cdot 503$. It will be recalled that the generation mean $\bar{P}_1 = 22 \cdot 9$ is based on 14 culture means. Hence the variance of the distribution of generation means is obtained as $0 \cdot 503/14 = 0 \cdot 0359$, i.e. $\hat{\sigma}^2 \bar{P}_1 = 0 \cdot 0359$. The sampling variances of $\bar{P}_2$ and $\bar{F}_1$ are obtained in the same way. Those of $\bar{F}_2$, $\overline{BX}_1$ and $\overline{XB}_2$ must be calculated on the Total *SS less* the Crosses *SS*. This is because we have averaged over crosses in order to obtain these family means, so that variation between

**Table 3.8** Sampling variance of generation mean bristle number

| | | | | | |
|---|---|---|---|---|---|
| O | $\hat{\sigma}^2 \bar{P}_1$ | = | 0·0359 | with | 13 *df* |
| C | $\hat{\sigma}_2 \bar{P}_2$ | = | 0·0669 | with | 13 *df* |
| $F_1$ | $\hat{\sigma}_2 \bar{F}_1$ | = | 0·0421 | with | 27 *df* |
| $F_2$ | $\hat{\sigma}_2 \bar{F}_2$ | = | 0·0610 | with | 27 *df* |
| $BX_1$ | $\hat{\sigma}_2 \overline{BX}_1$ | = | 0·0194 | with | 52 *df* |
| $BX_2$ | $\hat{\sigma}_2 \overline{BX}_2$ | = | 0·0191 | with | 52 *df* |

crosses within these families has become irrelevant, and if included, would spuriously inflate the sampling variance. The full set of six sampling variances is shown in table 3.8 and the tests of significance on the scaling tests in table 3.9. From these, it is clear that our genetical model accounts satisfactorily for the difference between family means.

**Table 3.9** Tests of significance on the scaling tests. (Since the degrees of freedom of $\sigma^2_A$, $\sigma^2_B$ and $\sigma^2_C$ are greater than 30, being 95, 95 and 80 respectively, we use the *z*-test criterion. A tabulation of *z* may be found on p. 434 of Steel & Torrie 1960)

| | | | | | | | | | | | |
|---|---|---|---|---|---|---|---|---|---|---|---|
| $\hat{A}$ | = | −0·4 | $\hat{\sigma}^2_A$ | = | 0·1556 | $\hat{\sigma}_A$ | = | 0·395 | $z$ | = | 1·01, $P$ = 0·16 |
| $\hat{B}$ | = | +0·2 | $\hat{\sigma}^2_B$ | = | 0·1854 | $\hat{\sigma}_B$ | = | 0·431 | $z$ | = | 0·46, $P$ = 0·32 |
| $\hat{C}$ | = | −0·4 | $\hat{\sigma}^2_C$ | = | 1·2472 | $\hat{\sigma}_C$ | = | 1·117 | $z$ | = | 0·36, $P$ = 0·36 |

# V. SECOND DEGREE STATISTICS: VARIANCE COMPONENTS

## (a) Expectations

In discussing the cause of quantitative variation earlier the total phenotypic variance of the $F_2$ family was partitioned into genetical and environmental components, i.e. $\sigma^2_{F_2} = G_{F_2} + E$.

By specifying gene effects in the manner employed in obtaining expectations of family means, namely, by using $m$, $[d]$ and $[h]$, it is possible to partition the genetical component, $G$, of both the $F_2$ and backcross family variance.

Thus, as previously noted, the genotypic frequencies and values in $F_2$ are:

| Genotype | $BB$ | $Bb$ | $bb$ | Mean |
|---|---|---|---|---|
| Genotype frequency | $\frac{1}{4}$ | $\frac{1}{2}$ | $\frac{1}{4}$ | |
| Genetic value | $m+d$ | $m+h$ | $m-d$ | $m+\frac{1}{2}h$ |

The genetical variance of $F_2$ individuals about their mean is then obtained as:

$$G_{F_2}=\tfrac{1}{4}d^2+\tfrac{1}{2}h^2+\tfrac{1}{4}d^2-(\tfrac{1}{2}h)^2$$
$$=\tfrac{1}{2}d^2+\tfrac{1}{4}h^2$$

— the $m$'s not entering into the calculations since they are constant. Now for many genes with independent distribution and effects, this expectation becomes,

$$G_{F_2}=\tfrac{1}{2}\Sigma_i d_i^2+\tfrac{1}{4}\Sigma_i h_i^2$$

or to use a more concise notation, writing

$$D=\Sigma_i d_i^2 \text{ and } H=\Sigma_i h_i^2$$

gives

$$G_{F_2}=\tfrac{1}{2}D+\tfrac{1}{4}H$$

and

$$\sigma^2{}_{F_2}=\tfrac{1}{2}D+\tfrac{1}{4}H+E.$$

For the backcross families we have similarly,

| | $BX_1$ | | | | $BX_2$ | | |
|---|---|---|---|---|---|---|---|
| Genotype | $Bb$ | $bb$ | Mean | | $BB$ | $Bb$ | Mean |
| Genotype frequency | $\frac{1}{2}$ | $\frac{1}{2}$ | | | $\frac{1}{2}$ | $\frac{1}{2}$ | |
| Genetic value | $m+h$ | $m-d$ | $m-\frac{1}{2}d+\frac{1}{2}h$ | | $m+d$ | $m+h$ | $m+\frac{1}{2}d+\frac{1}{2}h$ |

Hence the genetical variance of each backcross family is

$$G_{BX_1}=\tfrac{1}{2}h^2+\tfrac{1}{2}d^2-(-\tfrac{1}{2}d+\tfrac{1}{2}h)^2=\tfrac{1}{4}d^2+\tfrac{1}{4}h^2+\tfrac{1}{2}dh$$
$$G_{BX_2}=\tfrac{1}{2}d^2+\tfrac{1}{2}h^2-(\tfrac{1}{2}d+\tfrac{1}{2}h)^2 =\tfrac{1}{4}d^2+\tfrac{1}{4}h^2-\tfrac{1}{2}dh$$

The inconvenience of the terms in $dh$ can be removed by taking the sum of these variances, i.e.

$$G_{BX_1}+G_{BX_2}=\tfrac{1}{2}d^2+\tfrac{1}{2}h^2$$

Thus for many genes with independent distributions and effects

$$G_{BX_1}+G_{BX_2}=\tfrac{1}{2}D+\tfrac{1}{2}H$$

and

$$\sigma^2{}_{BX_1}+\sigma^2{}_{BX_2}=\tfrac{1}{2}D+\tfrac{1}{2}H+2E$$

—the coefficient of the environmental component being 2 because the sum of two independent variances is being considered.

### (b) Estimation

The full table of expectations can now be set out

| Family | Observed variance | | Expected variance | | |
|---|---|---|---|---|---|
| $P_1$ | 3·28 | | $\sigma^2_{P1}$ | $=$ | $E$ |
| $P_2$ | 3·59 | 4·20 | $\sigma^2_{P2}$ | $=$ | $E$ |
| $F_1$ | 4·96 | | $\sigma^2_{F1}$ | $=$ | $E$ |
| $F_2$ | 9·50 | | $\sigma^2_{F2}$ | $= \frac{1}{2}D + \frac{1}{4}H +$ | $E$ |
| $BX_1 + BX_2$ | 15·69 | | $\sigma^2_{B1} + \sigma^2_{B2}$ | $= \frac{1}{2}D + \frac{1}{2}H +$ | $2E$ |

As before one obtains a joint estimate of $E$ as the weighted average of the parental and $F_1$ family variances, i.e. $\frac{1}{4}\sigma^2_{P1}+\frac{1}{4}\sigma^2_{P2}+\frac{1}{2}\sigma^2_{F1}$.

Subtracting $E$ and $2E$ from $\sigma^2_{F2}$ and $\sigma^2_{BX1}+\sigma^2_{BX2}$ respectively we have:

$$\frac{1}{2}D+\frac{1}{4}H=5\cdot30 \qquad (1)$$

$$\frac{1}{2}D+\frac{1}{2}H=7\cdot29 \qquad (2)$$

From (2)—(1)

$$\frac{1}{4}H=1\cdot99 \text{ so that } H=7\cdot96.$$

And by substitution in (1)

$$\frac{1}{2}D=5\cdot30-1\cdot99=3\cdot31$$

so that $\qquad\qquad\qquad D=6\cdot62$

(As a check on these computations one can substitute these values in the expectations of $\sigma^2_{F2}$ and $\sigma^2_{BX1}+\sigma^2_{BX2}$ to recover the observed values of these statistics.)

### (c) Broad and narrow heritability

Earlier on heritability was estimated as the ratio of the genetical variance to the total phenotypic variance of the $F_2$ family:

$$h^2_{F2}=\frac{G_{F2}}{\sigma^2_{F2}}$$

but since it can now be written

$$G_{F2}=\frac{1}{2}D+\frac{1}{2}H$$

the heritability formula can be written in a rather more explicit form, namely,

$$h^2_{F2}=\frac{\frac{1}{2}D+\frac{1}{4}H}{\frac{1}{2}D+\frac{1}{4}H+E}$$

The ability to cast this formula in this form draws one's attention to two important points about the use of heritability estimates.

First, the heritability estimate calculated depends on the family in question. If one considers, for example the ratio of genetical variance to total variance in the backcross families one finds

$$\hat{h}^2{}_{(BX_1+BX_2)}=\frac{G_{(BX_1+BX_2)}}{\sigma^2{}_{BX_1}+\sigma^2{}_{BX_2}}=\frac{\frac{1}{2}D+\frac{1}{2}H_2}{\frac{1}{2}D+\frac{1}{2}H+E}=\frac{7\cdot29}{15\cdot69}=0\cdot465 \text{ or } 47 \text{ per cent,}$$

i.e. the heritability of the character is less than in the previous estimate. But since the composition of the genetical portion of the variance in backcross families is different from that of $F_2$ families, it is clear that these estimates of heritability are not comparable—a point frequently overlooked in discussions concerning heritabilities. There is no real difficulty here, provided heritability is estimated in a standard form, a task easily accomplished, given estimates of the genetical components $D$ and $H$. These may, of course, be obtained from a wide variety of generations. For inbreeding experiments of the type under discussion it is then possible to assemble these estimates so as to calculate a standard heritability on the basis of the expectation in respect of the $F_2$ generations.

The second point concerns the nature of heritability estimates and their use for the purpose of prediction. In principle, given estimates of $D$, $H$ and $E$, the phenotypic variance of any generation for which we have worked out an expectation may be predicted. In practice, the situation one would most like to predict is the response of the character of interest to selection. Now, on the single gene model, it is clear that the limit of response to selection will be reached when all individuals in the high line are $BB$, and all in the low line are $bb$. At this stage, therefore, differences between selection lines concern comparisons between homozygotes, which, as shown earlier are characterized by $d$ or additive genetical effects rather than $h$ or dominance effects, the latter being a property of heterozygotes. Thus for the purpose of predicting advance under selection the interest lies more in the proportion of additive genetical variance than in total genetical variance. An appropriate estimate of heritability for this purpose would thus be:

$$\hat{h}^2{}_{F_2}=\frac{\frac{1}{2}D}{\frac{1}{2}D+\frac{1}{4}H+E}=\frac{3\cdot31}{9\cdot50}=0\cdot348 \text{ or } 35 \text{ per cent.}$$

Defined in this way, *narrow* heritability has been estimated; the previous estimate was defined as *broad* heritability, so that the distinction between them is made clear.

### (d) Dominance and potence

These estimates of additive and dominance genetical variance are $D=6\cdot62$ and $H=7\cdot96$. Remembering that $D=\Sigma d^2$ and $H=\Sigma h^2$, these variance components provide an estimate of the magnitude of dominance effects relative to additive effects—that is, an estimate of dominance itself—which is

$$\sqrt{\frac{H}{D}}=\sqrt{\frac{7\cdot96}{6\cdot62}}=\sqrt{1\cdot20}=1\cdot09.$$

It was shown earlier that the first order components of generations means were $[d]=5\cdot4$ and $[h]=0\cdot8$. On these estimates, the magnitude of dominance is

$$\frac{[h]}{[d]}=\frac{0\cdot8}{5\cdot4}=0\cdot15.$$

On second degree statistics therefore, there is either complete ($1\cdot00$) or over-dominance ($> 1\cdot00$). On the components of family means, dominance is all but absent. What is the reason for this discrepancy?

First, it will be recalled that in general $[h]$ is a measure of the *net* dominance effects of the genes determining the character in question, whereas $H=\Sigma h^2$ is a measure of their gross effects. Suppose, for example, that bristle number is determined by a pair of genes $B$ and $C$. It can be supposed further that $B$ and $C$ are dominant to their respective alleles $b$ and $c$, and that this dominance is equal in sign and in magnitude, i.e. $h_1=h_2=1$. Then in these circumstances $[h]=2$ and $H=\Sigma h^2=2$ also. But suppose that the dominance relationship of the alleles at the second locus are reversed such that $c$ is dominant to $C$. In this case, while $h_1=1$ as before, $h_2=-1$, i.e. $h_1=-h_2=1$. Again, as before $H=2$, but now $[h]=0$. Thus it can be seen that while $H$ remains unaffected by the sign, that is to say, the direction of the dominance effects, this is not true of $[h]$, which, when dominance is ambi-directional, under-estimates its magnitude. For this reason, it is perhaps better to refer to $[h]$ as a measure of *potence*, rather than dominance.

Second, all estimates, whether they be from first- or second-order statistics are reliable only to the extent that the model used in deriving expectations is valid. The very good fit (on scaling tests) obtained from the estimates of $m$, $[d]$ and $[h]$ encourages one to believe that the right model has been used in predicting generation means. The adequacy of the model in respect of $D$, $H$ and $E$ cannot be tested because they are estimated as perfect fit solutions. Mather and Jinks (1971) describe methods for testing the adequacy of $D$, $H$ and $E$ models analogous to those used for generation means. They require, however, observations on more segregating generations than the $F_2$ and backcrosses described here.

## (e) Examining assumptions

In the present case, moreover, there is evidence that two of the assumptions that have been made in deriving our model are not strictly valid. The first is that the genes determining bristle number are all carried on the autosomal chromosomes, for the predictions that were made regarding the composition of the $F_2$ and backcross families assume this. It has been shown that there is good evidence that some of the genes determining this character are carried on the $X$-chromosome, thus causing sex-linked inheritance. The failure of this assumption is of little consequence in respect of family means, because reciprocal crosses were averaged. Sex-linkage will however make its full contribution to culture variances. Since there was no allowance for this type of effect in deriving expectations of family variance (though this can be done) it is clear that there is no expectation of the model fitting the data very well.

The second assumption which deserves scrutiny concerns the magnitude of the environmental component in relation to genotype. It has been assumed that the environmental effects on bristle number do not depend at all on the genotype of the family or individual. What is the evidence on point? The variance of the non-segregating families, together with their degrees of freedom are

| Family | Variance | Degrees of freedom |
|--------|----------|--------------------|
| $P_1$ | 3·28 | 126 |
| $P_2$ | 3·59 | 126 |
| $F_1$ | 4·96 | 252 |

The ratio of the largest to the smallest variance here is

$$F=\frac{4\cdot96}{3\cdot28}=1\cdot51$$

which for 252 and 126 degrees of freedom is certainly significant even allowing for the fact that the largest of three variances has been compared with the smallest. Thus the expression of $Bb$ heterozygotes appears to be more susceptible to environmental variation than that of either $BB$ or $bb$ homozygotes. Or, to put it another way, the reaction of an individual to a given environmental change depends on the genotype. Indeed, the phenomenon of *genotype–environment interaction* is frequently encountered in studies of this type and with more extensive data it is possible to allow for its effects in a way similar to those of the genotype. As with sex-linkage, the consequences of the failure of the assumption of no genotype–environment interaction fall more heavily on estimates from second-degree statistics.

A final point relevant to this discussion concerns the process of estimation itself in these circumstances. Estimation has been made of three quantities, $D$, $H$ and $E$ from just three equations, which as noted earlier, gives a perfect fit situation. If, however, due to an inadequate model, $D$, for instance, is under-estimated, $H$ will be over-estimated, i.e. the estimate of $D$ and $H$ are negatively correlated. Thus the ratio $H/D$, which was used earlier to estimate the dominance level for bristle number, is rather sensitive to small changes in $D$.

Taking all these considerations into account, it is clear that with the present experimental evidence it would be unwise to place too much reliance on these second-degree components. This, however, need not always be the case and indeed with more extensive data it has been shown that bristle number in these lines and their derivative crosses is in fact determined largely by additive gene effects.

## VI. EXPERIMENTS WITH OTHER ORGANISMS

Though *Drosophila* is in many ways a very convenient organism for the experiments that have been discussed, other organisms may be used. The chief requirement here is that inbred lines should be available. This requirement is met by several species, the most convenient of which are probably the small flowering plant, *Arabidopsis thaliana* and the mouse (chapter 2). Nevertheless, experiments with either of these species would take longer and could not be time-tabled in the manner possible with *Drosophila* experiments.

A comparable experiment with a pair of inbred lines of *Arabidopsis* would take six to seven weeks (p. 30). The earliest flowering lines of this species can flower within three weeks of the sowing date, provided that horticultural techniques are of a high standard and that natural daylight is supplemented with standard glasshouse mercury vapour lights (this is assumed to be necessary since class experiments are not usually performed during the best months of the year). This type of experiment can be sown as a miniature field experiment on a specially reinforced glasshouse bench covered with about 6 in of soil, the plants being sown at 2-in intervals on the square. The character days to flower or flowering time is a convenient one, but with this, scoring must be carried out each day whilst the experiment is in progress.

With mice, the main problem is one of maternal effects in which the dam contributes to many aspects of the progeny environment. With skill it is possible to carry out cross-fostering procedures. As with *Arabidopsis*, technical standards of hygiene and management must be high.

## VII. RANDOM MATING: EQUAL GENE FREQUENCIES

### EXPERIMENT 4

The analysis of families including and arising from a cross between a pair of inbred lines has already been discussed. The exact knowledge of the parental material in this situation allows a simple analysis of the families derived from them. This kind of experiment, therefore, is particularly suitable as an introduction to the study of quantitative inheritance. In many species, however, inbred lines are not available. What is required, therefore, are analyses of heritable variation which do not assume that the initial or parental material is inbred.

One rather simple breeding design concerns the relationship between parents mated at random and their offspring. A number of individuals are drawn at random from a population and then paired off at random. The data from this type of experiment consist of measurements on these parents and their progeny.

### (a) Expectations

In principle, no assumptions are needed about either the gene or genotypic frequencies of the parental population. It is convenient however to assume that each allele at each locus is as frequent as its alternative allele in this population. Such a situation is met with in the $F_2$ family discussed previously in which the frequency $u$ of $B$ is the same as that $v$ of $b$, both being $\frac{1}{2}$. The corresponding genotype frequencies (see also p. 297) are, of course

| Genotype | $BB$ | $Bb$ | $bb$ |
|---|---|---|---|
| Genotype frequency | $u^2$ | $2uv$ | $v^2$ |
| | $\frac{1}{4}$ | $\frac{1}{2}$ | $\frac{1}{4}$ |

If $F_2$ individuals are mated at random, the probability of a $BB$, mating with another $BB$ individual, is $\frac{1}{4}$, since $\frac{1}{4}$ of the offspring are of this type. Similarly the probability of a $BB$ fly mating with a $Bb$ and a $bb$ individual is $\frac{1}{2}$ and $\frac{1}{4}$ respectively. But since $BB$ individuals occur with a frequency of $\frac{1}{4}$, the probability of the matings $BB \times BB$, $BB \times Bb$, $BB \times bb$ is simply the product of their respective genotype frequencies, i.e. $1/16$, $2/16$ and $1/16$ respectively. The full list of matings, together with their frequencies (probabilities) and their progeny is shown in table 3.10.

It will be noticed that these progenies can be classified as to whether they are of $P_1$, $P_2$, $F_1$, $F_2$, $BX_1$ or $BX_2$ type, whose expectations have, of course, been obtained previously. These expectations can thus be rearranged in a form more suitable for present purposes (table 3.11).

A convenient measure of the resemblance between parents and their offspring is provided by the statistic, the *covariance* between parents and offspring, $\text{cov}_{o/p}$, whose expectation is:

$$\text{cov}_{o/p} = \tfrac{1}{16}d^2 + \tfrac{1}{4}(\tfrac{1}{2}(d+h))^2 + \tfrac{1}{4}(h)(\tfrac{1}{2}h) + \tfrac{1}{4}(\tfrac{1}{2}(-d+h))^2 + \tfrac{1}{16}d^2 - (\tfrac{1}{2}h)(\tfrac{1}{2}h)^2 = \tfrac{1}{4}d$$

For many genes with independent action and inheritance we may write $\text{cov}_{o/p} = \tfrac{1}{4}\sum_i d_i^2 = \tfrac{1}{4}D$. This statistic has no environmental component as parents and offspring are generally raised in different, uncorrelated environments.

**Table 3.10** Frequency of matings (top left hand corner of each cell in table), composition of progeny (middle) and type of progeny (top right) under random mating of $F_2$ individuals

| ♂ parent<br><br>♀ parent | $BB$<br>$\frac{1}{4}$ | $Bb$<br>$\frac{1}{2}$ | $bb$<br>$\frac{1}{4}$ |
|---|---|---|---|
| $BB$<br>$\frac{1}{4}$ | $\frac{1}{16}$    $P_1$<br>$BB$<br>$1$ | $\frac{2}{16}$    $BX_1$<br>$BB, Bb$<br>$\frac{1}{2}, \frac{1}{2}$ | $\frac{1}{16}$    $F_1$<br>$Bb$<br>$1$ |
| $Bb$<br>$\frac{1}{2}$ | $\frac{2}{16}$    $BX_1$<br>$BB, Bb$<br>$\frac{1}{2}, \frac{1}{2}$ | $\frac{4}{16}$    $F_2$<br>$BB, Bb, bb$<br>$\frac{1}{4}, \frac{1}{2}, \frac{1}{4}$ | $\frac{2}{16}$    $BX_2$<br>$Bb, bb$<br>$\frac{1}{2}, \frac{1}{2}$ |
| $bb$<br>$\frac{1}{4}$ | $\frac{1}{16}$    $F_1$<br>$Bb$<br>$1$ | $\frac{2}{16}$    $BX_2$<br>$Bb, bb$<br>$\frac{1}{2}, \frac{1}{2}$ | $\frac{1}{16}$    $P_2$<br>$bb$<br>$1$ |

**Table 3.11** The entries of table 3.10 rearranged showing expected parental and progeny means

| Mating | $BB \times BB$ | $BB \times Bb$ | $BB \times bb$ | $Bb \times Bb$ | $Bb \times bb$ | $bb \times bb$ |
|---|---|---|---|---|---|---|
| Type of family | $P_1$ | $BX_1$ | $F_1$ | $F_2$ | $BX_2$ | $P_2$ |
| Frequency | $\frac{1}{16}$ | $\frac{4}{16}$ | $\frac{2}{16}$ | $\frac{4}{16}$ | $\frac{4}{16}$ | $\frac{1}{16}$ |
| Parental mean | $d$ | $\frac{1}{2}(d+h)$ | $0$ | $h$ | $\frac{1}{2}(-d+h)$ | $-d$ |
| Progeny mean | $d$ | $\frac{1}{2}(d+h)$ | $h$ | $\frac{1}{2}h$ | $\frac{1}{2}(-d+h)$ | $-d$ |

Note too that the overall progeny mean is, at $m+\frac{1}{2}[h]$, identical with the $F_2$ or parental mean.

Now since $\text{cov}_{o/p}$ is just one half of the additive genetic variance of $F_2$ families, we may estimate narrow heritability as

$$h^2{}_{F_2}=\frac{2\ \text{cov}_{o/p}}{\sigma^2{}_{F_2}}=\frac{\frac{1}{2}D}{\frac{1}{2}D+\frac{1}{4}H+E}$$

The experiment discussed below concerns 16 $F_2$ and 16 random mating progenies, the $F_2$ family being the same as that of previous experiments (i.e. from $O \times C$). In each of these cultures, ten male and ten female flies were scored with respect to bristle number. Sixteen male and 16 female $F_2$ individuals were chosen at random from among those scored by drawing up a list prior to their being scored, using a convenient random permutation (Moses and Oakford 1963). As before, an excess number of matings is made up to ensure that 16 are available in the following generation. The cultures of the parental and offspring generations were randomized on a single shelf of an incubator set to a temperature of 25°C. The calculations of the culture means and sums of squares may again be left to the student (table 3.12).

### (b) Estimation

Previous evidence has shown that bristle number in these crosses is determined partly by genes carried on the $X$-chromosome. The model we have employed in the derivation of the expected composition of the covariance between parents and offspring, on the other hand, does not allow for sex-linked inheritance. A characteristic of sex-linked inheritance is that male progeny tend to resemble their maternal parent, and that female progeny resemble their paternal parent—so-called criss-cross inheritance. This effect can be accommodated by calculating separately the covariance of female offspring on to their male parents and male offspring on to their female parents, since both are affected to the same extent by sex-linkage (Mather and Jinks 1971).

Table 3.12 The data from experiment 4. The entries in respect of male and female progenies are the *sums* of the bristle numbers of the ten individuals scored in each progeny

| Progeny No. | 1 | 2 | 3 | 4 | 5 | 6 | 7 | 8 | 9 | 10 | 11 | 12 | 13 | 14 | 15 | 16 |
|---|---|---|---|---|---|---|---|---|---|---|---|---|---|---|---|---|
| Male parent | 30 | 36 | 27 | 31 | 27 | 26 | 28 | 32 | 34 | 33 | 23 | 26 | 30 | 32 | 32 | 40 |
| Female parent | 32 | 29 | 33 | 29 | 34 | 30 | 26 | 31 | 27 | 25 | 21 | 21 | 32 | 25 | 27 | 21 |
| Male progeny | 280 | 318 | 292 | 312 | 322 | 301 | 322 | 336 | 299 | 285 | 272 | 237 | 287 | 249 | 288 | 339 |
| Female progeny | 256 | 319 | 276 | 316 | 319 | 305 | 298 | 319 | 311 | 284 | 302 | 258 | 274 | 279 | 290 | 340 |

In estimating narrow heritability, the appropriate estimates of $\sigma^2_{F_2}$ are respectively, calculated on male and on female $F_2$ individuals. The values of the statistics required are (from table 3.12)

| | Female offspring/male parent | Male offspring/female parent |
|---|---|---|
| $\mathrm{cov}_{o/p}$ | 4·52 | 4·08 |
| $\sigma^2_{F_2}$ | 18·26 | 18·50 |

Thus the narrow heritabilities are:

$\female/\male\ \hat{h}^2_{F_2}=(2\times4{\cdot}52)/18{\cdot}26=0{\cdot}495$ or $\simeq$ 50 per cent

$\male/\female\ \hat{h}^2_{F_2}=(2\times4{\cdot}08)/18{\cdot}50=0{\cdot}441$ or $\simeq$ 44 per cent.

These estimates are reasonably consistent and are similar to that previously obtained from the $F_2$ and $BX$ experiment. Here, however, though not allowing explicitly for sex-linked effects, the covariances have been calculated in a way which accommodates them.

## VIII. RANDOM MATING WITH COMMON PARENTS

## EXPERIMENT 5

In the last experiment a general method of randomly mating a population, namely by pairing individuals of opposite sex at random, and its analysis was described. From the analytical viewpoint a more efficient method of randomly mating is to take a random sample of a population and cross the individuals in the sample in all possible combinations. This design, however, has practical limitations of two kinds. First, unless the sample is small, the number of matings that must be made and hence the number of progenies that must be raised, is prohibitive. Second, since each individual in the sample is included in a number of matings this mating programme can only be used where there are no severe restrictions on the number of times the same individual can be mated over a relatively short period of time. With animals, therefore, such a mating programme is virtually impossible because the female parents cannot usually be mated to each of the males in the sample in turn. In consequence, the use of this crossing programme is virtually confined to higher plants and fungi. While higher plants are ideal for this kind of experiment, they require facilities for growing up to 5000 plants under uniform conditions. A suitable subject, therefore, is the fungus *Schizophyllum commune* which has been used for many of the fungal experiments (chapter 5). Indeed, use can be made of the strains, media, techniques, and the character rate of growth, described in that chapter.

### (a) Mating programme

The procedure follows exactly that of experiment 7 of chapter 5 which are briefly summarized below.
(i)   The dikaryon is fruited on plates with SF medium.
(ii)   Monokaryon progeny are obtained by germinating samples of basidiospores.
(iii)   The incompatibility reactions of the progeny are determined.
(iv)   Six progeny are chosen at random from each of the four major compatibility groups and all possible compatible matings made between them. The design is summarized in table 3.13.

Three days after making the matings, inocula are transferred to growth tubes to determine the growth rates of the dikaryons on MT medium at 25°C as described in experiment 7, chapter 5.

**Table 3.13** The design of the experiments

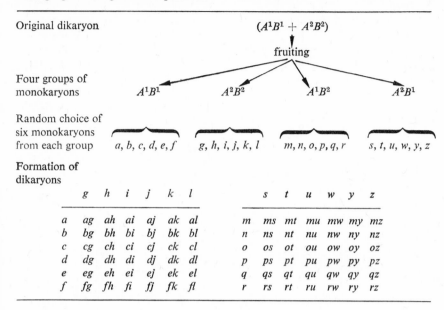

Original dikaryon $\qquad$ $(A^1B^1 + A^2B^2)$

fruiting

Four groups of
monokaryons $\qquad A^1B^1 \qquad A^2B^2 \qquad A^1B^2 \qquad A^2B^1$

Random choice of
six monokaryons
from each group $\qquad$ $a, b, c, d, e, f \qquad g, h, i, j, k, l \qquad m, n, o, p, q, r \qquad s, t, u, w, y, z$

Formation of
dikaryons

|   | g | h | i | j | k | l |   |   | s | t | u | w | y | z |
|---|---|---|---|---|---|---|---|---|---|---|---|---|---|---|
| a | ag | ah | ai | aj | ak | al |   | m | ms | mt | mu | mw | my | mz |
| b | bg | bh | bi | bj | bk | bl |   | n | ns | nt | nu | nw | ny | nz |
| c | cg | ch | ci | cj | ck | cl |   | o | os | ot | ou | ow | oy | oz |
| d | dg | dh | di | dj | dk | dl |   | p | ps | pt | pu | pw | py | pz |
| e | eg | eh | ei | ej | ek | el |   | q | qs | qt | qu | qw | qy | qz |
| f | fg | fh | fi | fj | fk | fl |   | r | rs | rt | ru | rw | ry | rz |

The parental dikaryon, the monokaryons chosen for crossing, and the dikaryons resulting from their mating are grown in the same experiment replicated four times, each replication containing 49 growth tubes randomized over a single shelf in an incubator.

### (b) Analysis

The analysis of the growth rates will be illustrated by reference to the experiment with dikaryon No. 1 described by Simchen and Jinks (1964). The results are summarized in table 3.14.

They are the growth in mm over a ten-day period summed over the four replicates for each dikaryon combination and each parental monokaryon. The data on the dikaryons is in the form of two $6 \times 6$ tables within each of which the total variation can be subdivided into three items corresponding with Rows (common parents $a$ to $f$ and $m$ to $r$), and Columns (common parents $g$ to $l$ and $s$ to $z$) and the interaction between Rows and Columns ($a$–$f \times g$–$l$ and $m$–$r \times s$–$t$). In addition there is a replicate error item derived from the replicate sum of squares for each dikaryon summed over all dikaryons. When, as in the illustrative example the replicates are arranged in four blocks, each block containing one replicate of each dikaryon and occupying a different shelf in the incubator, the replicates sum of squares can be broken down into two items, the block differences and the blocks $\times$ dikaryons

**Table 3.14** Ten days growth (in mm) of monokaryons (in italics) and resulting dikaryon combinations

| Monokaryon parents | *g* *295* | *h* *311* | *i* *340* | *j* *276* | *k* *287* | *l* *316* | Array means |
|---|---|---|---|---|---|---|---|
| *a 280* | 291 | 305 | 304 | 292 | 308 | 295 | 270·8333 |
| *b 275* | 299 | 318 | 305 | 290 | 278 | 271 | 293·5000 |
| *c 274* | 294 | 308 | 309 | 303 | 305 | 282 | 300·1667 |
| *d 296* | 314 | 339 | 326 | 323 | 325 | 313 | 323·3333 |
| *e 301* | 319 | 329 | 311 | 303 | 298 | 304 | 310·6667 |
| *f 288* | 308 | 310 | 312 | 289 | 301 | 281 | 300·1667 |
| Array means | 304·1667 | 318·1667 | 311·1667 | 300·0000 | 302·5000 | 291·0000 | |

| Monokaryon parents | *s* *279* | *t* *255* | *u* *279* | *w* *268* | *y* *292* | *z* *266* | Array means |
|---|---|---|---|---|---|---|---|
| *m 251* | 260 | 265 | 289 | 280 | 285 | 306 | 280·8333 |
| *n 301* | 311 | 297 | 295 | 296 | 302 | 298 | 299·8393 |
| *o 268* | 285 | 285 | 311 | 312 | 307 | 320 | 303·3333 |
| *p 272* | 291 | 297 | 302 | 299 | 281 | 307 | 296·1667 |
| *q 279* | 290 | 287 | 295 | 299 | 290 | 308 | 294·8333 |
| *r 287* | 277 | 308 | 300 | 298 | 302 | 315 | 300·0000 |
| Array means | 285·6667 | 289·8333 | 298·6667 | 297·3333 | 294·5000 | 309·0000 | |

interaction. The analysis of variance for the general case of $f$ monokaryons of one mating type crossed to $m$ monokaryons of another, compatible mating type grown as $n$ replicates one in each of $n$ blocks is as follows:

| Item | $df$ | Expected mean squares |
|---|---|---|
| 1. Between $f$ common parents | $(f-1)$ | $\sigma^2_E + n\sigma^2_I + nm\sigma^2_{P1}$ |
| 2. Between $m$ common parents | $(m-1)$ | $\sigma^2_E + n\sigma^2_I + nf\sigma^2_{P2}$ |
| 3. Interaction $f \times m$ | $(f-1)(m-1)$ | $\sigma^2_E + n\sigma^2_I$ |
| 4. Blocks | $(n-1)$ | $\sigma^2_E + fm\sigma^2_B$ |
| 5. Blocks $\times$ dikaryons | $(fm-1)(n-1)$ | $\sigma^2_E$ |

The expected mean squares in terms of components of variation for error variation ($\sigma^2_E$), block differences ($\sigma^2_B$), differences between the $f$ common parents ($\sigma^2_P$) the $m$ common parents ($\sigma^2_{P_2}$) and the interaction between them ($\sigma^2_I$) are given in the right hand column. It can be seen from these expectations that the estimates of the components are

$$\sigma^2_E = (5)$$
$$\sigma^2_B = [(4)-(5)]/fm$$
$$\sigma^2_{P1} = [(1)-(3)]/nm$$
$$\sigma^2_{P2} = [(2)-(3)]/nf$$
$$\sigma^2_I = [(3)-(5)]/n$$

In the special case under consideration $f=6$, $m=6$ and $n=4$.

The two analyses of variance along with their expected mean squares are as follows

| Item | df | MS | Item | df | MS |
|---|---|---|---|---|---|
| 1. Common parents $(a - f)$ | 5 | 173·92 | Common parents $(m - r)$ | 5 | 94·72 |
| 2. Common parents $(g - l)$ | 5 | 131·35 | Common parents $(s - z)$ | 5 | 97·43 |
| 3. Interaction | 25 | 16·14 | Interaction | 25 | 16·23 |
| 4. Blocks | 3 | 70·19 | Blocks | 3 | 121·21 |
| 5. Blocks × dikaryons | 103* | 8·05 | Blocks × dikaryons | 102* | 7·21 |

* Should be 105 but some replicates failed to grow.

All the items are significant when tested against the block × dikaryons mean squares, furthermore, the common parent mean squares are significant against the parental interactions. Hence, all the components of variation are significantly different from zero. The estimates of these components are as follows:

| Component | Estimates $(a - f)(g - l)$ | Estimates $(m - r)(s - z)$ |
|---|---|---|
| $\hat{\sigma}^2_{P1}$ | 6·57 | 3·27 |
| $\hat{\sigma}^2_{P2}$ | 4·80 | 3·38 |
| $\hat{\sigma}^2_{I}$ | 2·02 | 2·25 |
| $\hat{\sigma}^2_{B}$ | 1·72 | 3·17 |
| $\hat{\sigma}^2_{E}$ | 8·05 | 7·21 |

### (c) Fitting a model

The total variance among the dikaryotic combinations has now been partitioned among its significant components. Genetic contributions of two kinds have been recognized in this partitioning, namely, an additive difference among the sets of common parents $(\sigma^2_{P1}, \sigma^2_{P2})$ and the interaction between the sets $(\sigma^2_I)$. These have expectations in terms of the parameters of biometrical genetics. Thus assuming a model which allows for additive $(d)$ and dominance $(h)$ effects of the genes,

$$\sigma^2_{P1} = \sigma^2_{P2} = \tfrac{1}{4} D_R$$

and

$$\sigma^2_I = \tfrac{1}{4} H_R$$

where

$$D_R = \Sigma 4uv[d+(v-u)h]^2$$

and

$H_R = \Sigma 16u^2 v^2 h^2$, $u$ being the frequency of the increasing allele (e.g. $B$) and $v$ the frequency of decreasing allele (e.g. $b$) in the sample of parents.

$D_R$ and $H_R$ are the random mating equivalents of $D$ and $H$ for the case of unequal gene frequencies and they are equal to $D$ and $H$ in the special case of random mating where $u = v = \tfrac{1}{2}$ for all loci (see experiment 4).

It follows that the total genetical variation among the dikaryons $\sigma^2_G$ equals

$$\sigma^2_{P1} + \sigma^2_{P2} + \sigma^2_I = \tfrac{1}{2} D_R + \tfrac{1}{4} H_R$$

E

The environmental components of variation are the components for blocks and blocks × dikaryons components, namely $\sigma^2_B$ and $\sigma^2_E$. The latter is equivalent to the $E$ component of earlier experiments. Hence, the broad and narrow heritabilities of the dikaryotic populations can be estimated as

$$\text{Broad heritability} = \frac{\frac{1}{2}D_R + \frac{1}{4}H_R}{\frac{1}{2}D_R + \frac{1}{4}H_R + E} = \frac{\sigma^2_G}{\sigma^2_G + \sigma^2_E}$$

$$\text{Narrow heritability} = \frac{\frac{1}{2}D_R}{\frac{1}{2}D_R + \frac{1}{4}H_R + E} = \frac{\sigma^2_{P_1} + \sigma^2_{P_2}}{\sigma^2_G + \sigma^2_E}$$

The estimates of $\frac{1}{2}D_R$, $\frac{1}{4}H_R$, $E$ and the heritabilities for the two sets of dikaryons are as follows:

|  | $(a-f)$ | $(g-l)$ | $(m-r)$ | $(s-z)$ |
|---|---|---|---|---|
| $\frac{1}{2}D_R$ | 11·37 | (53·1%) | 6·65 | (41·3%) |
| $\frac{1}{4}H_R$ | 2·02 | (9·4%) | 2·25 | (14·0%) |
| $E_1$ | 8·05 | (37·5%) | 7·21 | (44·7%) |
| Broad heritability | 62·5% |  | 55·3% |  |
| Narrow heritability | 53·1% |  | 41·3% |  |

Both $D_R$ and $H_R$ are significantly greater than zero since the $\sigma^2$'s from which they are derived are significant. The fact that $D_R$ and $H_R$ can be separated, and their significance tested is one of the advantages of carrying out random mating by this mating design. It should be noted that $H_R/D_R$ does *not* estimate the dominance ratio unless, of course, $u = v$.

The results leave no doubt that heritable variation accounts for 55 to 62 per cent of the differences in the growth rate of dikaryons produced from the monokaryotic progeny of single dikaryons isolated from the wild.

### (d) Adequacy of model

In fitting the additive-dominance model to these data a number of assumptions have been made, the most important of which is that non-allelic interactions (epistasis) are absent. With the particular material and mating design used the validity of this assumption can be tested. Thus, two further statistics can be estimated, namely, the variances and covariances of members of an array; an array being defined as all the crosses (dikaryons) involving any one parental line (monokaryotic component). For example, the variance of the first array $Vr_a$ is (table 3.14 first row)

$$\frac{1}{5}\left[\frac{1}{4}(291^2+305^2+304^2+292^2+308^2+295^2) - \frac{1795^2}{24}\right] = 13\cdot54.$$

While the covariance on to the array means $Wr_a$ is

$$\frac{1}{5}\left[\frac{1}{4}(291\times304\cdot1667+305\times318\cdot1667+304\times311\cdot1667+292\times300\cdot0000+308\times\right.$$
$$\left.302\cdot5000+295\times291\cdot0000) - \frac{1795\times1827}{24}\right] = 9\cdot78.$$

In this way estimation can be made of six $Vr$'s and six corresponding $Wr$'s for each of the four sets of parents, namely, $a$–$f$, $g$–$l$, $m$–$r$ and $s$–$z$. If this additive-dominance model is correct then each set of $Wr$'s will give a linear regression when plotted against their $Vr$'s (see Simchen & Jinks 1964 for details). As an example the relationship between the six $Wr$'s and $Vr$'s for the set of parents $a$–$f$ will be analysed. The values of the $Wr$'s and $Vr$'s are as follows:

| Array | $Vr$ | $Wr$ |
|-------|------|------|
| a | 13·54 | 9·78 |
| b | 76·08 | 35·02 |
| c | 26·94 | 19·54 |
| d | 22·47 | 18·63 |
| e | 33·47 | 19·99 |
| f | 39·54 | 25·90 |

A regression analysis of $Wr$ on $Vr$ gives the following result:

| Item | $df$ | $MS$ | $F$ | $P$ |
|------|------|------|-----|-----|
| Linear regression | 1 | 5192·10 | 44·05 | 0·01 – 0·001 |
| Remainder | 4 | 117·87 | | |

The linear regression which has a coefficient of $b = 0\cdot3683 \pm 0\cdot0555$ is highly significant. Hence there is no reason to believe that the additive-dominance model is inadequate. Corresponding analyses lead to the same conclusion for the parental sets $g$–$l$, and $m$–$r$ but not for set $s$–$z$ which give a non-significant ($P = 0\cdot20$–$0\cdot10$) linear regression. There is, therefore, some suspicion as to whether or not an additive-dominance model is adequate for the $(m$–$r) \times (s$–$z)$ set of crosses. Nevertheless the estimates of the parameters from this set do not differ in any important respect from the $(a$–$f) \times (g$–$l)$ set of crosses where the model is adequate. If epistasis is present, therefore, its contribution to the total variation cannot be large.

### (e) Gene frequencies

The expected slope of the regressions ($b$) is

$$\frac{1}{2} - \frac{(u-v)h,}{2d}$$ that is, if $u = v$ $b = \frac{1}{2}$; if $u > v$ $b < \frac{1}{2}$

and if $u < v$ $b > \frac{1}{2}$ if $b$ is positive and the reverse if it is negative (Simchen and Jinks 1964). For the four sets of parents $b$ exceeds $\frac{1}{2}$ for the $m$–$r$ set only. But in all four sets the estimates of $b$ do not differ significantly from $\frac{1}{2}$. In the absence of differential viability one expects $u = v = \frac{1}{2}$ in the population of monokaryons from which the four small samples of six were chosen, but not necessarily within each of the small samples themselves.

### (f) Direction of dominance and gene distribution

The *Wr*, *Vr* linear relationship has a further useful property in that it allows us to arrange the parents within any set in order of the relative numbers of dominant : recessive alleles they possess. Thus the most dominant parent has the lowest values of *Wr* and *Vr* for its array and the most recessive parent the highest values. Reference to the values for the set (*a–f*) shows that parent *a* contains the most dominant alleles and *b* the most recessive. For the other three sets the corresponding parents are *i* and *l*, *n* and *m* and *z* and *t*, respectively. Reference to table 3·14 shows that there is no correlation between number of dominant: recessive alleles and rate of growth. Thus of the four parents (*a*, *i*, *n* and *z*) containing most dominants in their set, *a* produces slow growing dikaryons, *i* produces fast growing dikaryons and *n* and *z* dikaryons of intermediate growth rate. Hence, dominance is ambi-directional rather than directional.

## IX. GENOTYPE-ENVIRONMENTAL INTERACTION

In experiment 3 tests for the presence of genotype–environmental interactions were described and used to detect its presence in two inbred lines of *D. melanogaster* (O and C) and the $F_1$ between them. In this experiment the interaction arose from the differing sensitivities of sternopleural bristle production in the inbred and $F_1$ families to the uncontrolled environmental heterogeneity within the bottles in which they were raised. Such interactions, however, are more readily found and can be more informatively analysed if different genotypes are exposed to deliberately produced environmental differences. The choice of environmental differences will depend on the organism used. For example, with *Drosophila* three obvious differences are temperature, medium and density. With limited facilities the latter is the most easily achieved.

## EXPERIMENT 6

### Effect of density on yield in *D. melanogaster*

The environmental treatment consists of raising the progeny of single pair matings in either vials or bottles each containing the usual amount of the standard medium in the manner described in chapter 1. The simplest to score, and the most responsive character to this treatment, is the total number of adult flies per mating. This can be scored by counting the number of flies that emerge from each vial or bottle each day, after removing and etherizing them over a period of ten days.

The most useful genotypes to compare in this way are inbred lines and $F_1$'s produced by crosses between them. With such genotypes the interpretation is unambiguous since all single pair matings within an inbred line or within an $F_1$ should yield identical progenies apart from environmental effects. Since the total yield of adult flies is subject to many experimental errors and each mating provides only one estimate it is necessary to replicate each genotype in each treatment. Indeed as many replicates as possible should be set up.

An example has been kindly provided by Dr B.W.Barnes. This consists of single pair matings using two inbred lines, Oregon (O) and Samarkand (S) as follows:

eight single pair matings      O ♀ × O ♂, four in vials and four in bottles
eight single pair matings      S ♀ × S ♂, four in vials and four in bottles
eight single pair matings of the   $F_1$ O ♀ × S ♂, four in vials and four in bottles
and eight single pair matings of the   $F_1$ S ♀ × O ♂, four in vials and four in bottles

raised at 18°C, all bottles and vials being placed at random over the area they occupy during incubation.

The adult progeny are counted by removing them daily from each vial or bottle into an etherizer, counting and discarding on ten successive days after their first appearance. At this temperature some matings usually fail (see table 3.15) and it is advisable to set up more matings than it is intended to score. In which case, if more are successful than required, the appropriate number of matings should be chosen for scoring at random. Nevertheless, in this experiment less than four matings per treatment were scorable in three out of the eight treatments. However, a suboptimal temperature, i.e. below 25°C, is essential for observing genotype–environmental interactions with the strains and densities used. Where a suitable incubator is not available the experiment can be carried out at room-temperature but more replicates will be required to offset the loss of precision in order to detect the genotype–environmental interactions.

The scores from the illustrative experiment are listed in table 3.15, along with the means and standard errors of the eight treatments. Examination of the average number of flies emerging in vials and bottles shows that the smaller vials reduce the number by a factor of two, which is equivalent to an average reduction of 70 flies. That genotype–environmental interactions are

Table 3.15 The number of adult flies produced per mating in vials and bottles at 18°C

| Genotypes | Environments | | | |
| | Vials | Mean | Bottles | Mean |
|---|---|---|---|---|
| O | 50, 48, 65, 54 | 54·25 ± 6·57 | 92, 106, 94 | 97·33 ± 6·18 |
| $F_1$(O × S) | 73, 52, 67 | 64·00 ± 8·83 | 148, 190, 167, 235 | 185·00 ± 32·47 |
| $F_1$(S × O) | 87, 81, 74, 88 | 82·50 ± 5·59 | 184, 182, 139 | 168·33 ± 20·76 |
| S | 91, 77, 79, 68 | 78·75 ± 8·20 | 119, 97, 83, 141 | 110·00 ± 22·02 |

present is shown by the fact that this reduction differs markedly among the genotypes being as low as 31 for S and as high as 121 for (O × S), a difference which is many times the standard errors of the counts.

A simple analysis of these data yields the following results.

| Item | df | MS |
|------|-----|--------|
| Genotypes (matings) | 3 | 5055·22 |
| Environments (vials v bottles) | 1 | 36449·95 |
| Genotype × environment | 3 | 2137·93 |
| Replicates | 21 | 398·37 |

This analysis clearly shows a significant difference between matings and between environments and the presence of genotype–environmental interactions. The nature of the interactions can be pursued at a number of different levels. The simplest of these is to examine the relationship between the inbred parents and the $F_1$'s in the two environments since it appears that the inbreds have changed to a small and similar extent but the two $F_1$'s have changed much more between the treatments. This is made clear by reducing the data in table 3·15 down to four items which represent the mid-parent value and the mean of the reciprocal $F_1$'s in the two treatments.

|  | Vials | Bottles |
|------|-------|---------|
| Mid-parent | 66·50 | 103·67 |
| $F_1$ | 73·25 | 176·67 |

It will now be seen that the principal effect of the crowded conditions in the vials is to suppress the greater yielding capacity of the $F_1$ until it is not significantly greater than that of the parents. At the optimal temperature of 25°C this differential reduction of the $F_1$ yield does not occur, hence at this temperature there is no genotype–environmental interaction.

Alternatively, a biometrical genetical analysis of the treatment means for the four matings can be carried out. The general expectation for the mean phenotype of genotype $i$ in environment $j$ may be written as

$$\bar{P}_{ij} = \mu + g_i + e_j + g_{ij}$$

where $\mu$ is the overall mean

$g_i$ is the genetical contribution of line $i$

$e_j$ is the contribution of environment $j$

and $g_{ij}$ is the genotype–environmental interaction of the $i$th line in the $j$th environment.

For two inbred lines and the $F_1$ between them this can be rewritten as (see experiment 3, p. 101)

$$\bar{P}_{1j} = \mu + [d] + e_j + g_{dj}$$
$$\bar{P}_{2j} = \mu - [d] + e_j - g_{dj}$$
$$\bar{F}_{1j} = \mu + [h] + e_j + g_{hj}$$

The parameters in this model may be estimated as follows:

$$\mu=\sum_j \frac{\frac{1}{2}(\bar{P}_{1j}+\bar{P}_{2j})}{S} \quad \text{where } S=\text{the number of environments.}$$

$$[d]=\sum_j \frac{\frac{1}{2}(\bar{P}_{1j}-\bar{P}_{2j})}{S}$$

$$e_j=\frac{1}{2}(\bar{P}_{1j}+\bar{P}_{2j})-\mu$$

$$g_{aj}=\frac{1}{2}(\bar{P}_{1j}-\bar{P}_{2j})-[d]$$

$$[h]=\sum_j \frac{\bar{F}_{1j}-\frac{1}{2}(\bar{P}_{1j}+\bar{P}_{2j})}{S}=\sum_j \frac{\bar{F}_{1j}}{S}-\mu$$

$$g_{hj}=\bar{F}_{1j}-\frac{1}{2}(\bar{P}_{1j}+\bar{P}_{2j})-[h]$$

The value for $P_1$=Samarkand, $P_2$=Oregon and $S=2$ are

$$=85\cdot08$$
$$[\hat{d}]=9\cdot29$$
$$\hat{e}_j=\pm18\cdot58$$
$$\hat{g}_{aj}=\pm2\cdot96$$
$$[\hat{h}]=39\cdot88$$
$$\hat{g}_{hj}=\pm33\cdot13$$

where the $+$ is for the bottle environment and the $-$ for the vials.

In interpreting these estimates it should be borne in mind that two of them $[\hat{d}]$ and $[\hat{h}]$ relate to the average performance over both environments and $\hat{e}_j$, $\hat{g}_{aj}$, and $\hat{g}_{hj}$ to the differences in performance between the two environments. Overall, therefore, there is heterosis ($[h]>[d]$) the $F_1$ having a superior performance to the better parent. Equally, the $F_1$ reacts much more to the environmental difference than the parents ($\hat{g}_{hj}>\hat{g}_{aj}$). Indeed since $[\hat{h}]-[\hat{d}]\simeq[\hat{g}_{hj}]-[\hat{g}_{aj}]$ this greater reaction is sufficient to obliterate the heterosis in the poorer environment. Thus fitting the model has led to the same conclusion as examining the mid-parental and $F_1$ means in the two environments but it has also allowed us to quantify the relative contributions of genetic, environmental and genotype–environmental components to these means and to partition its genetic part between additive and dominance effects.

Since this demonstration of genotype–environmental interaction depends largely on the adverse effect of overcrowding on the higher yielding $F_1$ almost any pair on inbred lines of *Drosophila* could be used since crosses between most pairs of inbred lines lead to a higher yielding $F_1$.

## EXPERIMENT 7

### Effect of medium and temperature on *Schizophyllum commune*

Where the facilities are available fungi provide ideal material for demonstrating and analysing genotype–environmental interaction because of the range of controlled temperatures and media over which they can be successfully grown. Equally, isolates differing widely in easily scored characters such as rate of growth are readily obtainable both from the wild and from laboratory collections. One such experiment will be described, using monokaryons of *S. commune* growing at three temperatures 20, 25 and 30°C and two media, 1 and 2 per cent malt extract. However, the design and analysis are applicable to any combination of two environmental variables in any relative numbers.

The two strains used were monokaryons from the progeny of a wild dikaryon selected for their high and low rates of growth at 25°C on 2 per cent malt extract. Comparable strains could, of course, be chosen from the monokaryotic progeny of experiment 7, chapter 5. In this particular experiment six replications were used. The techniques for preparing the growth tubes and media, inoculating and randomizing the tubes and scoring the character are all as described in experiment 7, chapter 5.

In table 3.16 are given the growth in mm over a nine-day period for each of the two strains in each of the six environments, summed over the six replications. The high growth rate selection has clearly grown more than the low selection both overall and in every one of the environments. The amount of growth increases markedly with increasing temperature but the effect of

Table 3.16 Growth in mm over a nine-day period for high and low growth rate strains of *S. commune* summed over six replicates when grown in six different environments

| % malt | Temp. | High | Low | Total growth in environments |
|---|---|---|---|---|
| 1 | 20°C | 322 | 192 | 514 |
| 2 | 20°C | 303 | 169 | 472 |
| 1 | 25°C | 516 | 330 | 846 |
| 2 | 25°C | 511 | 298 | 809 |
| 1 | 30°C | 601 | 479 | 1080 |
| 2 | 30°C | 627 | 456 | 1083 |
| Total growth of strains | | 2880 | 1924 | 4804 |

the medium is small, growth being slightly higher on 1 per cent than on 2 per cent malt at the lower temperatures but not at the highest temperature. The effect of genotype–environmental interaction is revealed by examining the difference in amount of growth between the high and low strains (H–L) for each of the environments. This difference varies markedly from one environment to another being lowest for 1 per cent malt at 30°C, and the highest for 2 per cent malt at 25°C. Overall the difference is highest at 25°C and always higher on 2 per cent malt than on 1 per cent malt at all temperatures.

A simple analysis of variance confirms that the genotypes (high and low strains) and environments differ significantly and that there is significant genotype–environmental interaction. The error variance is obtained from the variation among the replicates.

| Item | df | MS |
|---|---|---|
| Genotypes | 1 | 12693·56 |
| Environments | 5 | 5834·38 |
| Genotypes × environments | 5 | 110·39 |
| Error | 60 | 5·83 |

Since media and temperatures are orthogonal treatments we can partition both environments and the interactions into three items as follows.

| | | df | MS |
|---|---|---|---|
| Environments | Temperature | 2 | 14520·51 |
| | Medium | 1 | 81·71 |
| | Temperature × medium | 2 | 24·60 |
| Genotype × environments | Genotype × temperature | 2 | 211·91 |
| | Genotype × medium | 1 | 90·46 |
| | Genotype × temperature × medium | 2 | 18·83 |

This shows, as expected, that the temperature has a considerably greater influence on growth and interacts more with the genotypes than the medium.

One could of course fit a biometrical model to these data of the kind used with the *Drosophila* data (expt. 6), but this requires a more complex treatment of the orthogonal environmental parameters ($e_j$'s) and the corresponding interaction parameters ($g_j$'s). However, it involves no new principle so it will not be considered further here. Instead the analysis will be pursued as if the environments were six treatments that could not be specified in terms of orthogonal combinations of temperature and medium. We can estimate $\mu$ and $[d]$ as previously and six $e_j$'s and six $g_j$'s corresponding with the six treatments as

$$e_j = \tfrac{1}{2}(\bar{H}_j + \bar{L}_j) - \mu$$

$$g_j = \tfrac{1}{2}(\bar{H}_j - \bar{L}_j) - [d]$$

These estimates, which must be calculated from the means of the replicates are as follows

$$\hat{\mu}=66\cdot72 \qquad\qquad [\hat{d}]=13\cdot28$$

| $j=$ | 1 | 2 | 3 | 4 | 5 | 6 |
|---|---|---|---|---|---|---|
| $e_j$ | $-28\cdot33$ | $-27\cdot39$ | $3\cdot78$ | $0\cdot69$ | $23\cdot28$ | $23\cdot53$ |
| $g_j$ | $-2\cdot44$ | $-2\cdot11$ | $2\cdot22$ | $4\cdot47$ | $-3\cdot11$ | $0\cdot97$ |

It is frequently observed that the genotype–environmental component is a linear function of the environmental component, that is

$$g_j = \beta e_j$$

where $\beta$ is a linear regression coefficient.

An analysis of the regression of the six values of $g_j$ on the corresponding $e_j$ values yields the following

| Item | SS | df | MS |
|---|---|---|---|
| Linear regression | 30·16 | 1 | 30·16 |
| Remainder | 521·78 | 4 | 130·45 |
| Total | 551·94 | 60 | 5·83 |

The first point to note is that if a correction is made for the fact that the regression analysis is being prepared from components derived from the means of six replicates, whereas the earlier analyses were based on the totals, one finds that the linear regression *SS* and the regression remainder *SS*, sum to the genotype–environmental interaction *SS* of the earlier analyses. Second, while the linear regression accounts for a significant portion of the interaction as shown by its significance against the replicate error, most of the interaction is not accounted for in this way (remainder *SS*). Hence, one cannot substitute $\beta e_j$ in the equations for $g_j$ in this particular case.

Examination of the $g_j$ and $e_j$ values shows why this is so. The largest $g_j$ values occur in the intermediate environments (3 and 4), i.e. there is a curvilinear relationship between $g_j$ and $e_j$. In other words the greatest difference in growth rate between the high and low strains occurs in that environment in which they were selected for their high and low growth rates.

## X. NUMBER OF GENES (EFFECTIVE FACTORS)

By making certain assumptions estimates of the number of genes controlling the variation for a quantitative character can be obtained by carrying out experiments of the kind described under experiments 3 and 6 of this chapter and (experiment 7 of chapter 5). The assumptions are

(1) The $d$'s are equal for all loci, i.e. $d_a = d_b = d_c = \ldots d_k = d$.

(2) Two inbred lines are available which have all the genes in association,

i.e. the higher scoring line has all the increasing genes (the $+d$'s) and the lower scoring line all the decreasing genes (the $-d$'s).

Under these circumstances two inbred lines, $P_1$ and $P_2$ differing at $k$ loci controlling the variation will have the expected mean phenotypes.

$$\bar{P}_1 = m + kd$$
$$\bar{P}_2 = m - kd$$

Hence $\frac{1}{2}(\bar{P}_1 - \bar{P}_2) = kd$ instead of the usual $\Sigma d$.

The two lines can be crossed and sufficient generations raised to estimate $D$, i.e. for experiment 3 and 6, $P_1 P_2 F_1 F_2 BX_1$ and $BX_2$ and for experiment 7, chapter 5 the $P_1 P_2$ and $F_1$ generations. If the assumptions hold then:

$$D = kd^2 \text{ instead of the usual } \Sigma d^2$$

From which it follows that:

$$\frac{\frac{1}{2}(\bar{P}_1 - \bar{P}_2)^2}{D} = \frac{(kd)^2}{kd^2} = k$$

### (a) Estimate of $k$ in *Drosophila melanogaster*

Oregon and 6CL are two lines which are relatively the low and high extremes for the character number of sternopleural bristles. Oregon presumably contains most of the decreasing genes and 6CL most of the increasing genes. Hence, one of the conditions for estimating $k$ is satisfied. However, while the experiments using O and C described in this chapter do not allow any estimation of the relative magnitudes of the $d$'s for the different genes, it is known from other experiments that they are not identical. Hence the estimate will be biased by the inequality.

From experiment 3

$$\frac{1}{2}(\bar{P}_2 - \bar{P}_1) = [d] = 5 \cdot 4.$$

Similarly from experiment 6

$$\hat{D} = 6 \cdot 62.$$

Hence

$$\hat{k} = \frac{(5 \cdot 4)^2}{6 \cdot 62} = 4 \cdot 4.$$

Since the failure of either assumption reduces the value of $k$ below its true value there are at least four or five genes controlling the difference in sternopleural bristle number between O and C. Since sex-linkage can be demonstrated for the character in this cross (experiment 1) one can locate at least one of these genes on the $X$ chromosome.

The 'genes' to be counted are units of segregation in that if two genes are closely linked they will be counted as one by this method. Hence, they are frequently referred to as 'effective factors' rather than genes. In view of this the number estimated is as much a function of the number of chromosomes and the chiasma frequency as it is of the number of genes. Thus if there were three chromosomes, each carrying two genes between which chiasma rarely, if ever, are formed, one would estimate three genes; if, however one chiasma always forms between each pair one would estimate six genes.

### (b) Estimate of $k$ in *Schizophyllum commune*

The effect of the failure of the assumption that all the increasing genes are present in one inbred line and all the decreasers in the other can be illustrated by reference to the character rate of growth in *Schizophyllum* (experiment 7, chapter 5). Reference to the data obtained from that experiment shows that, $\bar{P}_1 = 58 \cdot 5$ and $\bar{P}_2 = 58 \cdot 0$ hence $\frac{1}{2}(\bar{P}_1 - \bar{P}_2) = 0 \cdot 25$, from the segregation of the monokaryotic progeny $\sigma^2_G = \dot{D} = 10 \cdot 566$

$$\text{therefore } \hat{k} = \frac{0 \cdot 25^2}{10 \cdot 566} = 0 \cdot 0059.$$

This is clearly a ridiculous result in view of the highly significant variation among the monokaryotic progeny of the cross $P_1 \times P_2$, and it could only result from the failure of the assumptions on which the estimates are based. Reference to fig. 5.3 shows that the variation is continuous, the distribution being similar to that observed for sternopleural bristles (fig. 3.2) where, as we have seen, many genes are involved.

In *Schizophyllum* the haploid progeny of a cross provide a further opportunity for estimating $k$ because they are in fact themselves 'inbred' lines. Furthermore, we can come closer to satisfying the assumption of association of like genes by selecting the opposite extreme segregants present in the progeny and determining their rate of growth.

Examination of the data shows that these two extremes have growth rates of $49 \cdot 5$ and $69 \cdot 0$ mm, respectively (table 5.8, p. 195), hence

$$\tfrac{1}{2}(\bar{P}_1 - \bar{P}_2) = 9 \cdot 75.$$

$D$ remains the same, therefore

$$\hat{k} = \frac{9 \cdot 75^2}{10 \cdot 566} = 9 \cdot 05$$

Clearly, the previous estimate based on the original parental monokaryons was a gross underestimate because increasing and decreasing genes were equally distributed between them, i.e. they were dispersed. Because of this, first generation selections differ 39 times more than the parents in rate of growth. This contrasts with the situation in *Drosophila* where after four generations of selection the selections did not differ more than the original parents for sternopleural bristles (fig. 3.3).

## (c) Mean size of gene effects

An assumption underlying the estimates of $k$ is that all gene effects are equal. The estimates of $k$ can be used to calculate the magnitude of the effect of each gene assuming this to be true.

Thus $\qquad \frac{1}{2}(\bar{P}_1 - \bar{P}_2) = kd.$

Hence $\qquad\qquad d = \frac{\frac{1}{2}(\bar{P}_1 - \bar{P}_2)}{k}$

For *Drosophila* $\hat{d} = \frac{5\cdot4}{4\cdot4} = 1\cdot23$ bristles.

Therefore, on average each gene controls the production of $1\cdot23$ bristles.

For *Schizophyllum* $\hat{d} = \frac{9\cdot75}{9\cdot05} = 1\cdot08$ mm of growth.

Hence, on average each gene controls the production of $1\cdot08$ mm of growth.

## XI. REFERENCES AND FURTHER READING

Moses, Lincoln E. & Oakford, Robert V. (1963) *Tables of random permutations.* Allen and Unwin, London.

Simchen, G. & Jinks, J.L. (1964) The determination of dikaryotic growth rate in the Basidiomycete *Schizophyllum commune*: a biometrical analysis. *Heredity*, **19,** 629–649.

Two useful statistical text-books

Snedecor, George W. & Cochran, William G. (1967) *Statistical Methods.* Iowa State University Press, Ames, Iowa.

Steel, Robert G.D. & Torrie James H. (1960) *Principles and Procedures of Statistics.* McGraw-Hill, New York.

Two indispensible books of statistical tables

Fisher, Sir Ronald A. & Yates, Frank (1963) *Statistical Tables.* Oliver and Boyd, Edinburgh.

Lindley, D.V. & Miller, J.C.P. (1966) *Cambridge Elementary Statistical Tables.* Cambridge University Press.

Advanced treatments of the subject of biometrical genetics

Mather, K. (1949) *Biometrical Genetics.* Methuen, London.

Mather, K. & Jinks, J.L. (1971) *Biometrical Genetics*, 2nd Edn. Chapman Hall, London.

# CHAPTER 4

# CYTOGENETICS

## S. WALKER

## I. INTRODUCTION

As the study of chromosomes forms the basis of cytogenetics, it is important that practical methods for their display be available for all levels of teaching. It would be an impossible task to produce a comprehensive section on techniques in a limited number of pages, but it is intended here to suggest quick and simple techniques which can be used to provide preparations for chromosome investigation, particularly suitable for school, college and university courses.

The basic techniques for making chromosome preparations are usually a selection of procedures ranging from pre-treatment of the tissues to finally making the slide preparation permanent. Methods of pre-treatment, fixation, staining and slide preparation will therefore be described. These are succeeded by specific schedules for dealing with suggested material and finally an appendix giving details for the preparation of stains and chemicals. Unfortunately space does not permit any chromosome theory or morphology to be discussed and the reader must turn to other appropriate books for this (White 1961; Swanson 1958; De Robertis *et al.* 1954). The primary aim here is to present practical methods for the student to study mitosis, meiosis, chromosome mutations and aberrations, and the morphology of special types of chromosome. As the procedure for embedding material prior to sectioning and staining is a long one and only rarely used in cytogenetic studies, this has been omitted. If required, details of this technique can be found in Darlington & La Cour (1962).

## 1. EQUIPMENT

The range of chemicals and stains can be found in the detailed schedules and stain preparation sections. Apart from these, other requirements may be divided into glassware, instruments, general equipment and optical equipment.

## Glassware

(a)  Microscope slides—75 mm × 25 mm glass (thickness preferably between 0·8 mm–1·1 mm).
(b)  Cover slips—22 mm square no. 1½ gauge (No. 0 for larger working distance on oil immersion).
(c)  Specimen tubes—50 mm × 13 mm, glass or plastic with corks or caps (a tight seal is essential) are the most useful. If larger sizes are necessary screw-capped bottles ranging from 30 ml universals upwards may be used.
(d)  Dropping bottles—for reagents and stains. Those with ground-in pipette and teat or with a glass rod are to be preferred.
(e)  Staining troughs, etc.—corked tubes 100 mm × 30 mm may be used for two slides (back to back) but grooved troughs are preferable. The simplest, though not the cheapest, are stainless steel racks which are transferable together with the slides, from one dish to the next. The racks may carry up to 20 slides.
(f)  Mountant bottle—with glass rod and loose fitting glass cover.
(g)  Watch glasses—for staining or ease of handling material.
(h)  Spirit lamp.

## Instruments

(a)  Dissecting instruments—to include fine and coarse forceps, fine scissors, scalpels, mounted needles, single-edged razor blade.
(b)  Mounted needle, flattened at the tip and preferably with two cutting edges, for cutting small tissues, smearing and squashing.

## General equipment

(a)  Water bath or oven at temperature of 60°C for hydrolysis. One also at 37°C is useful for mammalian tissue work.
(b)  Hot-plate for drying permanent mounts, though this is not essential.
(c)  Slide storage trays or boxes.
(d)  Filter paper, for use in squashes.

## Optical equipment

(a)  A magnifying lens, hand or mounted, is useful. The microscope is, however, the essential apparatus. Chromosome work requires high resolution and hence the better the instrument, the more information will be gained. Essential components are three objectives (× 10, × 40, × 100 oil), a sub-stage condenser, eyepiece ($\times$ 10) and a source of illumination. The limiting factor of resolution is the numerical aperture (N.A.) of either the objective or the sub-stage condenser. Maximum useful magnification (above which resolution tends to diminish) may be regarded for simplicity as 1000 N.A. (the lowest N.A. in use at the time being the limiting one). Hence an eyepiece magnification of $\times$ 10 is usually ideal. If a binocular microscope is used whereby the magnification of the head may be $\times$ 1·3 or $\times$ 1·5, then $\times$ 8 eyepieces are recommended. The simplest light source is a 60 w bulb with a means of diffusing the light, but lamps fitted with a source condenser and diaphragm are preferable together with a means of varying their intensity. These may be either external to the microscope or an integral part of it. Phase contrast is extremely useful for critical work, being essential for unstained or lightly stained material.

## II. GENERAL TECHNIQUES

### 1. PRE-TREATMENT

A variety of treatment methods are available, any of which, when applied to the tissue prior to fixation, is an aid to the study of chromosome morphology. This involves exposure of the material to chemicals for reasons such as arresting a larger number of cells in metaphase, separation of the chromatid arms, contraction of the chromosomes, enhancement of constriction regions and heterochromatin, demonstration of coiling, and production of aberrations. Maceration treatment can be given after fixation to aid spreading of the tissues.

#### (a) Morphology and metaphase arrest

A series of chemicals which are used for metaphase arrest, by inhibition of spindle formation, also bring about chromosome contraction and separation of the chromatid arms. The replication of the chromosomes is not affected, nor is their splitting into chromatids. The chromatids remain attached at the centromere, the division of which is delayed for several hours. The inhibition of spindle formation and the chromosome contraction facilitate chromosome spreading. These chemicals are therefore most useful for general morphological study of the chromosomes and for detailed study of the karyotype. Treatment is followed immediately by fixation. Those considered as the simplest to use, being effective at room temperature, are as follows:

#### (i) Colchicine

This alkaloid is most commonly used because of its high effectiveness over a wide range of concentrations on both plant and animal tissues (Levan 1938;

Barber & Callan 1943). It is water soluble and the duration of treatment is inversely proportional to the concentration. The favoured concentrations range from 0·05 per cent to 1 per cent aqueous solution, these being used for time periods between 1–4 hours. Tissues are placed directly into the colchicine solution, or for tissue cultures it is added to the medium. Both mitotic and meiotic tissues respond well to colchicine treatment. Actively dividing mammalian cells can be treated *in vivo*, but as this requires injection of the alkaloid into the animal (permissible only by special licence in Britain), this is not included as a technique here. Colchicine is a poison and hence should be handled with care. Although expensive, its effectiveness at low concentrations offsets this.

### (ii) Para-dichlorobenzene (PDB)

This chemical is most effective for plant tissues though it may also be used for animal tissues (Meyer 1945). It is probably the simplest and cheapest to use. Whereas after colchicine treatment the chromatid arms in mitosis lie well apart, with PDB sister chromatids lie close together. Treatment ranges from 1–4 hours using a saturated aqueous solution.

### (iii) α-monobromonaphthalene

Also very useful for plant tissues. Between 2–4 hours treatment is necessary in a saturated aqueous solution (O'Mara 1948).

### (b) Induction of polyploidy

The production of polyploid cells in plant tissue is effected most readily by prolonged pre-treatment of actively dividing cells in colchicine. When spindle function is inhibited, nuclear and cell division are also, as a rule. The chromatids separate but are included within a common nuclear envelope so that the number of chromosomes per mitotic cycle in each cell is doubled. In order to demonstrate polyploid induction, seeds, after initiation of germination in water, or seedlings can be bathed in an aqueous colchicine solution of 0·05 per cent to 1 per cent concentration for 1–3 days. Although polyploid cells are formed by this treatment over a period of time, selection against such cells may cause the reversion of the meristematic zone to a diploid condition. (Root or shoot apices can be cytologically investigated after a period of growth to ascertain whether polyploidy has been maintained.)

### (c) Chromosome coiling

Shock treatments such as exposure of the tissues to heat, cold, acid or alkali tend to uncoil the chromosome threads so that their structure is more readily visible. The most reliable method is by treatment with ammonia, either as a vapour (Kuwada & Nakamura 1934) or by one or two drops of concentrated ammonia in 10 ml dilute alcohol (20 per cent–30 per cent) (Sax & Humphrey 1934; Creighton 1938). There are however, many variables involved so that trial and error is essential, but the ammonia/alcohol mixture is considered more reliable (see schedule 18). Meiotic tissues (e.g. pollen mother cells or testicular material) respond most readily to this pre-treatment although somatic tissues may also be utilized.

### (d) Heterochromatin

Many heterochromatic regions may be seen without pre-treatment. Primary and secondary constrictions are enhanced by pre-treatment with any of the metaphase-arrest chemicals (see section (a) above). Additional regions of heterochromatin can be seen also in some animal and plant tissues by special cold treatment immediately prior to fixation (Callan 1942; La Cour *et al.* 1956; Evans 1956). The plant (e.g. rhizome of *Trillium*) or animal (e.g. *Triturus*, the newt) is kept at temperatures of 0–3°C for 3–5 days; the requisite tissues are then fixed and stained, preferably by the Feulgen technique (see staining techniques p. 140). Cold pre-treatment can also inhibit spindle formation in some tissues (as in *Trillium* and *Triturus*), thus having a similar effect to that of the chemicals used for metaphase arrest.

### (e) DNA synthesis

Sites of the incorporation of radioactive tracers into tissues may be located by autoradiographic techniques. Handling of such radioactive chemicals, even those which have a minute degree of penetrance, requires special permits. Hence these techniques would be limited to laboratories with the necessary facilities and they are probably only suitable for advanced students. Autoradiographic study of DNA synthesis and chromosome replication in *Vicia faba* was carried out by Taylor *et al.* (1957) since when similar studies have been carried out by numerous workers on many other tissues (Taylor 1958, 1960; Prescott & Bender 1963).

The basis of autoradiography is the exposure of a photographic emulsion, which has been brought into contact with the isotope-incorporated tissue, to the β-rays emitted by the radioactive chemical (e.g. $^{32}P$, $^{14}C$, $^{3}H$). It is essential for chromosome work to use a radioactive isotope which will be utilized in DNA synthesis and one which emits electrons over a very limited range. Thymidine being a compound incorporated specifically into DNA, and tritium being an isotope which emits β-radiation only to a maximum range of 3 μ, the use of tritiated thymidine in DNA synthesis experiments is ideal. Exposure of the film emulsion to the radiation produces, on development, a pattern of black dots over the sites of thymidine incorporation. Background in autoradiographs (development of silver grains in the emulsion not due to the experimental source) can often be severe unless strict precautions are taken in handling the film. Causes of background are numerous but the most frequent are excess exposure to the safelight, over-development, occurrence of static when using stripping film, and pressure on the emulsion surface.

The tritiated thymidine is added either to the medium in which the tissue grows, or it can be injected into the tissue. For student purposes addition to the medium is the most practical. Hence plant roots and plant or animal tissue cultures can be allowed to grow in the isotope-medium. As incorporation of thymidine occurs in the S-period (phase of DNA synthesis), the time of 'labelling' can be adjusted for set periods, depending on the experimental requirements. The exposure may be over the whole S-period, or a mere fraction of it as in so-called 'pulse-labelling' (Bender & Prescott 1962; Evans 1964).

Fixation is best carried out in acetic-alcohol (one part glacial acetic acid to three parts absolute alcohol) and hydrolysis, if necessary, in HCl. Staining of medium intensity with aceto-carmine, acetic-orcein or by the Feulgen method after squashing or air-drying techniques is most useful. If staining is omitted until after photographic processing, then toluidene blue is most useful for staining through the film. The slides are then covered by Kodak stripping film AR 10 (Pelc 1956) or coated in liquid emulsion (Kopriwa & Leblond 1962) and exposed in complete darkness and dryness (preferably packed with a desiccant) in a refrigerator at 4°C for a period of 1–4 weeks. Time of exposure varies with the degree of incorporation of tritiated thymidine. Development in Kodak D-19 is carried out at 18°C, and fixing and drying in a dust-free atmosphere (see schedule 19).

## (f) Induction of aberrations

Radiation and chemicals are well known as sources of damage to the hereditary material. Their effects on chromosomes are primarily to cause breakage with subsequent reunion and/or loss of fragments. Such damage may bring about an aberration leading to inviability of the cell, or the outcome may be one of chromosome mutation (inversion, translocation, etc.) which can be perpetuated in polymorphic states (e.g. inversions in *Drosophila*, *Chironomus*) or as permanent heterozygotes (e.g. translocations in *Oenothera*).

Aberrations can readily be induced by ionizing radiation such as X-rays, first shown by Muller (1927) in *Drosophila* and Stadler (1928) in maize, α-, β-, as is evident in autoradiographic experiments, and γ-rays. They may also be induced by non-ionizing radiation such as ultraviolet light. Both radiation and chemical agents are used to bring about fragmentation of chromosomes and hence cell death in the treatment of cancer. Some chemicals, however, are very specific in their action. Whereas ionizing radiation brings about chromosome breakage throughout the cell cycle, most chemical mutagens appear to act after the onset of the S-period. Many are also temperature and pH sensitive and may be sensitive to oxygen concentration.

The following agents are recommended :

### (i) X-ray

If a source of X-rays is available, it provides one of the simplest methods for aberration induction. The material can be exposed directly to the X-ray source, the dosage being determined best by a fixed output and variation of exposure time (Darlington & La Cour 1945; Bishop 1950; Fox 1966). The effects on mitotic cells are time dependent. There is a primary effect causing chromosome stickiness and anaphase breaks. This is followed by a depression of mitotic frequency brought about by a delay in the onset of mitosis. Breaks and reunions which had occurred in the resting stage nuclei then appear in later divisions. Resting seeds are delayed in germination. Example exposure times are, for seeds (preferably soaked), 10,000–50,000 rad; root-tips, pollen grains and anthers, 25–250 rad; animal embryo 25–400 rad; testis in adult insect (e.g. locust) 500 rad (see schedules 20 and 21).

### (ii) Maleic hydrazide

This is a most effective agent with *Vicia faba* (Darlington & McLeish 1951). A large proportion of the induced aberrations tend to be localized in the heterochromatic segment, close to the nucleolar arm of the $M$-chromosome (McLeish 1953). The frequency of aberrations increases with temperature and with a decrease in pH. Root-tips can be treated with $10^{-4}$M maleic hydrazide at 20°C for 2 hours. As cells so treated are delayed in their development they are best fixed for preparations approximately 24 hours later.

### (iii) Potassium cyanide

Produces breakage in *Vicia* root-tips after treatment for 1 hour with $4 \times 10^{-4}$M KCN (Lilly & Thoday 1956). The effect is again delayed and best results are found approximately 24 hours later. It is not temperature sensitive but aberration frequency increases with oxygen concentration.

### (iv) Nitrogen mustard (HN2)

This alkylating agent provides another relatively simple class method for producing chromosome aberrations. Ford (1949) studied the effect of HN2 on the root-tips of *Vicia*. Although breaks do not occur at random they are not so localized as with maleic hydrazide. The recommended treatment is for half an hour with $10^{-5}$M solution of HN2. Further growth for 12–24 hours yields numerous structural aberrations.

## (g) Aids in spreading of chromosomes

Satisfactory separation of cells for squashing can only be obtained for somatic tissues of plants by dissolving the pectic salt of the middle lamella. This may be carried out prior to staining but after fixation by maceration of the tissue by either HCl or ammonium oxalate or enzymes. A saturated aqueous ammonium oxalate solution is particularly useful for plant meristematic tissues; treatment is from 2–5 min (C.E.Ford, unpublished, referred to in Darlington & La Cour 1962). Gut cytase, obtained from the stomach extract of snails (*Helix pomatia*) was used as an overnight treatment at room temperature by Fabergé (1945), and is useful for a wide variety of plant tissues. Pectinase (5 per cent in 1 per cent aqueous solution of peptone) has also been used for treatment periods of 4–16 hours (Chayen & Miles 1954). HCl is however the most readily available and most commonly used method because of its adaptability. When used prior to staining it also brings about hydrolysis of the nucleic acids, releasing aldehydes so that the Feulgen reaction, specific for DNA, can occur. It is most frequently used as N.HCl at 60°C but this can be varied via a range of concentrations and temperatures to 5N.HCl at 20°C. If other stains such as aceto-carmine or preferably acetic-orcein are to be used, then satisfactory maceration can be obtained by mixing N.HCl with the stain in approximately 1 : 9 ratio and heating the tissues in this acidified stain. Hard roots, such as those from grasses, should be treated overnight with gut cytase, and then with HCl.

For animal cells, treatment with a hypotonic solution prior to fixation effectively increases the volume of the cells and permits much improved chromosome spreading. Hypotonic solutions which have been used with very satisfactory results in mammals are distilled water, dilute Hanks (see appendix), sodium citrate and potassium chloride solutions (see schedules 6, 7, 8 and 13).

## 2. FIXATION

The prime purpose of fixation is to coagulate the cell contents in such a way as to retain their shape, structure and position. It is therefore important to use a fixation fluid that will penetrate the tissues rapidly. Fixation should also prepare the surface of the chromosomes so that they readily take up suitable stains and it should soften the tissue when required for squash techniques. Absolute alcohol or acetic acid alone will bring about coagulation of the cell contents, but alcohol tends to harden the tissues and does not penetrate them rapidly enough whilst acetic acid, which penetrates speedily, tends to swell the protoplasm and causes excessive softening. Combined, these two chemicals provide an excellent fixative for materials prior to squashing and they are used most frequently as a mixture of three parts absolute alcohol : one part glacial acetic acid. Propionic acid may be used advantageously as a replacement for acetic acid in some tissues. It is not so rapid in penetration, but does not cause swelling of the protoplasm to such an extent. The combination of chloroform with absolute alcohol and acetic acid was first used by Carnoy (1886) as a mixture of six parts absolute alcohol : three parts chloroform : one part glacial acetic acid. Chloroform permits rapid penetration and also acts as a solvent for fats. It is therefore very useful for plant or animal tissues (e.g. testes) which contain or are surrounded by fatty substances. When fixing plant material in bulk for classes a simplified Carnoy fixative is particularly useful. For this purpose the quantity of chloroform is reduced to proportions of six alcohol : two acetic acid : one $CHCl_3$, or three alcohol : one acetic acid: one $CHCl_3$. When material is to be left in fixative for a long period, the fixative should be changed preferably after 1–3 hours or certainly before cold storage (see below).

If the material is left in the acetic/alcohol fixatives for long at room temperature it will become brittle and be unsuitable for making squash preparations. The hardening will however be retarded at lower temperatures. Hence the material should not be allowed to stand in the fixative for more than three days at room temperature or more than two weeks in a refrigerator at 4°C. If kept in a deep-freeze at −20°C, however, the material will keep satisfactorily almost indefinitely. Should facilities for low temperatures not be available, then the material is best transferred from the fixative, via absolute alcohol to 70 per cent alcohol where it can remain at room temperature without hardening. Fixation and staining gradually deteriorate, however, and preparations should be made as soon as possible.

As an alternative, and often for convenience, the fixation and staining may be carried out in a combined operation (as in aceto-carmine, -orcein) where the acetic acid acts as the fixative. This is ideal for class work if the living tissues are available.

Should staining with the basic dyes prove to be inadequate, then premordanting may help. This can be done in the fixation stage by the addition of a few drops of ferric acetate or ferric chloride to the fixative.


## 3. SLIDE PREPARATION

For most cytogenetic work slides are prepared by direct smear or squash techniques, or by air-drying if the fixed cells are in suspension, as after tissue culture. These techniques are ideal because of their ease and speed in execution, coupled with a high efficiency.

## (a) Smear and squash

The importance of both these techniques is to make an even spread of cells over the surface of the slide, thereby permitting observations on single cells and a better spread of the chromsomes due to the release of outside pressures from surrounding tissue.

Smear methods are to be preferred with soft tissues such as anthers and testes for meiotic studies or buccal mucosa tissue for nuclear chromatin (see schedules 1 and 9). The contents of an anther or of the tubules of a testis are smeared with the aid of a scalpel, or similar flat instrument, direct on to a dry slide. The material can then either be fixed by placing the slide in fixative (e.g. 3 : 1 alcohol-acetic) prior to staining, or it can be fixed and stained in a combined operation by the addition of an acetic stain. It is essential not to delay fixation of either kind or the material will dry on the slide causing the contents of the cells to clump.

Squash methods incorporating the combined stain-fixation of an acetic stain have become the most widely used cytological techniques for chromosome studies. The first of these was described by Belling (1926) using iron-aceto-carmine for the study of pollen mother cells. It can be used, however, for almost any animal or plant tissue that can be teased out with a needle and which is adequately soft to allow squashing and spreading of the cells. Root tips require prior maceration treatment. The material is mounted directly into a drop of acetic stain (e.g. carmine, orcein) and teased out. Carmine staining is often much improved by an iron mordant but this is not required for orcein (see section (c) below). Teasing out is best effected by a mounted steel needle with a flattened end (e.g. a small needle-knife). The material can be broken up with the flattened needle which helps to separate the cells without undue damage. All unwanted debris (e.g. anther wall) should be removed to facilitate efficient spreading of the remaining cells by the squash method. A cover slip is lowered gently on to the drop of stain containing the teased out material (the drop of stain should be just sufficient to cover the area of the cover slip), care being taken to avoid the production of air bubbles. Further teasing out can be accomplished at this stage by gentle tapping of the cover slip with a blunt-ended instrument (e.g. a rubber-tipped pencil).

The slide is then heated gently over a spirit lamp. This is a crucial stage which is perfected often only after much trial and error. Heating helps the adhering of the cells to the slide, the spreading of the chromosomes and the differential staining. Too much heat will bring about shrinkage of the cells and clumping of the contents as well as an excess of stain in the cytoplasm. The heating should be gradual by moving the slide in, out and over the flame, and should not normally exceed 60°C. A good estimate of this can be obtained by testing the heat of the slide on the back of the hand; one should stop heating when the slide is too hot to be held in contact with the hand. The material can then be squashed by placing the slide and cover slip between filter paper and using vertical thumb pressure against a really flat surface such as a sheet of plate glass. Heavy pressure on even a slightly uneven surface may lead to breakage of the slide. Care should be taken not to move the cover slip sideways at this stage or the squash will be ruined. The amount of pressure, ranging from nil to the maximum possible, depends on the fragility of the material and can only be judged by experience. It should, however, never be violent, but gentle to begin with, steadily increasing pressure if required. Perhaps even more than with most laboratory techniques, the squash method is one in which the user will find that only experience brings regular success, and this is not gained quickly. Final results, however, can be superb and worthy of perseverance.

Material which has been hydrolysed for Feulgen staining can also be squashed by the above method. After staining, the tissue should be mounted in a drop of 45 per cent acetic acid, teased out, covered by a cover slip and squashed without heating. If desired, additional contrast can be obtained by mounting in an acetic stain in place of the 45 per cent acetic acid, when the cytoplasm (normally unstained by Feulgen's method) will take up a pink colouration and the stain in the chromosomes will be intensified.

### (b) Air drying

This technique has gained considerable popularity in the past decade following the improvement of mammalian chromosome techniques. It was first introduced by Rothfels & Siminovitch (1958) and has since been modified by different workers. It is used if dealing with cells in suspension (e.g. mammalian blood, marrow, etc.). After fixation the cells are suspended in either a 3 : 1 alcohol-acetic mixture or a 60 per cent acetic acid medium, pipetted on to a slide, preferably pre-cooled, and allowed to dry completely. Drying may be hastened either by blowing over the surface of the slide, by gentle heating over a spirit lamp, or even by igniting the fixative and allowing to burn dry. Slides can be stored in this dried state if desired before staining (see schedules 7, 8 and 22). This method is usually preceded by colchicine and hypotonic treatment. The spreading of the chromosomes can be excellent.

### (c) Staining

This is principally a process of dye adsorption on the surface of the chromosomes. The chromatic material is acidic and hence staining is effected by basic dyes. Some of these may require a mordant to enhance their adherance to the chromosome surface (e.g. carmine) whereas Feulgen staining requires acid hydrolysis, for this is dependent on a chemical reaction with the resultant liberated aldehydes. Recently fluorescent dyes such as quinacrine mustard and quinacrine hydrochloride, or the use of stains such as Giemsa or Leishman after special post-fixation treatment, has enabled the display of specific banding patterns in chromosomes (see schedule 22 for further detail and references).

Unfortunately, stained material frequently tends to lose its colouration with time if the preparations are made permanent. This fading is due to a combination of factors such as exposure to light (principally ultraviolet) and the acidity of the mountant. Both aqueous and acetic stains can be used, the following being considered best for the techniques described. Tissues can be stained in bulk or following smear/squash/air-drying techniques.

### (i) Aceto-carmine

Carmine is probably the most commonly used stain for all chromosomes. It is primarily used for plant material but can be equally useful for animal chromosomes. It is prepared as a solution in 45 per cent acetic acid, often at a concentration of 1 per cent as first used by Belling in 1921. It is preferable however, to use a saturated solution, more in the region of 2 per cent (see appendix for stain preparation). The staining is improved by an iron mordant which can be added to the stock solution of the stain as a few drops of ferric acetate. Too much iron brings about precipitation of the stain; if this occurs the precipitate must be filtered off, for it hinders squashing of the material when slide-making and also produces a very dirty background to the preparation. Addition of iron to the stain is unnecessary if an iron pre-mordant has been used or if unplated steel needles are used for teasing out the tissue in the stain when making squashes. The steel will react with the acetic acid to provide the iron acetate necessary for mordant action. If iron has been added to the stain then stainless steel needles should be used during the teasing out process to prevent stain precipitation.

If too much stain is taken up by the cytoplasm, the excess can be removed by differentiation using 45 per cent acetic acid.

### (ii) Acetic-orcein

This is a more selective stain than carmine and preferably is used for most animal chromosomes, particularly in salivary gland cells, and occasionally in plants if carmine proves inefficient. No mordanting is necessary. It was first recommended as a chromosome stain by La Cour (1941) and is used most frequently as a 1 or 2 per cent solution in 45 per cent acetic acid.

Both carmine and orcein may be used in a 9 : 1 mixture of stain : N.HCl as a combined stain/hydrolysis technique (see schedule 2).

### (iii) Propiono-carmine and propionic-orcein

Either may be used as a 2 per cent solution after fixation in propionic acid-alcohol (1 : 3). They tend to give better differentiation than aceto-carmine and acetic-orcein and hence are useful for small chromosomes. Propionic-orcein does not precipitate so readily at room temperature as acetic-orcein.

### (iv) Lactic-acetic-orcein

If staining is of necessity prolonged, either for satisfactory adsorption of the dye or when the stain is being used as a temporary mountant, then the introduction of lactic acid prevents rapid drying out. Lactic-acetic-orcein is used at various concentrations from 0·25 per cent to 2 per cent in a solution of equal parts lactic acid and glacial acetic acid. It is particularly useful for salivary gland and other animal chromosomes.

### (v) Leuco-basic fuchsin (Feulgen)

This stain, at its best, surpasses all others for it is specific for DNA, is translucent and requires no differentiation. The method is wide in its application but unfortunately is subject to failure at times. It is dependent on Schiff's reaction for aldehydes and requires adequate pre-hydrolysis which liberates the aldehyde groups of the nucleic acid. Subsequent chemical reaction between the aldehydes and the stain gives the chromosomes a magenta colouration.

Hydrolysis is usually carried out in N.HCl at 60°C but both concentration of the acid and the temperature can be varied reciprocally (e.g. 5N.HCl at 20°C). Timing of hydrolysis is important particularly at the higher temperature and for 60°C a suitable duration may range from 3–15 min depending on the tissue. The stain is prepared from basic fuchsin and decolourized to leuco-basic fuchsin by the release of $SO_2$ into the solution (see appendix for details). Preparation of the stain is critical and should always be carried out with the utmost care. The stain is unstable to air or light exposure and hence storage of the stain and staining of material should be carried out in an airtight container in the dark. The occasional failures which do occur are frequently due to incorrect hydrolysis time or to an inefficiently prepared stain solution.

### (vi) Toluidene blue

This is a very effective stain for chromosomes, but as it stains both RNA and DNA the cytoplasm becomes too heavily stained. It is therefore necessary to hydrolyse the preparation previously, at 60°C with N.HCl or at room temperature with 5N.HCl for a period of 2–5 min. This removes the ribose nucleic acid and the chromosomes become selectively stained, using a 0·05 per cent aqueous solution of toluidene blue. Staining is rapid, half to 1 min being adequate.

This stain is particularly useful in the technique for semi-permanent stain mounting (see section (d) below).

### (vii) Giemsa

May be used as a general chromosome stain but is recommended for the staining of mammalian chromosomes after treatment to display specific banding patterns. It is used as a 1 : 50 solution of Giemsa : pH 6·8 buffer (see schedule 22a).

### (viii) Leishman

May be used as a general chromosome stain but it has been recommended specifically for the display of banding in mammalian chromosomes following trypsin treatment (see schedule 22b). It is used as a 1 : 3 solution of 0·2 per cent Leishman : pH 6·8 buffer.

### (ix) Fast green

This is a simple counter-stain suitable for use for nucleoli and cytoplasm after Feulgen staining. It may be used as a 0·1 per cent aqueous or alcoholic solution.

### (d) Mounting

It is often convenient or necessary to keep preparations for future reference. Such preparations may be kept up to a week if treated as a temporary mount; for longer than this period, it is essential to treat them as semi-permanent or permanent mounts.

### (i) Temporary mounts

The simplest method for making a temporary mount is to leave the preparation as made in its acetic stain or 45 per cent acetic acid (after Feulgen staining) and to seal round the edges of the cover slip so that drying out of the mountant cannot occur. Preparations which are to be treated in this way should not contain air bubbles beneath the cover slip. Sealing is best carried out by an application of either rubber solution or nail varnish (preferably coloured); in either case a second application, after drying of the first, is advisable to ensure a good air-tight seal. The seal can be removed at a later date and the slide made permanent. Rubber solution is simple to remove; raising at one point will allow it to be stripped off as a complete ring. Nail varnish also strips off with ease, particularly if immersed briefly in acetone.

The main disadvantages of temporary mounts are due to inefficient sealing (thereby allowing a gradual drying out), to fading of the stain (often in Feulgen preparations due to diffusion into the mountant), or to precipitation and becoming overstained in the acetic dyes. There is less danger of drying out if a lactic-acetic stain is used. Removal of any precipitate and differentiation of overstained preparations can be carried out in a relatively strong concentration of acetic acid, preferably 75 per cent. This acts quickly and leaves adequate stain still in the chromosomes. The slide can then be made permanent if desired.

### (ii) Semi-permanent mounts

For semi-permanent mounts a technique which is well recommended makes use of a stain-mountant resin (C.E.Ford, personal communication). A saturated aqueous solution of the resin, dimethyl hydantoin formaldehyde, is obtained by adding resin little by little to a flask of distilled water over a period of several days and shaking periodically. The solution is filtered and a 1 per cent aqueous solution of toluidene blue is added until the mixture stains satisfactorily. Previous hydrolysis of the preparation is advisable, from 2–5 min in 5N.HCl being recommended, to prevent excessive staining of the cytoplasm. The slide is then passed rapidly through 50 per cent and 90 per cent alcohol, washed in absolute alcohol and dried in air before the stain-mountant is added. The chromosomes take up an intense blue stain immediately but the stain is not permanent, fading after 2–8 weeks. The slides can often be re-stained satisfactorily by dissolving the mountant by exposure to warm water ($c.$ 50°C) for 3–4 days if necessary, before passing through the alcohols and re-drying.

### (iii) Permanent mounts

For permanent mounts the mounting medium should be optically sound by possessing a refractive index equivalent to that of glass or slightly higher so as to approach that of the tissue. Preferably it should also be of neutral pH, for acidity brings about fading of the stain. There are numerous mounting media available but those most commonly used are Canada Balsam and Euparal. The former is the better optically but tends to be acidic and cause fading. Euparal has the advantage that the tissue can be mounted in it directly from alcohol, thereby omitting the xylene stage essential for Canada Balsam. Either of these can be recommended following the smear/squash techniques. For preparations made by the air-drying method, the quick drying neutral Xam mountant (supplied by G.T.Gurr) is very useful. This does not cause fading to the extent of either Canada Balsam or Euparal, but for squash preparations it may cause excessive shrinkage of the tissue. As when using Canada Balsam the tissue must be passed through a xylene stage before mounting.

The most important part of permanent slide making from smear/squash preparations is the successful separation of the cover slip from the slide without the loss of the material. Most authors will recommend the use of albumenized cover slips whereby the cover slip is smeared very thinly with Mayer's albumen (see appendix) and allowed to dry before making the preparation. This certainly aids adhesion of the material to the cover glass but if one is not careful it can provide a less clean background and to some extent hinder squashing. With experience the use of albumen can be avoided. Slides and cover slips, as received direct from the manufacturers, are wiped clean with a fine dust-free cloth and the preparation made in the normal manner. (Slides and cover slips which have previously been cleaned in detergents are not recommended to be used without albumen.) Separation of the slide and cover slip can be effected either by placing the prepared slide in a dish of solvent (e.g. 45 per cent acetic acid) or by the quick-freeze method introduced by Conger & Fairchild (1953). The latter method is the more rapid and once mastered is very successful. There is perhaps a greater tendency for some cells to collapse after this method however.

To separate the slide and cover slip in solvent the preparation is submerged in 45 per cent acetic acid in a Petri dish, until the cover slip glides away freely without physical disturbance. In this method the adherance of the tissue to the slide and/or cover slip is entirely dependent on the gentle heating during the slide preparation and the degree of 'drying down' allowed before the preparation is submerged in the solvent. The 'drying down' process should be just sufficient to see the stain mountant beginning to pull away from the edge of the cover slip. After separating the slide and cover slip, the material is dehydrated by taking both the slide and cover slip through 95 per cent alcohol and absolute alcohol (two changes), leaving in each of the first two alcohols for approximately 2 min and the last alcohol for 5 min. The slide and cover slip can then be reunited using Euparal or they may be taken into one part xylene : 1 part absolute alcohol before mounting in Canada Balsam.

With the quick-freeze method the preparation is either placed in contact with a flat surface of dry ice (solid $CO_2$) or frozen by a jet of $CO_2$ from a liquid $CO_2$ cylinder. Rapid freezing is important and the slide should be kept in close contact with the dry ice by slight pressure with a blunt-ended instrument for at least 30 seconds. After the preparation has been frozen for about 1 min (leaving longer has no detrimental effect) the cover slip is separated from the slide with the aid of a sharp scalpel or preferably a razor blade. This is run along the edges of the cover slip and then a little under one edge until the cover slip is free. Both slide and cover slip are placed immediately into 95 per cent alcohol; it is at this stage that most damage to the preparation can occur and under no circumstances should thawing be allowed before placing in the alcohol. The slide and cover slip are then transferred through absolute alcohol (two changes) and either mounted in Euparal or after passing through one part xylene : one part absolute alcohol mounted in Canada Balsam. Cloudiness in the mountant indicates insufficient dehydration. The slide and cover slip should be drained of any excess of the final dehydrating liquid by tilting them sideways on a piece of filter paper before uniting them with a drop of mountant. This should be done as quickly as possible. Some workers prefer to take the preparation into pure xylene before mounting in Canada Balsam; this is not necessary but it eliminates any danger of the absolute alcohol, present in the final dehydrating medium and left on the slide, readily attracting moisture from the atmosphere as the mountant is added. Slides take about three days to dry at room temperature and hence are best warmed to approximately 40°C either in a drying oven or on a warming plate. Euparal tends to be slower drying than Canada Balsam.

One objection to the quick-freeze method for permanent slides is the excess of cytoplasmic and background stain left in the preparation. This is particularly noticeable with aceto-carmine, but it can be counteracted by placing the slide and cover slip, immediately after separation, into 95 per cent alcohol to which acetic acid has been added (in the proportion of three parts alcohol : 1 part acetic acid). Differentiation can thus be carried out before transferring through 95 per cent alcohol into the absolute alcohols as described previously.

## III. SUGGESTED MATERIAL AND SPECIFIC SCHEDULES

### 1. SEX CHROMATIN

The presence of a distinct chromatin mass has been observed in the nuclei of many normal female mammals, but not in normal males (Moore & Barr 1954). Sex chromatin has proved on further investigation to be a marker which indicates the number of $X$ chromosomes present. Hence the technique has been widely used in the investigation of abnormal sexual development and other aspects of medical cytology. Although a simple technique it is one which should only be done under careful supervision, due to the method of obtaining the sample of tissue. The sex chromatin mass, which stains readily with basic dyes, is seen in interphase nuclei and is usually found against the inner surface of the nuclear membrane (see plate 4.1*).

The most readily available cells which can be used for investigating sex chromatin are obtained from the buccal mucosa of man as follows:

**Schedule 1. Sex chromatin mass**

(i)   Draw the edge of a sterile metal spatula (as used in analytical weighing) firmly over the inside surface of the cheek. As the first sample of tissue usually

* Plates in this chapter follow p. 148.

includes many dead cells this should be discarded and a second gentle scrape made.

*Either:*

(ii)   Smear on a slide, add a drop of acetic-orcein and cover with a cover glass.

(iii)   Gently heat and squash. (Direct observations can then be made, but as it is better to use an oil-immersion objective, a more permanent preparation is recommended.)

*Or:*

(ii)   Spread tissue on an albumenized slide (very thinly coated with Mayer's glycerin/albumen) making a relatively thick preparation.

(iii)   Fix immediately in 95 per cent alcohol (15–30 min).

(iv)   Stain in acetic-orcein (15–30 min), rinse in 45 per cent acetic acid, dehydrate through 95 per cent and absolute alcohol and mount in Euparal (see section on mounting, p. 144).

An alternative staining method giving better contrast may be used after fixation in 95 per cent alcohol (as in step (iii) above).

(iv)   Wash in distilled water for 5 min.

(v)   Hydrolyse in N.HCl at 60°C for 6 min (or 5N.HCl at room temperature for 25 min).

(vi)   Rinse well in distilled water.

(vii)   Stain in leuco-basic fuchsin (1 h).

(viii)   Rinse in two changes of $SO_2$ water (see appendix).

(ix)   Wash in tap water (10 min).

(x)   Counterstain in 0·1 per cent Fast Green in 95 per cent alcohol (few seconds only).

(xi)   Dehydrate in absolute alcohol and mount in Euparal, or clear in xylene and mount in Xam.

## 2. CHROMOSOMES IN MITOSIS

There are many plant and animal tissues which provide suitable material for the study of mitotic chromosomes. Root tip tissue is the simplest and most readily available, but techniques of varying complexity will produce excellent preparations from pollen grains, pollen tubes, insect embryos, mammalian eyes, marrow and blood.

### (a) Root tips

A good supply of root tips with rewarding mitoses can be obtained from a selection of germinated seeds, *Vicia faba* ($2n=12$, $14$), *Pisum sativum* ($2n= 14$), *Haplopappus gracilis* ($2n=4$), from bulbs, *Allium cepa* ($2n=16$), *Crocus balansae* ($2n=6$), *Endymion non-scripta* ($2n=16$), *Hyacinthus orientalis* ($2n=16$, $24$, etc.), and numerous growing plants such as *Ranunculus* spp.

($2n=14$, 16, 32) (see plate 4.2), *Crepis* spp. ($2n=6$), *Nothoschordum fragrans* ($2n=16$, 18, 20), *Melandrium* spp. ($2n=24$, 48).

In all cases it is essential to use actively growing roots. These are obtained from bulbs such as *Allium* and *Hyacinthus* by merely growing them with their bases almost in contact with water in the dark, whereas *Vicia* and *Crocus* are best planted in a box of moist vermiculite. The primary root or secondary lateral roots may be used in *Vicia*. Secondary laterals can be stimulated to grow by removing the meristematic tip of the primary root when the root is about 25–35 mm long and replanting the seedling in the vermiculite. Roots of approximately 10–25 mm in length should be used. Pot-grown plants produce good roots for cytological investigation particularly if kept moist from below by standing the pot in wet shingle. Pot-bound plants should be re-potted and allowed to develop new root growth. Plant roots are best obtained by removing the pot to expose growing roots at the base. The following schedule is recommended:

**Schedule 2. Root tip squash preparation**

(i)   Remove approximately 10 mm length of root tips by means of fine-pointed forceps.

(ii)   Immerse in pre-treatment solution, saturated aqueous paradichloro-benzene for 4 h at room temperature. (This stage can be omitted or other pre-treatment methods may be used—see section on pre-treatment, p. 132.)

(iii)   Fix in fresh mixture of 3 : 1 absolute alcohol : glacial acetic acid for a minimum of 15 min. Fixed material will remain workable for long periods (see section on fixation, p. 137).

(iv)   Stain by means of one of the following techniques. (Hard roots, such as those of grasses, should first be washed free of fixative and treated overnight with snail gut cytase at room temperature in a stoppered bottle prior to hydrolysis with HCl.)

*Feulgen technique*

(a)   Hydrolyse in N.HCl at 60°C for 5–15 min, being dependent on the material (e.g. 6 min for *Crocus balansae*; 10 min for *Allium, Hyacinthus*).

(b)   Rinse in water.

(c)   Stain in leuco-basic fuchsin for $\frac{1}{2}$–3 h in dark.

(d)   Mount on a slide in a small drop of 45 per cent acetic acid, or an acetic stain (orcein or carmine), removing all tissue except the dense meristematic tip (*c.* 1–2 mm).

(e)   Break up with needle knife to separate the cells as well as possible.

(f)   Place cover slip in position and tap with blunt instrument to spread tissue.

(g)   Heat very gently, over small flame, or not at all.

(h)   Squash under blotting (filter) paper.

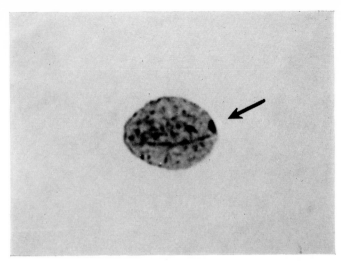

**Plate 4.1.** Sex chromatin mass (arrowed) stained in leuco-basic fuchsin (schedule 1) in a human female buccal mucosa cell.

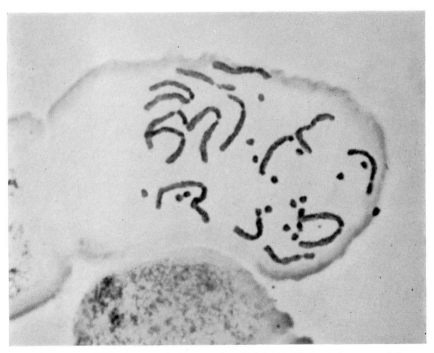

**Plate 4.2.** Metaphase of mitosis in a root tip cell of *Ranunculus acris* ($2n = 14 + 18\beta$). after pretreatment in colchicine and stained in aceto-carmine (schedule 2). (Courtesy of Dr N.M.Gregson.)

[*Facing p.* 148]

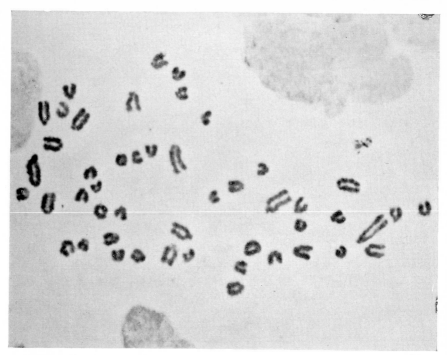

**Plate 4.3.** Mitotic metaphase from cornea of *Apodemus sylvaticus* ($2n=48$). Prepared and stained in acetic-orcein (schedule 6). (Courtesy of W.H.Jones.)

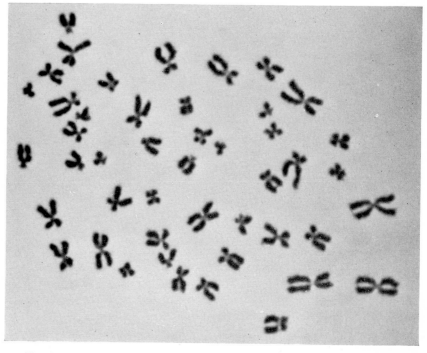

**Plate 4.4.** Metaphase in mitosis in female human leukocyte ($2n=46$). Stained in lactic-acetic-orcein (schedule 8).

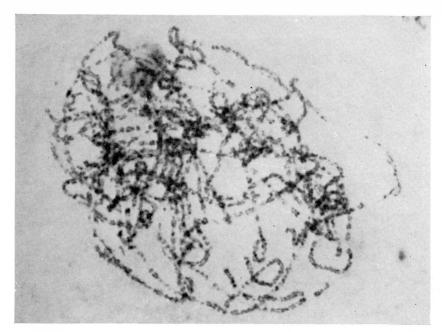

**Plate 4.5.** Zygotene stage of meiosis in *Osmunda regalis*. Stained in aceto-carmine (schedule 10).

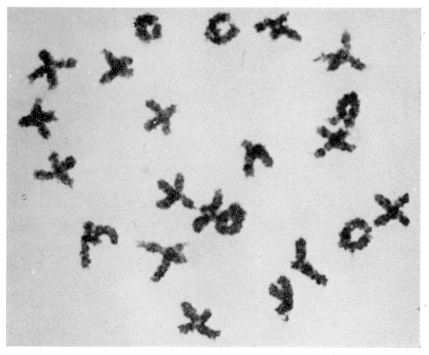

**Plate 4.6.** Diakinesis stage of meiosis in *Osmunda regalis* ($n = 22$) (schedule 10).

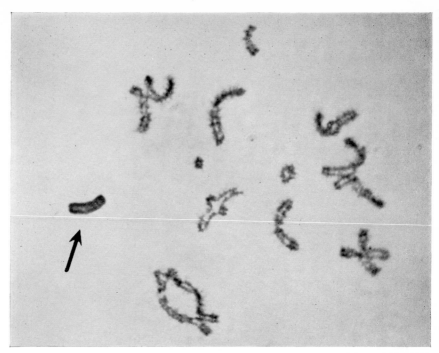

**Plate 4.7.** Diakinesis stage of meiosis in male *Schistocerca gregaria* showing 11 bivalents+a single heterochromatic X chromosome (arrowed). Stained in acetic-orcein (schedule 11). (Courtesy of Dr J.J.B.Gill.)

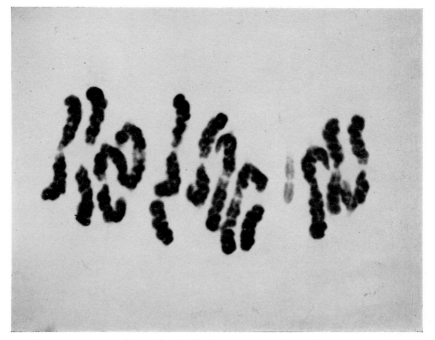

**Plate 4.8.** 1st metaphase of meiosis in autotetraploid *Tradescantia virginiana* ($n=12+\beta$) showing multivalent formation, bivalents, univalents and a pair of heterochromatic accessory chromosomes (schedule 9).

*Aceto-carmine or acetic-orcein technique*

(a)   Heat root tips gently over small flame, in 9 : 1 mixture of acetic stain : N.HCl until steaming. Do not boil. (Previous fixation can be omitted for this is a combined fixation, hydrolysis and staining stage.) Leave preferably for 5–10 min.

(b)   Mount on a slide in a drop of acetic stain (without HCl), removing all tissue except the dense meristematic tip.

(c)   Break up with the aid of a needle knife to separate the cells.

(d)   Place cover slip in position and tap with a blunt instrument to aid spreading of cells.

(e)   Heat slowly to about 60°C over a spirit lamp (see in smear and squash method, p. 139).

(f)   Squash between filter paper by vertical thumb pressure.


(v)   Examine immediately. (Do not let material dry out. If necessary add small drops of mountant, 45 per cent acetic acid, or stain, at the edge of cover slip.)

(vi)   Make temporary or permanent mounts if desired by methods described in the section on mounting (p. 143).


### (b) Pollen grains and pollen tubes

Mitosis in pollen grains and of the generative nucleus in pollen tubes offers the simplest means for investigating the basic haploid set of chromosomes and often provides superb chromosome morphology. The first mitotic division within the grain can be seen by a direct smear and squash technique (see schedule 3). *Allium ursinum* and *Tradescantia* spp. are specially recommended; pollen tube mitosis requires prior stimulation of tube growth by artificial culture methods, simulating conditions on the stigma, and usually requiring a special medium together with an optimum temperature of about 20°C and a high humidity. The culture media used can be a 3–30 per cent solution of sucrose in water, a 2 per cent solution of agar or gelatin in water, or an extract of the style and/or stigma. Several methods are available, including the hanging drop technique (as described in Darlington & La Cour 1962), the coated slide method (Conger 1953), the celloidin membrane method (Savage 1957) and the floating cellophane method (La Cour & Fabergé 1943). Only the last will be described here (see schedule 4). The concentrations of sucrose solutions required for germination of pollen differ considerably with and within the species. Hence only by trial can the optimum concentration be found. Colchicine (0·05 per cent) may be added to the medium to produce metaphase arrest (see p. 132).

F

**Schedule 3. Pollen grain mitosis**

*Either:*
(i)   Break up anther in acetic-orcein stain on slide.
(ii)  Add cover slip and squash.
*Or:*
(i)   Smear contents of anther on slide.
(ii)  Fix in 3 : 1 absolute alcohol : glacial acetic acid for not less than 15 min.
(iii) Hydrolyse in N.HCl at 60°C for 6–7 min.
(iv)  Stain in leuco-basic fuchsin (1–3 h).
(v)   Rinse in water and mount in 45 per cent acetic acid.
(vi)  Squash.

**Schedule 4. Pollen tube mitosis**

(i)   Cellophane (non-waterproof) of size approximately 20 mm × 20 mm should be soaked in the sucrose solution (e.g. 14 per cent for *Tulipa*, 8 per cent for *Tradescantia*) of strengths suitable to effect germination of the pollen.
(ii)  Remove excess by blotting.
(iii) Float cellophane on a large drop of sugar solution in a Petri dish. (This can contain 0·05 per cent colchicine for metaphase arrest if desired.)
(iv)  Scatter ripe pollen on the surface of the cellophane square, cover the Petri dish and incubate at *c.* 20°C.
(v)   Germination will probably commence in a few hours and mitotic division of the generative nuclei in 12–24 h (can be less in some species).
(vi)  Mount the cellophane, pollen side uppermost, on a slide.
(vii) *Either* fix and stain by flooding with acetic-orcein.
*Or* fix in 3 : 1 absolute alcohol : glacial acetic acid (minimum of 1 h), hydrolyse in N.HCl at 60°C for about 6 min, stain in leuco-basic fuchsin ($\frac{1}{2}$–3 h) and mount in 45 per cent acetic acid.
(viii) Make temporary or permanent slides as described in section on mounting (p. 143).

### (c) Tadpole tails mitons

The various stages of mitosis can be seen readily in the tails of urodele larvae such as *Triturus*, the newt ($2n=24$) or *Rana*, the frog ($2n=26$). Full-grown tails give best results. Unless the larva is killed prior to cutting off the tail and fixing in 3 : 1 absolute alcohol : glacial acetic acid for a minimum of 2 hours, then a special licence for vivisection is necessary in Great Britain. Squash preparations can then be made in acetic-orcein or in leuco-basic fuchsin after hydrolysis. Pretreatment with 0·05 per cent colchicine is advantageous (see p. 132).

## (d) Embryos

Early embryo development is an obvious source of mitosis but most young embryos are relatively inaccessible. The desert locust, *Schistocerca gregaria*, is however excellent for this purpose. (Specimens can be obtained on request from the Anti-Locust Research Centre, London W.8.) Locusts breed all the year round and, under conditions of culture as used by the Anti-Locust Research Centre, embryos of 3–6 days old provide excellent mitoses.

### Schedule 5. Locust embryo mitosis

(i)   Take a fertilized egg of desert locust after 3–6 days' culture and puncture under insect saline (0·67 per cent NaCl) with a sharp needle. The embryo and yellow yolk will flow out, but the embryo is readily distinguished for it is white and transparent.

(ii)   Mount direct into acetic-orcein and squash. (Prior fixation in 3 : 1 absolute alcohol : glacial acetic acid can be incorporated if it is desired to accumulate embryos before preparations are made.)

### (e) Mammalian eye (cornea)

A good yield of epithelial cells undergoing mitosis can be obtained from the cornea of mammals. The most readily obtained mammal will probably be the mouse, but the technique will suffice for any available mammal (see plate 4.3). The schedule recommended is a modification of that used by Fredga (1964).

### Schedule 6. Corneal mitosis

(i)   Kill animal (preferably by cervical dislocation).

(ii)   Snip corners of eye socket with scissors and remove the whole eye by the optic nerve.

(iii)   Place in pre-warmed isotonic citrate/colchicine solution (0·05 per cent colchicine in 3·8 per cent sodium citrate) for 30 min at 37°C.

(iv)   Transfer to pre-warmed hypotonic citrate/colchicine solution (0·05 per cent colchicine in 0·7 per cent sodium citrate) for a further 25 min at 37°C.

(v)   Fix in a mixture of 50 per cent acetic acid and N.HCl (ratio 9 : 1) for 5–30 min.

(vi)   Stain in 2 per cent acetic-orcein (2–3 min).

(vii)   Gently rub the corneal part of the eye on a clean slide so as to remove some epithelial cells together with a little stain.

(viii)   Cover the cells with a cover glass, warm gently and squash.

An alternative to the latter part of the above schedule is recommended to give a more even distribution of cells and better squashing. After fixation in part (v) of the schedule, proceed:

(vi)   Tap the fixed eye on a slide together with a drop of acetic-orcein until all the cells of the cornea have been effectively released.
(vii)   Take up the cells in the stain with the aid of a micropipette.
(viii)   Deliver to a series of fresh slides and squash.

### (f) Bone marrow

Marrow is the best somatic tissue available in a small mammal. The technique however requires the use of a centrifuge and careful manipulation but the basic procedure is relatively simple. The mouse, Chinese hamster, etc., are suitable animals to use.

### Schedule 7. Marrow mitosis

(i)   Kill animal.
(ii)   Quickly dissect out the femurs, cut off the epiphyses, and wash out the marrow into a small centrifuge tube with a suitable syringe and needle using about 2–3 ml of medium. The medium comprises TC 199 (see appendix) together with colchicine to a final concentration of 0·0004 per cent. Gently draw into the syringe and expel several times to break up any clumps of cells (avoid bubbling).
(iii)   Leave at room temperature for 1–2 h (the medium may be pre-warmed to 37°C but this is not necessary).
(iv)   Spin down at about 500 rev/min (with a centrifuge arm radius of approximately 200 mm) for 5 min.
(v)   Remove supernatant and resuspend in a hypotonic solution of either 0·7 per cent sodium citrate for 15–30 min (room temperature), or preferably 0·45 per cent KCl for 5 min.
(vi)   Spin down again and fix in approximately 1 ml cold 3 : 1 methanol : glacial acetic acid by gently adding the fixative down the tube and then flicking the cells into suspension (2–3 min).
(vii)   Spin down as before and resuspend in fresh fixative. Repeat this once more and leave in final fixative for at least 15 min.
(viii)   The fixed cells may then be used direct or kept in a refrigerator (preferably at −20°C) until required.
(ix)   Spin down and resuspend in fresh fixative. Take a little of the cell suspension in a narrow-bore Pasteur pipette and allow one droplet to fall on to a well-cleaned slide (grease-free). Let the droplet expand and aid drying by gentle blowing over the surface of the slide.
(x)   Repeat with one or two more droplets, depending on cell concentration.
(xi)   Allow to air-dry thoroughly (the preparation can then be stored at this stage indefinitely).
(xii)   Stain in 2 per cent lactic-acetic-orcein ($\frac{1}{2}$–2 h), or by any other standard procedure.

(xiii)   Dehydrate and mount in Euparal or Xam (after finally passing through xylene).

### (g) Peripheral blood

The technique for the display of mitosis in blood cells has evolved rapidly during the past decade. Stimulated by the success on human chromosomes, the techniques have been adapted to a wide range of animals. Peripheral blood is much more easily obtained than bone marrow, but unfortunately division in the cells cannot be obtained without a culture period before harvesting.

The technique for human peripheral blood culture as described by Moorhead *et al.* (1960) has formed the basis for techniques now widely used for producing mitosis in leukocytes. A mitotic stimulant, phytohaemagglutinin, is necessary in blood culture and the cells are harvested after two or three days. Several modifications to the original technique have since been developed, especially in utilizing small quantities of blood. Owing to the interest in human chromosomes and the excellent results which can now be obtained by relatively simple techniques (see plate 4. 4), the following schedule is included here (see also special staining, schedule 22). It must be appreciated however that it incorporates sterile techniques and utilizes a variety of laboratory facilities and equipment; hence it can only be undertaken by more advanced classes and under supervision.

Although the best blood samples are obtained by venopuncture, this is not practicable for general class or student use and the following micro-method using blood from a finger-stab has been found to give very satisfactory results even if the number of dividing cells obtained is relatively small. As well as suitable culture tubes and special chemicals, an incubator or water bath maintained at 37°C and a centrifuge are essential.

### Schedule 8. Human chromosome mitosis

(i)   Squeeze end of thumb (or finger) to accumulate blood and swab surface with ether to sterilize. (Ether is preferable to alcohol for it evaporates and dries more quickly.)

(ii)   Prick thumb with sharp *sterile* instrument (e.g. a hagedorn needle).

(iii)   Collect blood (0·1 to 0·2 ml) in a sterile micropipette and discharge quickly into the sterile culture tube (suitable tubes are available from Stayne Laboratories Ltd, High Wycombe, Bucks) containing the following culture medium: 4·5 ml TC 199 (see appendix); 1·5 ml human AB or foetal calf serum; 0·05 ml phytohaemagglutinin (see appendix). Provided the blood is discharged into the medium without delay there should be no danger of clotting and heparin is not essential in the medium for these small amounts of blood.

(iv)   Incubate at 37°C for 69 h (cap of tube must be airtight).

(v)   Add 0·1 ml of colchicine (0·02 per cent) and continue incubation for a further 3 h.

(vi)   Centrifuge tubes for 5 min at 500 rev/min (radius of centrifuge approximately 200 mm). Pour off supernatant with care so as not to disturb cells at base of tube, and resuspend cells in the remaining drop.

(vii)   Add 4 ml of 0·56 per cent KCl and incubate for 3 min. (This acts as a hypotonic solution to swell the cells.)

(viii)   Centrifuge for 5 min at 500 rev/min. (Timing of this and section (vii) is rather critical for satisfactory hypotonic treatment. Too long will cause excessive bursting of the cells.)

(ix)   Pour off supernatant and resuspend cells in remaining drop.

(x)   Fix in 3 : 1 methanol : glacial acetic acid. Add fixative a few drops at a time, mixing well between each addition, using about 4 ml of fixative in all.

(xi)   Centrifuge again immediately at 500 rev/min for 5 min. Pour off supernatant, resuspend cells and add fresh fixative.

(xii)   Repeat with another change of fixative and then either store at −20°C, until required, or make the preparations as follows:

(xiii)   Centrifuge at 500 rev/min for 5 min. Pour off supernatant, resuspend in remaining drop and add one or two drops of fresh fixative to give a reasonably thick suspension of cells.

(xiv)   Using a Pasteur pipette allow 2–3 drops of the cell suspension to fall onto a dry slide (this can have been chilled in iced water to give better spreading, if necessary) and allow to air-dry.

(xv)   Stain for $\frac{1}{2}$–1 h in 1 per cent lactic-acetic-orcein (see also special staining, schedule 22).

(xvi)   Wash briefly (few seconds) in 45 per cent acetic acid. Dehydrate through 70 per cent alcohol, into absolute alcohol (2–5 min in each).

(xvii)   Pass through 1 : 1 absolute alcohol : xylene (2 min) and mount in Xam or Canada Balsam.

## 3. CHROMOSOMES IN MEIOSIS

The majority of meiotic tissues is even more simple to handle than mitotic tissues for the quick smear/squash technique, and can be used without prior hydrolysis being necessary (unless Feulgen staining is required). Meiosis is readily obtained from pollen mother cells of plants, or testes of animals. Where possible the organism should be used for both mitosis and meiosis.

### (a) In plants

The timing of meiosis is critical, often occurring only over a relatively short period of time during the season. It is the rule in flowering plants to select buds for investigation which have not begun to show any bursting of the perianth or well before the anthers begin to yellow. In cases where there are numerous buds aggregated into an inflorescence it is convenient to test the oldest and the youngest to ensure that all are neither too young nor too old; then systematic investigation of buds within the inflorescence can be carried out, thereby obtaining all stages of meiosis. Examples of flowering plants excellent for meiotic study are as follows (see schedule 9).

*Allium* spp.: Readily available in the wild or cultivation. Usually in meiosis in May ($n=7, 8, 9$).

*Endymion non-scripta:* Ready when flower still protected within shoot well below ground. Dig up plants in January. If winter has been extreme the inflorescence may still be too young. Keep bulbs in a polythene bag in a refrigerator at 4°C until ready (often only a few days) ($n=8$).

*Hyacinthus* (garden varieties): Undergo meiosis about November when the inflorescence is still in the bulb. Bulbs can be bought direct from the nursery and cut open to test for meiosis. If too young allow bulbs to be kept at room temperature and test one or two every few days ($n=8$, etc.).

*Nothoschordum fragrans:* Available most of the year round ($n=8, 9$).

*Paeonia* spp.: In meiosis about April when buds are still tightly shut and about 10 mm in diameter. The numerous stamens provide an excellent gradation of maturity, the oldest on the periphery. The chromosomes are large and have a low chiasma frequency. Hence they are very suitable for chiasma analysis and quantitative work ($n=5, 10$).

*Papaver orientale:* In flower during the summer months and ideal for its many stamens ($n=14, 21$).

*Secale* spp. (Rye): Easy to grow and provides good normal meiosis ($n=7$). This material is also useful for exhibition of accessory chromosomes and chromosome mutations (see p. 163 and p. 159).

*Tradescantia paludosa:* An all-year-round greenhouse species and hence of special merit. The inflorescence is of suitable age when the oldest bud is just beginning to open. Very large chromosomes with clear major coils ($n=6$).

*Vicia faba:* Useful class material, along with *Allium* and *Hyacinthus* in that they can be obtained so simply and provide good mitotic and meiotic preparations for comparison within the same material. In meiosis about May ($n=6, 7$).

One of the most rewarding plants for the observation of meiosis is the Royal fern, *Osmunda regalis* ($n=22$) (see plates 4.5 and 4.6). It provides beautiful illustrations of all stages of meiosis and most often on a single slide (see schedule 10). The fern is relatively rare in the wild and hence it is important that material is not collected in quantity except from plants in cultivation. There are many of these, particularly in botanic gardens. Meiosis occurs usually in April when the fronds are from 75–300 mm high with the pinnae and terminal sporangia still tightly coiled.

### Schedule 9. Meiosis in flowering plants

Material may be fixed singly or in quantity in 3 : 1 absolute alcohol:glacial acetic acid or modified Carnoy's fixative (see section on fixation, p. 137), but this is not essential if the smear/squash technique using acetic stains is used.

(i)   Dissect out a single anther from bud and place on a clean slide.

(ii)   Either smear the anther on the slide with a flattened needle-knife and add a drop of aceto-carmine, or break up the anther in a drop of stain.

(iii)   Carefully remove all debris and large pieces of material leaving the contents of the anther as a cell suspension in the drop of stain.

(iv)   Cover with a cover glass, heat gently in a spirit lamp and squash (see smear and squash technique in slide preparation section, p. 139).

(v)   Make temporary or permanent mounts (see appropriate section, p. 143).

### Schedule 10. Meiosis in *Osmunda*

(i)   Fix the terminal portion of the coiled frond, which is modified for sporangial production, in 3 : 1 absolute alcohol:glacial acetic acid. (It is not often possible to use the material fresh, but bulk fixation should not be attempted until sample fixations have been checked for meiosis.)

(ii)   The sporangia are oldest lower down each modified pinnule or pinna. Test individual pinnae, starting from the outermost part of the coiled frond tip, by checking the development in old and young sporangia as follows:

(iii)   Mount a few sporangia (collectively about the size of a pin head) in a drop of aceto-carmine and break up well with a needle-knife.

(iv)   Remove all debris, leaving contents of the sporangia in suspension.

(v)   Cover with a cover glass and heat gently until the preparation almost boils. (Bubbles will form in the stain and pass rapidly to the edge of the cover glass.)

(vi)   Squash immediately between filter paper with hard pressure.

(vii)   Make temporary or permanent as previously recommended (see appropriate section, p. 143).

### (b) In animals

Of the animal testes which are suitable for demonstrating meiosis, the simplest and best are those of the adult grasshopper (*Chorthippus*, $2n=17$) and desert locust (*Schistocerca*, $2n=23$). The odd somatic chromosome number in male locust and grasshopper is due to the sex-determining mechanism whereby a single $X$ chromosome is present in the male (see plate 4.7) and two in the female (see later section on special chromosomes, p. 162). Good preparations can also be obtained from the newt (*Triturus*, $2n=24$) and many mammalian spp., such as the mouse (*Mus*, $2n=40$). Mammalian testes however require a more complex technique for good results and hence such material is handled better by more advanced students (see schedule 13).

### Schedule 11. Meiosis in grasshopper/locust

Locusts have the advantage of being easy to breed in captivity and available all the year round, whereas grasshoppers are suitable during July/September only. Both provide excellent preparations of all stages of meiosis, and are noteworthy for demonstrating chiasmata at diplotene.

(i) Kill insect by placing in a container with a pad soaked in chloroform.

(ii) Dissect under insect saline (0·67 per cent NaCl) using a mid-dorsal abdominal incision. The testes lie in the anterior third of the abdomen.

(iii) Remove the testes comprising a series of follicles invested by fat body and whilst still in insect saline, remove the fat body as much as possible (for this interferes with the squash technique) by means of fine needles.

(iv) Fix the follicles in 3 : 1 absolute alcohol : glacial acetic acid for at least half an hour. (They may be stored in the fixative for long periods at low temperature—see fixation section, p. 137.)

(v) Mount in a drop of acetic-orcein (2 per cent) and squash gently. The preparation can be sealed or made permanent as previously described (p. 143).

### Schedule 12. Meiosis in newt (*Triturus*)

The testes of the newt will show stages of meiosis during June/July.

(i) Dissect out testes of freshly killed newt.

(ii) Cut into small piecesa nd fix in 3 : 1 absolute alcohol : glacial acetic acid (preferably overnight).

(iii) Mount a small piece in a drop of aceto-carmine or orcein stain and tap out with a needle-knife.

(iv) Cover with cover glass and squash gently, sufficient to separate the cells.

### Schedule 13. Meiosis in small mammals (e.g. mouse)

This is a technique requiring extra care and skill but will produce rewarding results. It has been used on testes from a variety of small mammals with success. Pre-treatment with colchicine is useful to accumulate cells in metaphase but this can only be done by injection and hence under special licence in Great Britain. This pre-treatment is not essential however.

(i) Optimal pre-treatment—inject animal intraperitoneally with pre-warmed (37°C) 0·05 per cent colchicine in 2·2 per cent sodium citrate (0·25 ml solution per 10 g body weight) 2 h before animal is killed.

(ii) Kill animal by cervical dislocation.

(iii) Dissect out testes, pierce the tunica and put tubules in 1 per cent sodium citrate (10–15 min). Ensure tubules are separated.

(iv) Fix the tubules in 3 : 1 absolute alcohol : glacial acetic acid. (Store after this stage at −20°C if required.)

(v)   After 10–20 min transfer a few tubules to 2 ml of 50 per cent acetic acid in a centrifuge tube. Flick tube with fingers to disperse the cells.

To remove most of the sperm and also to aid subsequent spreading, centrifuge material at 500 rev/min, centrifuge radius approximately 200 mm for 5 min. Discard the supernatant and resuspend in 50 per cent acetic acid.

(vi)   Pipette one drop of solution onto a clean slide. After 30 s warm gently over a spirit lamp and air-dry.

(vii)   Stain in 1 per cent lactic-acetic-orcein for 15–30 min.

(viii)   Wash briefly (few seconds) in 45 per cent acetic acid and dehydrate via 70 per cent alcohol and absolute alcohol (2–5 min in each).

(ix)   Transfer to 1 : 1 absolute alcohol : xylene and mount in Xam or Canada Balsam.

### (c) In plant hybrids

Examination of hybrids is of considerable importance to illustrate the degree of non-homology between chromosomes of different species and to reveal chromosome evolution within a genus. It is of course possible to find hybrids in the wild but such material is not often readily available.

Hybrid meiosis can be seen in *Narcissus 'Geranium'* ($2n=17$) for this plant has evolved a system which maintains it permanently hybrid. The parents of this hybrid are *N. poeticus* ($2n=14$) and *N. tazetta* ($2n=20$). Lack of homology determines the presence of univalents but a few bivalents, which are unequal (heteromorphic), can be seen. Meiosis in this hybrid is probably unique and difficult to appreciate except by the more advanced student. First anaphase is absent and a restitution nucleus is formed in which, during second division, the chromosomes can still be seen paired, but uncoiled. Separation of chromatids and/or chromosomes and segregation usually occurs at second anaphase, though a restitution nucleus may give rise to pollen grains with 34 chromosomes. Hence, as a numerical and structural hybrid this material is an excellent example, illustrating abnormal meiosis usually in about September. Preparations can be made as per schedule 9, above.

It is a good exercise for advanced students to synthesize hybrids. Unfortunately there are relatively few species with different chromosome numbers, which hybridize easily and have large enough chromosomes for appreciating hybrid meiosis. Two species of *Ranunculus*, *R. marginatus* ($2n=32$) and *R. muricatus* ($2n=48$) can be recommended for this purpose. Seed can be obtained from botanic garden sources if necessary.

Hydridization can be effected firstly by destamination of the female parent flower some 2–3 days before opening and before the pollen is ripe. Protection of the destaminated flower from other pollen can be effected by a small nylon bag tied by thin wire. When the flower is opened the stigmas are dusted with pollen from the male parent. This should be repeated daily until the achenes set (approximately 3–7 days). Germination of hybrid seed is good. Preparations of meiosis in the hybrid plants will show typical paired and unpaired chromosomes, indicating a degree of chromosome homology.

# 4. CHROMOSOME MUTATIONS

Changes in chromosome number (e.g. aneuploidy and polyploidy) and in structural rearrangement (e.g. inversions and translocations) have been an important evolutionary mechanism within the plant and animal kingdoms.

## (a) Aneuploidy and polyploidy

Changes in chromosome number are more readily tolerated by plants and there is a host of examples (Darlington and Wylie 1955).

Suitable for showing both aneuploidy and polyploidy are *Hyacinthus orientalis* (garden varieties), ranging from $2n=14-31$, and *Secale* ($2n=14, 21, 28$). Diploids ($2n=16$), triploids and tetraploids can be found within *Ranunculus ficaria* and the chromosomes are of a size suitable for all classes of students. A whole range of polyploids can be found within the genus *Chrysanthemum* (with a base number of $x=9$).

*Tradescantia* however is perhaps the best material for polyploid meiosis. The common garden variety *T. virginiana* ($2n=24$) is an autotetraploid showing multivalent formation of rings and chains with terminal chiasmata (see plate 4·8). Comparison should be made with the diploid species *T. paludosa* and *T. bracteata*. (In Great Britain these diploid species are usually grown in a greenhouse, but the latter can be grown outdoors.) Triploid plants are also available and all are easy to handle.

Preparation of mitosis and meiosis in aneuploids and polyploids should be made as per schedules 2 and 9. The production of polyploids by colchicine treatment (see in pre-treatment section p. 133) may also be investigated.

## (b) Inversions and translocations

It is an advantage that students become familiar with a basic set of chromosomes so they are able to appreciate and understand any structural changes which have occurred. For this reason rye (*Secale*) is an ideal teaching material for it illustrates haploidy, diploidy, triploidy, tetraploidy, inversions, translocations and accessory chromosomes (see p. 163).

Multivalent ring formation can be demonstrated readily in the plant *Rhoeo discolor* ($2n=12$) and also in many *Oenothera* spp. ($2n=14$) but the latter is not easy material to work with. In animals translocations can be seen in some populations of *Periplaneta* (see schedule 14, below). Inversions are difficult to detect except by the consequences of meiotic segregation. They are however readily displayed in *Chironomus* and *Drosophila* by investigating the salivary gland chromosomes (see next section on Special Chromosomes).

### Schedule 14. Translocations in *Periplaneta* (meiosis)

(i)  Inject males intra-abdominally with 0·05 per cent colchicine in insect saline (0·67 per cent NaCl) so as to distend the abdomen. Leave overnight with food and water.

(ii)  Kill insect by means of chloroform pad.

(iii)  Dissect out testes, removing any fat body, and fix in 3 : 1 absolute alcohol : glacial acetic acid for at least half an hour.

(iv)  Take a small piece of a testis and mount on a clean slide in a drop of acetic-orcein.

(v)  Tap out in the drop of stain, cover with a cover glass and squash gently.

(vi)  Make permanent by any of the techniques described previously (p. 144).

## 5. SPECIAL CHROMOSOMES

In this category are placed those of a special significance such as the giant chromosomes in the salivary glands and other secretory cells of some Dipteran species, and the lamp-brush chromosomes seen in developing oocytes of vertebrates. Sex chromosomes are included in that they can be recognized only in heteromorphic complements and accessory (or β) chromosomes for their unusual occurrence and behaviour in segregation.

### (a) Salivary gland chromosomes

These are of particular importance, not only for their size and polytene structure, but for the amount of cytological evidence they have provided in genetics. They are about 100 times as long as the normal metaphase chromosomes and show a distinct and specific chromatin banding. Chromosome homologues are permanently paired together as in meiosis and hence chromosome mutations, particularly of the structural kind and in the heterozygous condition, can be seen readily.

For observation the simplest are found in the larvae of *Chironomus* ($2n=8$). Here the chromosomes are in four distinct pairs, whereas in *Drosophila melanogaster* ($2n=8$) the four pairs are all held together at the chromocentre, from which radiate six paired chromosome arms, and as a consequence it is more difficult to obtain well-spread nuclei.

### Schedule 15. S.G. chromosomes in *Chironomus*

The larvae can be readily found in almost stagnant, fresh or brackish water, particularly in early spring.

(i)   Cut off the head and anterior segment of the larval body (in 0·67 per cent. NaCl or in a drop of stain).

(ii)   Squeeze the remaining part of the body gently with a dissecting needle and the salivary glands will float out, attached to the gut (see fig. 4.1).

(iii)   Place the glands in a drop of 2 per cent orcein in 45 per cent acetic acid on a clean slide.

(iv)   Cover with a cover glass and squash gently (very little pressure required).

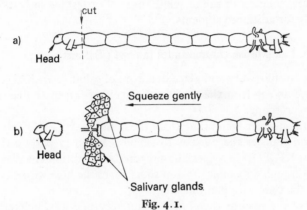

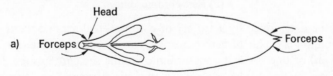

Fig. 4.1.

## Schedule 16. S.G. chromosomes in *Drosophila*

The larvae should be well fed.

(i)   Place the larva in a drop of saline solution or stain on a glass slide. Take hold of the head of the larva with a pair of fine forceps and the rear end of the body with another pair. Pull gently and the head end will break away with the salivary glands attached behind it (see fig. 4.2).

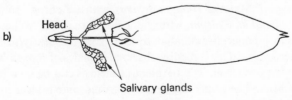

Fig. 4.2.

(ii)   Mount the glands in a drop of 2 per cent orcein in 70 per cent acetic acid under a cover slip and squash. (Spreading can be assisted by prior warming of the glands gently in N.HCl for 1 min.)

### (b) Lamp-brush chromosomes

These can be seen at a stage of meiosis in the ovarian oocyte, corresponding to diplotene. Because of their fragility they are best observed by phase contrast after isolation in saline (Gall 1954). The technique is difficult and suitable only for advanced students.

### Schedule 17. Lamp-brush chromosomes in newt (*Triturus*)

(i)   Dissect out the ovary and place in a drop of saline (0·1M KCl) on a slide.
(ii)   Remove an egg from the ovary with the aid of forceps, a fine needle and dissecting microscope.
(iii)   Free the nuclear membrane from the adhering yolk and cytoplasm by pipetting and transfer the nucleus to saline solution (7 parts 0·1M KCl : 3 parts 0·001M $KH_2PO_4$) in a specially prepared observation chamber. (Bore a 6 mm hole through a 75 mm × 25 mm slide, cover the hole with a cover glass and seal to the slide with paraffin wax.)
(iv)   Remove the nuclear membrane using very fine watchmaker's forceps and a tungsten needle pointed by dipping in molten sodium nitrite. The chromosomes are liberated and will fall on to the cover slip at the base of the chamber.
(v)   Cover the top of the chamber with a second cover glass and seal to prevent evaporation.
(vi)   View by means of an inverted phase microscope.

### (c) Sex chromosomes

Sex may be determined by a variety of chromosomal arrangements, but the most common is a pair of heteromorphic chromosomes in one sex (e.g. $XY$ mechanism) or one sex being monosomic for the sex chromosome (e.g. $XO$ mechanism). Mitotic and particularly meiotic arrangements of these sex chromosomes are of interest. Their display can be made using schedules 2, 9, 11 or 13 and the following material is recommended.

(a)   $XY$ *mechanism:* Males of *Humulus lupulus*, with a $Y$ chromosome smaller than the $X$; *Melandrium dioicum*, with the $Y$ larger than the $X$ chromosome. Meiosis occurs in summer. Mammalian testes (e.g. *Apodemus sylvaticus*).
(b)   $XY_1Y_2$ *mechanism:* This is readily observed in males of *Rumex hastatulus*, though the flowers are small. The trivalent at meiosis can be seen in May.
(c)   $XO$ *mechanism:* The single unpaired $X$ chromosome is clear in males of grasshopper or locust (see plate 4.7).

## (d) Accessory chromosomes
## (β chromosomes, supernumerary chromosomes)

These chromosomes are extra to the basic complement, apparently non-homologous with them and therefore different from normal extra chromosomes found in polysomic individuals. They are usually, but not always, smaller than the normal chromosomes and sometimes they are heteropycnotic. Accessory chromosomes tend to exist in polymorphic states within populations. They are present in varying numbers in many species and can be seen readily in preparations of mitosis or meiosis, using schedules 2, 9 and 11 as appropriate. A number of genera containing species which exhibit accessory chromosomes in some populations can be found by scanning through the *Chromosome Atlas* (Darlington & Wylie 1955), particularly in the Ranunculaceae, Cruciferae, Polemoniaceae, Amaryllidaceae, Commelinaceae, Liliaceae and Gramineae. Recommended material is:

(a)  Plants: Rye (*Secale cereale*), maize (*Zea mays*), *Ranunculus acris* (see plate 4.2), *R. ficaria, Tradescantia virginiana* (see plate 4.8).
(b)  Animals: grasshoppers (*Chorthippus parallelus* and *Myrmeleotettix maculatus*), particularly from populations in the southern half of Britain.

# 6. SPECIAL TECHNIQUES

The following schedules on chromosome coiling, autoradiography and induced chromosome aberrations are given to supplement some parts of the general discussion given in the sections on pre-treatment (p. 132) and slide preparation (p. 138), to which reference should also be made.

### Schedule 18. Chromosome coiling (spiral structure)

Plant tissue undergoing meiosis is considered the best material for trial. Concentrate initially on the first meiotic division, this being by far the easiest to handle. This technique makes use of ammoniated alcohol. A few ml of the ammoniated alcohol must be made up fresh immediately before use by putting one or more drops of concentrated ammonia into a few ml of 20 per cent or 30 per cent alcohol. As the strength of ammonia varies with time the exact amounts are difficult to specify and in practice one must proceed by trial and error. A convenient method is to take a small vessel (e.g. a 35 mm Petri dish), half fill with alcohol and add one drop of the ammonia. In working (see below) it may be necessary to dilute with additional alcohol or add a second drop of ammonia according to the way in which the material is responding. (One should be able to smell the ammonia with the nose close to the mixture, but it should not be strong enough to cause sneezing.)

(i)  Check the anthers by the squash technique (schedule 9), using aceto-carmine so as to ensure the bud is at the correct stage.

(ii)   Take a fresh anther, or part of one. Place on a clean slide, break it and smear rapidly with a flat-ended needle.

(iii)   Add immediately, by means of a glass rod, a drop of ammoniated alcohol (see above) to the surface of the slide covered by the smear. (This must be done as rapidly as possible so that the cells do not suffer from exposure to the air.)

(iv)   The length of time required for the pre-treatment of the live cells with ammoniated alcohol varies with the material and surroundings, such as temperature. It is therefore useful to try several lengths of time unless satisfied by inspection that the treatment has been successful (see note (vii) below). The treatment periods can usefully range from 10–30 s.

(v)   Within a second or two of the end of treatment time, drain away the alcohol mixture by means of filter paper and as rapidly as possible add a drop of aceto-carmine. Immediately drain away the first drop of aceto-carmine and replace by another so as to remove all traces of ammoniated alcohol.

(vi)   Cover the slide with a Petri-dish lid whilst waiting for the stain to be taken up by the cells, or if dealing with other anthers in rapid succession.

(vii)   Make a direct microscope inspection by means of a dry lens to see whether spirals are showing. The ideal is to have a bright colour in the chromosomes and spirals distinctly visible. If the chromosomes are brightly stained but without a trace of spirals, the ammonia treatment has been too little and should be increased by either a longer time or adding a further drop of concentrated ammonia to the alcohol mixture. Too much ammonia treatment gives a preparation with very faint staining in the chromosomes, which also appear rather diffuse. If all cells appear like this shorten the time of treatment or dilute the ammoniated alcohol with additional 20 per cent alcohol. The most satisfactory preparations will either show spirals on this inspection or contain a graduation of stages between intense staining and pale staining of the chromosomes.

(viii)   Before the aceto-carmine drop starts to dry up, blot away as much as possible with filter paper, without touching the cells and transfer the slide into 2 : 1 absolute alcohol : glacial acetic acid (2–15 min).

(ix)   Transfer to 3 : 1 absolute alcohol : glacial acetic acid fixative to harden (30 min–overnight).

(x)   Dry the back of the slide and drain as much fixative as possible from the front by means of filter paper without touching the cells. Add a drop of aceto-carmine (or if the stain has previously gone into the cytoplasm intensely, add a drop of aceto-carmine diluted with 45 per cent acetic acid) and cover with a cover slip.

(xi)   Warm gently and squash between filter paper using light thumb pressure sufficient to flatten the cells and spread the chromosomes more or less in one plane.

(xii)   Make temporary or permanent mounts as previously recommended (see appropriate section, p. 143).

**Schedule 19. Autoradiography of roots (for DNA synthesis)**

(i)  Culture roots in a water solution of tritiated thymidine (see appendix) at a concentration of 2 $\mu$C/ml (specific activity 3·0 C/mM) for 30 min.

(ii)  Transfer to a non-isotope solution for varying periods of time before fixation to observe isotope incorporation at different stages of the DNA synthesis period. The times are dependent on the duration of the DNA synthesis period (S period) and the period between the end of S and beginning of mitotic prophase ($G_2$). Example times are for 5–12 h for *Vicia faba* (S= 7·5 h, $G_2$=5 h) and 3–13 h for *Tradescantia paludosa* (S=10·5 h, $G_2$=2·5 h).

(iii)  Fix in 3 : 1 absolute alcohol : glacial acetic acid for 30 min or more. (N.B. A longer period of exposure to the isotope may be given initially and colchicine to a final concentration of 0·05 per cent in the non-isotope solution may be added 2 h before fixation to accumulate cells in metaphase.)

(iv)  Hydrolyse in N.HCl at 60°C (10 min for above-mentioned roots).

(v)  Stain by Feulgen method and squash (see schedule 2).

(vi)  Remove cover slip by quick-freeze method (see section on mounting, p. 145).

(vii)  Transfer to absolute alcohol and then via 70 per cent alcohol to water (2–5 min in each).

(viii)  *Either* coat slide with Kodak NTB or Ilford L4 liquid emulsion (see appendix) by dipping (in the dark room using red safelight with Kodak Wratten series 2 filter). Dry for 30 min and then apply a second coat. (Storage of slides coated with liquid emulsion is limited to approximately two months before background fogging becomes a problem.)

*Or* apply Kodak stripping film AR 10, in the dark room with red safelight only, as follows (to ensure that film will not slip, once applied, the original squash may be made on to a slide, which has been dipped into a 0·5 per cent aqueous solution of gelatine and 0·1 per cent chrome alum. Such slides should be air-dried for 48 h before use):

(a)  With a sharp scalpel cut through the film on the glass plate to produce pieces of approximately 40 mm × 40 mm.

(b)  Peel one of the squares of film slowly from the glass (fast stripping may cause fogging).

(c)  Turn it over and float in a dish of distilled water (24°C) so that the emulsion side faces downwards. Leave about 2 min for piece of film to expand.

(d)  Place the specimen slide (cells uppermost) in the water beneath the film. Lift slide and film together from the water (avoiding air pockets). Drain and dry in an air-stream from a fan (this must not produce dust or light as fogging will result).

(ix)  Store slides in trays in a black polythene bag. Place in a box with silica gel as a dehydrating agent, seal to ensure light-tight and keep at approximately 4°C in a refrigerator for 2–4 weeks.

(x)   Develop in Kodak D 19B for 2–5 min at 18°C. The shorter development times give smaller silver grains. (The temperature is important and if necessary water should be cooled with ice cubes to attain 18°C.)

(xi)   Dip into water (15 s) and fix in acid fixer at 18°C for twice the time taken for the film to clear (usually 2–3 min).

(xii)   Wash in tap water (18°C) for 10–15 min, with a final rinse in distilled water, and dry in front of a fan.

(xiii)   Pass through absolute alcohol and mount in Euparal or via xylene and mount in Xam.

### Schedule 20. X-ray effects on roots and inflorescences

(i)   Expose inflorescences of a pot-growing plant or roots of seedlings in a Petri dish to 250 rad.

(ii)   *Either* allow the inflorescence to continue development. Make preparations at daily intervals for investigation of meiosis as per schedule 9 and of mitosis in the pollen grain or pollen tube as per schedules 3 and 4.

*Or:* Transfer the seedlings to moist vermiculite or nutrient medium and make root tip squashes as per schedule 2. For the primary effects of X-rays fix at half-hourly intervals, and at daily intervals for later effects.

### Schedule 21. X-ray effects on locust embryos

(i)   Expose 3–6 day old eggs of *Schistocerca gregaria*, developed at 28°C, to 150 rad.

(ii)   For primary X-ray damage, dissect out the embryo under insect saline (schedule 5), at intervals of 15 min up to 1·5 h after irradiation.

(iii)   Fix in 3 : 1 absolute alcohol:glacial acetic acid (30 min or more).

(iv)   Squash in acetic-orcein stain and make permanent by the quick-freeze method (see section on mounting, p. 145).

### Schedule 22. Banding patterns in mammalian chromosomes

Characteristic fluorescent banding patterns have been shown in plant and animal metaphase chromosomes using quinacrine mustard and quinacrine hydrochloride (Caspersson *et al.* 1968; Caspersson *et al.* 1970). These techniques however require the use of an ultraviolet fluorescence microscope and are not therefore of general use in laboratories considered here.

Similar banding patterns have been shown by staining procedures using Giemsa or Leishman stain following post-fixation treatment of the chromosomes with either warmed saline solution (Sumner *et al.* 1971) or trypsin, which is a more rapid method (Seabright 1971). Both techniques are variable in their results dependent on several factors, probably associated with cell physiology, fixation methods, etc. Results can be obtained, however, which are spectacular and provide a basis for the recognition of individual chromosomes and chromosome aberrations. Results often vary between cultures, between slides and even between cells on the same slide. Hence trial and error is often the rule with either technique. Banding is seen at its best on chromosomes which are not too contracted and with their sister chromatids lying close together.

The schedules given below are for the staining of chromosome preparations as described by Sumner *et al.* (1971) and Seabright (1971, 1972). They may be adapted however for other chromosomes in many animals and plants.

### Schedule 22a. Banding by the acetic saline Giemsa (ASG) method

(i)  Human blood cultures (see schedule 8) are fixed in 3 : 1 methanol : glacial acetic acid (three changes) and chromosome spreading effected by the air drying technique.

(ii)  Slides are incubated for 1 h at 60°C in saline sodium citrate solution (0·3M sodium chloride + 0·03M trisodium citrate).

(iii)  Rinse briefly with distilled water.

(iv)  Stain in Giemsa, using a 1 : 50 solution of Giemsa : pH 6·8 buffer (see appendix) for $1\frac{1}{2}$ h.

(v)  Rinse in distilled water and allow to dry thoroughly after soaking up the excess water with filter paper.

(vi)  Transfer to xylene for 2 min and make permanent by mounting in a neutral medium (e.g. Xam).

### Schedule 22b. Banding by the trypsin method

(i)  Human blood cultures (see schedule 8) are fixed in 3 : 1 methanol : glacial acetic acid (three changes) and chromosome spreads made by the air-drying technique.

(ii)  A slide is placed horizontally and flooded with 1 : 10 trypsin (Difco) solution (see appendix) in isotonic saline (0·9 per cent NaCl) usually for 10–15 s, but this time may be extended up to 1 min dependent on the material and batch of trypsin.

(iii)  Rinse twice with isotonic saline. (At this point the preparation may be examined by phase contrast microscopy to assess the action of the enzyme; the chromosomes appear swollen and rather indistinct at the edges when suitable for staining.)

(iv)  Stain with a 1 : 3 solution of Leishman (B.D.H.) : pH 6·8 buffer (see appendix) for 3–5 min.

(v)  Rinse well in pH 6·8 buffer and allow to air-dry after blotting excess buffer with filter paper.

(vi)   Transfer to xylene for 2 min and mount in a neutral medium (e.g. Xam).

On a suitable preparation some nuclei show chromosomes with banding and others show chromosomes with an indistinct outline. If banding patterns cannot be seen and all the chromosomes are indistinct then the trypsin treatment was too long. If the chromosomes remain clear and evenly stained then the trypsin treatment should be extended; such a slide may be subjected to further treatment by removing the cover slip in xylene, de-staining in 3 : 1 methanol : glacial acetic acid and air-drying prior to trypsin treatment. The slide can then be re-stained.

## APPENDIX

Additional data on some of the chemicals and stains cited in the text are given below, particularly with regard to the preparation of some and recommended source (in Great Britain) of others.

## CHEMICALS

Foetal calf serum—obtainable from Biocult Laboratories Limited, Washington Road, Paisley, Scotland; or Flow Laboratories Limited, Heatherhouse Road, Irvine, Scotland.

Hanks's solution—as there is a tendency for the precipitation of bicarbonates of magnesium and calcium if all the required chemicals are mixed and autoclaved together, this balanced salt solution should be prepared as follows:

Solution A—$NaHCO_3$, 0·35 g in 10 ml dist. $H_2O$. Autoclave at 121°C for 15 min.

Solution B—$CaCl_2$, 0·14 g; $MgSO_4.7H_2O$, 0·10 g; Glucose, 1·00 g; NaCl, 8·00 g; KCl, 0·40 g; $KH_2PO_4$, 0·06 g; $Na_2HPO_4.2H_2O$, 0·06 g; Phenol Red, 0·02 g.

Add all to 990 ml of distilled $H_2O$ and autoclave at 121°C for 15 min. Mix solutions A and B together (to make 1 litre in all) under sterile conditions.

Mayer's albumen/glycerine—may be bought commercially or prepared by mixing the following and filtering before use: 25 ml albumen (white of egg); 25 ml glycerine; 0·5 g sodium salicylate.

α-Mono-bromonaphthalene—as a pre-treatment liquid for metaphase arrest it is required as a saturated aqueous solution. Mix excess with water and shake well to effect a saturated solution.

Para-dichlorobenzene—as a pre-treatment liquid for metaphase arrest this is required as a saturated aqueous solution. Mix excess with water and warm to 60°C. Shake well and allow to cool.

Photographic emulsions—for autoradiographic technique.

*Ilford L.4* can be stored up to two months at 4°C. Melt at 45°C and use in a 1 : 1 dilution with $H_2O$ at this temperature. Do not heat the same batch several times as this produces a high background. The required amount of emulsion should be taken from the stock for each occasion.

*Kodak NTB* can be stored up to two months at 4°C. Melt and use at 45°C. The same batch of emulsion may be melted and used several times. Not usually diluted for use.

Phytohaemagglutinin—obtainable from Burroughs Wellcome, Wellcome Research Laboratories, Beckenham, Kent.

$SO_2$ water—required in Feulgen staining procedure: 5 ml N.HCl; 5 ml 10 per cent $K_2S_2O_5$; 100 ml distilled $H_2O$.

TC 199—tissue culture medium, obtainable from BDH Chemicals Limited, Poole, Dorset.

Tritiated thymidine—for autoradiography, obtainable from Radiochemical Centre, Amersham, Buckinghamshire.

Trypsin—a temporary stock solution can be prepared by adding 10 ml isotonic saline (0·9 per cent NaCl) to a vial of trypsin (Difco catalogue no. 0153–60) and kept at +4°C. For use in the banding technique freshly make a further 1 : 10 dilution of this stock solution in isotonic saline as required.

## STAINS

Acetic-orcein—the standard 1 per cent solution in 45 per cent acetic acid tends to deteriorate relatively quickly. It is best prepared as required from a more stable stock solution of 2·2 per cent orcein in glacial acetic acid (45 ml stock solution to 55 ml distilled $H_2O$). To prepare the stock solution dissolve 2·2 g orcein in 100 ml glacial acetic acid by gentle boiling. Cool and filter.

For a stronger solution (as recommended in the text) add 2 g orcein to 45 ml glacial acetic acid and 55 ml distilled $H_2O$. Boil gently. Cool and filter.

Aceto-carmine—a saturated solution of carmine is recommended. Add excess carmine (2 g) to 45 ml glacial acetic acid and 55 ml distilled $H_2O$. Boil gently in a reflux condenser for 10 min. Cool and re-boil for a further 5 min. Shake well, cool and filter.

More dilute solutions can be used if required or a drop or two of ferric acetate may be added to act as a mordant. Too much of the ferric acetate will precipitate the carmine.

Fast green—use as a 0·1 per cent solution (aqueous or 95 per cent alcohol).

Giemsa—use as a 1 : 50 solution of stain (Giemsa R.66, G.T.Gurr, London. N.W.9) : pH 6·8 buffer (buffer tablets, G.T.Gurr).

Leishman—prepare a stock solution of 0·2 per cent in methanol. Warm if necessary to dissolve the stain. Use for banding technique as a 1 : 3 solution of 0·2 per cent Leishman (B.D.H.) : pH 6·8 buffer (buffer tablets, G.T.Gurr).

Leuco-basic fuchsin (Feulgen)—dissolve 1 g basic fuchsin in 200 ml boiling distilled H₂O. Shake well and cool to 50°C. Filter and add 30 ml N.HCl to the filtrate. Add 3 g K₂S₂O₅ and keep in an air-tight bottle for 24 h to allow the solution to bleach. The solution should then be a pale straw colour. To remove remaining colour, though this is not essential, add 0·5 g decolourizing carbon, shake well and filter rapidly under suction. Store in an air-tight bottle in the dark, preferably at 4°C.

Lactic-acetic-orcein—dissolve the appropriate amount of orcein (0·25–2·0 g) according to strength required in a mixture of 50 ml lactic acid and 50 ml glacial acetic acid.

Propionic orcein—dissolve 2 g orcein in 45 ml propionic acid and 55 ml distilled H₂O. Boil gently. Cool and filter.

Toluidene blue—dissolve 50 mg toluidene blue in 100 ml distilled H₂O for a working strength solution of 0·05 per cent.

## IV. ACKNOWLEDGEMENTS

The author wishes to thank the following for their most helpful advice and criticism during the preparation of the manuscript: Dr C.E.Ford, F.R.S., Dr J.J.B.Gill, Prof. B.John, Dr J.D.Lovis and Dr T.G.Walker.

## V. REFERENCES

BARBER H.N. & CALLAN H.G. (1943) The effects of cold and colchicine on mitosis in the newt. *Proc. R. Soc. B*, **131** : 258–271.

BELLING J. (1921) On counting chromosomes in pollen-mother cells. *Am. Nat.*, **55** : 573–574.

BELLING J. (1926) The iron-aceto-carmine method of fixing and staining chromosomes. *Biol. Bull.*, **50** : 160–162.

BENDER M.A. & PRESCOTT D.M. (1962) DNA synthesis and mitosis in cultures of human peripheral leukocytes. *Expl. Cell Res*, **27** : 221–229.

BISHOP C.J. (1950) Differential X-ray sensitivity of *Tradescantia* chromosomes during the mitotic cycle. *Genetics*, **35** : 175–187.

CALLAN H.G. (1942) Heterochromatin in *Triton. Proc. R. Soc. B*, **130** : 324–335.

CARNOY J.B. (1886) Les globules polaires de L'*Ascaris clavata* (Appendice de la Conférence donnée à la Societé Belge de Mircoscopie, 1887). *La Cellule*, **3** : 276.

CASPERSSON T., FARBER S., FOLEY G.E., KUDYNOWSKI J., MODEST E.J., SIMONSSON E., WAGH U. & ZECH L. (1968) Chemical differentiation along metaphase chromosomes. *Expl. Cell Res*, **49** : 219–222.

CASPERSSON T., ZECH L. & JOHANSSON C. (1970) Analysis of the human metaphase chromosome set by aid of DNA-binding fluorescent agents. *Expl. Cell Res*, **62** : 490–492.

CHAYEN J. & MILES U.J. (1954) Preparing plant root tip cells for microscopical examination. *Stain Tech.*, **29** : 33–39.

CONGER A.D. (1953) Culture of pollen tubes for chromosomal analysis at the pollen tube division. *Stain Tech.*, **28** : 289–293.

CONGER A.D. & FAIRCHILD L.M. (1953) A quick-freeze method for making smear slides permanent. *Stain Tech.*, **28** : 281–283.

CREIGHTON M. (1938) Chromosome structure in *Amblystoma punctatum*. *Cytologia* **8** : 497–504.

DARLINGTON C.D. & LA COUR L.F. (1945) Chromosome breakage and the nucleic acid cycle. *J. Genet.*, **46** : 180–267.

DARLINGTON C.D. & LA COUR L.F. (1962) *The Handling of Chromosomes*, 4th Edn. Allen and Unwin, London.

DARLINGTON C.D. & MCLEISH J. (1951) Action of maleic hydrazide on the cell. *Nature, Lond.*, **167** : 407–408.

DARLINGTON C.D. & WYLIE A.P. (1955) *Chromosome Atlas of Flowering Plants*. Allen and Unwin, London.

DE ROBERTIS E.D.P., NOWINSKI W.W. & SAEZ F.A. (1954) *General Cytology*. Saunders, London.

EVANS H.J. (1964) Uptake of $^3$H-thymidine and patterns of DNA replication in nuclei and chromosomes of *Vicia faba*. *Expl. Cell Res.*, **35** : 381–393.

EVANS W.L. (1956) The effect of cold treatment on the DNA content in the cells of selected plants and animals. *Cytologia*, **21** : 417–432.

FABERGÉ A.C. (1945) Snail stomach cytase, a new reagent for plant cytology. *Stain Tech.*, **20** : 1–4.

FORD, C.E. (1949) Chromosome breakage in nitrogen mustard treated *Vicia faba* root tip cells. *Proc. 8th Int. Congr. Genet.* (*Hereditas* Suppl. to Vol. **35** : 570–571).

FOX D.P. (1966) The effects of X-rays on the chromosomes of locust embryos I. The early responses. *Chromosoma (Berl.)*, **19** : 300–316.

FREDGA K. (1964) A simple technique for demonstration of the chromosomes and mitotic stages in a mammal. Chromosomes from cornea. *Hereditas*, **51** : 268–273.

GALL J.G. (1954) Lampbrush chromosomes from oocyte nuclei of the newt. *J. Morph.*, **94** : 283–351.

KOPRIWA B.M. & LEBLOND C.P. (1962) Improvements in the coating technique of autoradiography. *J. Histochem. Cytochem.*, **10** : 269–284.

KUWADA Y. & NAKAMURA T. (1934) Behaviour of chromonemata in mitosis, II. *Cytologia*, **5** : 244–247.

LA COUR L.F. (1941) Acetic-orcein. *Stain Tech.*, **16** : 169–174.

LA COUR L.F., DEELEY E.M. & CHAYEN J. (1956) Variations in the amount of Feulgen stain in nuclei of plants grown at different temperatures. *Nature, Lond.*, **177** : 272–273.

LA COUR L.F. & FABERGÉ A.C. (1943) The use of cellophane in pollen tube technique. *Stain Tech.*, **18** : 196.

LEVAN A. (1938) The effect of colchicine on root mitoses in *Allium*. *Hereditas*, **24** : 471–486.

LILLY L.J. & THODAY J.M. (1956) Effects of cyanide on the roots of *Vicia faba*. *Expl. Cell Res.*, **14** : 257–267.

MCLEISH J. (1953) The action of maleic hydrazide in *Vicia*. *Heredity*, **6** (Suppl.) : 125–147.

MEYER J.R. (1945) Prefixing with paradichlorobenzene to facilitate chromosome study. *Stain Tech.*, **20** : 121–125.

MOORE K.L. & BARR M.L. (1954) Nuclear morphology, according to sex, in human tissues. *Acta. Anat.*, **21** : 197–208.

MOORHEAD P.S., NOWELL P.W., MELLMAN W.J., BATTIPS D.M. & HUNGERFORD D.A. (1960) Chromosome preparations of leucocytes cultured from human peripheral blood. *Expl. Cell Res.*, **20** : 613–616.

MULLER H.J. (1927) Artificial transmutation of the gene. *Science*, **66** : 84–87.

O'MARA J.G. (1948) Acetic acid methods for chromosome studies at prophase and metaphase in meristems. *Stain Tech.*, **23** : 201–204.

PELC S.R. (1956) The stripping- film technique of autoradiography. *J. App. Rad. Isotopes* **1** : 172–177.

PRESCOTT D.M. & BENDER M.A. (1963) Autoradiographic study of chromatid distribution of labelled DNA in two types of mammalian cells *in vitro*. *Expl. Cell Res.*, **29** : 430–442.

ROTHFELS K.H. & SIMINOVITCH L. (1958) An air-drying technique for flattening chromosomes in mammalian cells grown *in vitro*. *Stain Tech.*, **33** : 73–77.

SAVAGE J.R.K. (1957) Celloidin membranes in pollen tube technique. *Stain Tech.*, **32** : 283–285.

SAX K. & HUMPHREY L.M. (1934) Structure of meiotic chromosomes in microsporogenesis of *Tradescantia*. *Bot. Gaz.*, **96** : 353–361.

SEABRIGHT M. (1971) A rapid banding technique for human chromosomes. *Lancet*, **ii** : 971–972.

SEABRIGHT M. (1972) The use of proteolytic enzymes for the mapping of structural rearrangements in the chromosomes of man. *Chromosoma (Berl.)*, **36** : 204–210.

STADLER L.J. (1928) The rate of induced mutation in relation to dormancy, temperature and dosage. *Anat. Rec.*, **41** : 97.

SUMNER A.T., EVANS H.J. & BUCKLAND R.A. (1971) New technique for distinguishing between human chromosomes. *Nature New Biol.*, **232** : 31–32.

SWANSON C.P. (1958) *Cytology and Cytogenetics*. Macmillan, London.

TAYLOR J.H. (1958) The mode of chromosome duplication in *Crepis capillaris*. *Expl. Cell Res.*, **15** : 350–357.

TAYLOR J.H. (1960) In *Cell Physiology of Neoplasia* (M.D.Anderson Hosp. and Tumor Inst.), Univ. Texas Press, Austin, pp. 547–575.

TAYLOR J.H., WOODS P.S. & HUGHES W.L. (1957) The organisation and duplication of chromosomes as revealed by autoradiographic studies using tritium-labelled thymidine *Proc. Natn. Acad. Sci.*, **43** : 122–128.

WHITE M.J.D. (1961) *The Chromosomes*, 5th Edn. Monog. on Biolog. Subjects. Methuen, London.

# CHAPTER 5

# GENETICAL EXPERIMENTS WITH FUNGI

## J. CROFT AND J. L. JINKS

## I. INTRODUCTION

The short life cycles and ease with which the growth, development and reproduction of fungi can be controlled make them ideal material for use in genetical studies. Whether one wishes to study simple Mendelian inheritance, tetrad analysis, linkage and fine structure of genes, biosynthetic systems and gene function, differentiation, parasexuality, mutagenesis, host–parasite relationships, extra-chromosomal inheritance or biometrical and population genetics, suitable material and appropriate techniques for their investigation are available from past and present researches on fungi. The experiments described in this chapter illustrate this versatility of fungi as experimental organisms but attention has been confined to those uses where fungi have a clear advantage, particularly for teaching purposes. Further experiments with fungi are described in chapter 3.

Some knowledge of mycology and of genetics is assumed, but for general background information reference should be made to Alexopoulos (1962) and to Fincham & Day (1971). With a few exceptions the experiments described in this chapter do not require very complicated equipment. The basic equipment required for sterilizing media and glassware is of course essential, and the only other expensive item which is required is a stereoscopic microscope capable of magnifications of up to at least $\times 40$. Although a rigid schedule is given for each experiment described, most are easily adaptable to different requirements, for example to large or to small classes and so on. The size of experiment given should perhaps be regarded as that required in order to illustrate adequately the principles involved in that particular experiment. Time scales are also readily modified to suit class time-tables since most experiments can be stopped temporarily at almost any stage by placing the cultures in the refrigerator at about $5°C$. Finally, this is by no means intended as a complete practical course on fungal genetics, but more as a collection of possible class experiments, all of which are known to work well under class conditions.

## II. GENERAL TECHNIQUES

Throughout the descriptions of the experiments a knowledge of general sterile culture techniques will be assumed. However, a description of some of the basic methods required for the culture of the fungi used is given here. Any special culture techniques will be given separately as required in the description of each experiment.

### 1. MASS TRANSFERS

Mass hyphal transfers, that is the use of small groups of hyphae as inocula, can be made from colonies growing on a solid medium. With a microspatula (made from nichrome wire hammered flat and filed to about 1 mm width) a small agar block carrying the hyphae is cut out from the colony and used as the inoculum. A maximum size of 1 to 2 mm square is suggested for inocula of this sort though in the case of *Phytophthora infestans*, which is not easy to keep in culture, inocula measuring 3 to 4 mm square will give better results. Non-sporulating mycelia such as those of *Schizophyllum commune* present few problems and inocula can be cut from the whole colony. The *Aspergillus* species produce many conidia and greater care is required. The conidia become very powdery and easily scattered with age and colonies less than one week old should be used. The inocula should be taken from the edges of the colonies only, thus avoiding the mature conidia in the centre. Because of the ease with which the light powdery conidia of *Neurospora crassa* can be scattered, mass hyphal transfers of this species should not normally be used. However, if a medium containing sorbose as the main carbon source is used, a compact growth and delayed conidiation results enabling mass hyphal inocula to be taken safely.

As an alternative to mass hyphal inocula, mass transfers of conidia can be used. In the case of *Aspergillus* the spores can be taken directly from the growing colony using a fine rounded glass needle moistened by dipping it into a clean agar plate. In the case of *Neurospora crassa*, again to reduce risk of contamination, suspensions of conidia should always be used, the transfers being made by means of a wire loop or a Pasteur pipette. (The wire loop, about 3 to 4 mm diameter, can be fashioned from nichrome wire.) Mass transfers of cells of *Saccharomyces cerevisiae* can be made from a liquid culture using a loop or a Pasteur pipette, or directly from a colony growing on a solid medium using a loop. In all cases great care and the use of the smallest inoculum possible will give the best results.

### 2. SPREADING OF SPORE SUSPENSIONS

When it is required to examine segregations among conidia of the *Aspergillus* species or of *Neurospora crassa*, among cells of a culture of *Saccharomyces cerevisiae* or among the sexual spores of *Aspergillus nidulans* or of *Schizo-phyllum commune* or to grow colonies from single spores or cells for any other reason, the method of dilution and spreading of spore or cell suspensions on solid media in Petri dishes should be used. Conidia of *Aspergillus* and of *Neurospora crassa* are difficult to 'wet' and suspensions are best made in a $10^{-5}$ dilution of the substance known commercially as 'Tween 80' (poly-oxyethylene sorbitan mono-oleate). In most other cases a 0·85 per cent NaCl solution should be used as the suspension medium.

Conidia of the *Aspergillus* species can be picked directly off colonies growing on a solid medium with a moistened wire loop which can then be washed off in the dilute 'Tween 80' solution. In the case of *Neurospora crassa* extra care should be taken and the following procedure is suggested. The strain should be grown on an agar slope in a test tube. When the conidia are mature the cotton-wool bung is removed and a pipette containing a suitable volume of 'Tween 80' solution and surrounded by a cotton-wool collar is inserted into the tube. The cotton-wool collar is then fitted firmly into the neck of the tube to form a bung with the pipette passing through it. The 'Tween 80' solution is run into the tube which is shaken in order to suspend the conidia. The suspension is then taken up into the pipette which can be withdrawn leaving the culture tube safely bunged. Suspensions of cells of *Saccharomyces cerevisiae* are simply made by growing the strain in a liquid medium to the stationary phase. The making of suspensions of sexual spores of *Aspergillus nidulans* and of *Schizophyllum commune* and of sporangia or zoospores of *Phytophthora infestans* is discussed in the descriptions of the respective experiments. It is best if suspensions are washed by centrifugation and resuspension in fresh suspension medium. This is necessary for the hand-ling of *Saccharomyces cerevisiae* liquid cultures, but most of the experiments described can be carried out without washing the suspensions if no centrifuge is available.

The number of spores or cells per ml of suspension, or of a suitable dilution of the suspension, can be estimated by the use of the haemocytometer slide. This is a graduated chamber slide upon which the number of particles in a known volume of liquid can be counted microscopically. On the basis of this estimation a dilution series can then be made so that in the final dilution the suspension contains the same number of spores or cells per 0·1 ml as the required number of colonies per plate. Aliquots (0·1 ml in volume) are then placed on each plate of solid medium and spread over the entire surface using a spreader suitably fashioned from a glass rod. It is best if plates to be used in this way are dried at 35°C for one to two days after pouring so that the liquid content of each 0·1 ml aliquot of suspension is easily absorbed into the medium upon spreading.

## 3. REPLICA PLATING

Two types of replicator will be referred to in these experiments. The first is the velveteen pad replicator which was first developed for use with bacteria. This consists of a wood or metal cylindrical block with a diameter a little less than that of the inside of a Petri dish. The sterile velveteen pad is placed over the flat end of the block and held in place firmly by a metal ring or rubber band. The master plate of colonies is then pressed lightly on this pad so that inoculum from each colony on the master plate is transferred on to the pad. This master impression can now be used to inoculate a number of plates by pressing them lightly on the pad. This type of replicator can be used unmodified for *Saccharomyces cerevisiae*. It may also be used for *Aspergillus nidulans* where the conidia adhere more firmly if the velveteen pad is used damp. Also, the *A. nidulans* colony as normally grown has a large surface area and provides too much inoculum. This difficulty is overcome if the medium upon which the master colonies are grown contains sodium deoxycholate at a concentration of 0·8 mg per ml. This results in colonies with a small diameter and a restricted, but ample, conidial development. (The concentration of deoxycholate required may vary somewhat from one source of this substance to another.)

The second type of replicator is used in these experiments with *A. nidulans*. It consists of a perspex or metal base plate into which needles have been set in an equally spaced 4 × 4 or 5 × 5 arrangement. The master plates for use with this replicator must be hand inoculated with each colony carefully positioned to correspond to the positions of the needles on the replicator. The replicator is readily sterilized by dipping the needles into alcohol and flaming.

## 4. GROWTH TUBES

When it is necessary to measure the rate of growth of a mycelium this is often best done in tubes. In these experiments growth tubes are used for *Schizo-*

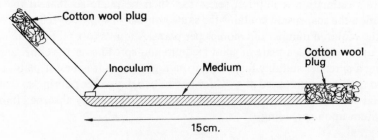

Fig. 5.1. Longitudinal section of growth tube.

*phyllum commune* and the following pattern is suitable. They are made from glass tubing (interior diameter 12 mm) cut into 250 mm lengths and with one end bent up as shown in figure 5.1. Both ends are plugged with non-absorbent cotton-wool. The plugged tubes are autoclaved, dried, and fastened together in groups of five with adhesive tape. The medium (about 6 to 7 ml) is then poured into each tube and allowed to set. The inoculum is placed on the medium at the bend of the tube and incubated until it starts to grow. The growing front is then marked and incubation continued for the required length of time. The distance from the mark to the growing front is then measured.

## III. MEDIA

The following media are referred to during the descriptions of the experiments. In all cases the final composition is given, though concentrated stock solutions could be prepared in advance for dilution as required. Separate stock solutions of mineral salts, trace elements, and the carbon source should be kept and these may be preserved by adding a small crystal of thymol tied in a dialysing membrane and storing at about 5°C. It is often necessary to autoclave the phosphate and sugar solutions separately from the other components in order to prevent precipitation or 'charring'. In all cases sterilization at about 120°C for 15 min is adequate. The media listed here are all based on, though not necessarily identical to, the media commonly referred to in the research literature.

### 1. MEDIA FOR THE *ASPERGILLUS* SPP.

#### Malt Extract (MT)

| | |
|---|---|
| Malt extract | 20·0 g |
| Agar (Oxoid No. 3) | 20·0 g |
| Water | Make up to 1 litre |

This simple complete medium is adequate for most strains. A richer complete medium suitable for *A. nidulans* is medium CM which is listed with the media for *Saccharomyces cerevisiae*. In both cases the addition of riboflavine and *p*-aminobenzioc acid (10 μg per ml) is often necessary for strains requiring these vitamins.

## Minimal Medium (M)

| | |
|---|---|
| NaNO$_3$ | 6·0 g |
| KCl | 0·52 g |
| MgSO$_4$.7H$_2$O | 0·52 g |
| FeSO$_4$.7H$_2$O | 0·01 g |
| ZnSO$_4$.7H$_2$O | 0·01 g |
| K$_2$HPO$_4$ | 0·75 g |
| KH$_2$PO$_4$ | 0·75 g |
| Glucose | 10·0 g |
| Agar (Oxoid No. 1 or No. 3) | 15·0 g |
| Water | Make up to 1 litre |

## 2. MEDIA FOR *NEUROSPORA CRASSA*

### Vogel's Minimal Medium (VM)

| | |
|---|---|
| Na$_3$ Citrate.2H$_2$O | 3·0 g |
| KH$_2$PO$_4$ | 5·0 g |
| NH$_4$NO$_3$ | 2·0 g |
| MgSO$_4$.7H$_2$O | 0·2 g |
| CaCl$_2$.2H$_2$O | 0·1 g |
| Trace element solution (see below) | 1·0 ml |
| Biotin solution (10 μg per ml) | 1·0 ml |
| Sucrose | 20·0 g |
| Agar (Oxoid No. 1) | 15·0 g |
| Water | Make up to 1 litre |

For a medium which restricts the rapid growth of *N. crassa* replace the sucrose in the above medium with 2 g gluclose, 2 g glycerol and 15 g sorbose.

### Westergaard's Crossing Medium (W)

| | |
|---|---|
| KNO$_3$ | 1·0 g |
| KH$_2$PO$_4$ | 1·0 g |
| MgSO$_4$.7H$_2$O | 0·5 g |
| NaCl | 0·1 g |
| CaCl$_2$.2H$_2$O | 0·1 g |
| Trace element solution (see below) | 1·0 ml |
| Biotin solution (10 μg per ml) | 1·0 ml |
| Sucrose | 20·0 g |
| Agar (Oxoid No. 3) | 15·0 g |
| Water | Make up to 1 litre |

Trace elements solution for VM and W:

| | |
|---|---|
| Citric acid.$H_2O$ | 500 mg |
| $ZnSO_4.7H_2O$ | 500 mg |
| $Fe(NH_4)_2(SO_4)_2.6H_2O$ | 100 mg |
| $CuSO_4.5H_2O$ | 25 mg |
| $MnSO_4.4H_2O$ | 5 mg |
| $H_3BO_3$ | 5 mg |
| $Na_2MoO_4.2H_2O$ | 5 mg |
| Water | Make up to 100 ml |

## 3. MEDIA FOR SCHIZOPHYLLUM COMMUNE

### Complete Medium (SCM)

| | |
|---|---|
| Peptone | 5·0 g |
| $MgSO_4.7H_2O$ | 0·5 g |
| $KH_2PO_4$ | 0·46 g |
| $K_2HPO_4$ | 1·0 g |
| Thiamine HCl | 0·12 g |
| Glucose | 2·0 g |
| Agar | 20·0 g |
| Water | Make up to 1 litre |

### Fruiting Medium (SF)

| | |
|---|---|
| Peptone | 2·0 g |
| Yeast extract | 2·0 g |
| $MgSO_4.7H_2O$ | 0·5 g |
| $KH_2PO_4$ | 1·0 g |
| $K_2HPO_4$ | 1·0 g |
| Glucose | 20·0 g |
| Agar | 20·0 g |
| Water | Make up to 1 litre |

MT. 2 per cent malt extract (see under *Aspergillus nidulans*) is also used in one experiment.

## 4. MEDIA FOR *SACCHAROMYCES CEREVISIAE*

### Complete Medium (YPG)

| | |
|---|---|
| Yeast extract | 10·0 g |
| Peptone | 10·0 g |
| Glucose | 20·0 g |
| Agar (Oxoid No. 3) | 20·0 g |
| Water | Make up to 1 litre |

### Respiratory Differential Medium (YPD)

Replace the glucose in YPG with 20 g glycerol and 1·0 g glucose.

### Alternative Complete Medium (CM)

| | |
|---|---|
| Yeast extract | 5·0 g |
| Peptone | 3·0 g |
| Hydrolysed casein | 5·0 g |
| Vitamin solution (see below) | 2·5 ml |
| Glucose | 40·0 g |
| Agar (Oxoid No. 3) | 20·0 g |
| Water | Make up to 1 litre |

### Minimal Medium (YM)

| | |
|---|---|
| $NH_4H_2PO_4$ | 6·0 g |
| $MgSO_4.7H_2O$ | 0·5 g |
| $(NH_4)_2SO_4$ | 2·0 g |
| $KH_2PO_4$ | 1·0 g |
| NaCl | 0·1 g |
| $CaCl_2.2H_2O$ | 0·1 g |
| Trace element solution (see below) | 1·0 ml |
| Vitamin solution (see below) | 5·0 ml |
| Glucose | 20·0 g |
| Agar (Oxoid No. 1) | 20·0 g |
| Water | Make up to 1 litre |

Trace elements solution for YM:

| | |
|---|---|
| $H_3BO_3$ | 50 mg |
| $FeSO_4.7H_2O$ | 40 mg |
| $MnSO_4.4H_2O$ | 40 mg |
| $ZnSO_4.7H_2O$ | 40 mg |
| $Na_2MoO_4.2H_2O$ | 20 mg |
| KI | 10 mg |
| $CuSO_4.5H_2O$ | 4 mg |
| Water | Make up to 100 ml |

Vitamin solution for CM and YM:

| | |
|---|---|
| Calcium pantothenate | 40·0 mg |
| Thiamin HCl | 40·0 mg |
| Inositol | 40·0 mg |
| Pyridoxine | 40·0 mg |
| Nicotinic acid | 10·0 mg |
| Biotin | 0·5 mg |
| Water | Make up to 100 ml |

## Presporulation Medium (PSP)

| | |
|---|---|
| Yeast extract | 0·5 g |
| $(NH_4)_2SO_4$ | 0·2 g |
| $KH_2PO_4$ | 0·2 g |
| Glucose | 2·0 g |
| Water | Make up to 100 ml |

## Sporulation Medium (SP)

| | |
|---|---|
| Yeast extract | 0·25 g |
| Potassium acetate | 0·98 g |
| Glucose | 0·10 g |
| Water | Make up to 100 ml |

G

## 5. MEDIUM FOR *PHYTOPHTHORA INFESTANS*

**Rye Extract Medium (R)**

| | |
|---|---|
| Rye grain | 60·0 g |
| Sucrose | 20·0 g |
| Agar (Oxoid No. 3) | 15·0 g |
| Water | Make up to 1 litre |

To prepare this medium soak the grain overnight in a part of the water. Heat for 2 hours in a boiling water bath and filter through gauze. Add the sucrose to the filtrate. Finally add this solution to the agar and make up the volume to 1 litre with water.

## IV. STOCK MAINTENANCE

### 1. *ASPERGILLUS* SPP.

Keep as conidiating colonies on slopes of medium MT in screw cap bijou bottles. Store at about 10°C. Freezing or excess condensation in the bottle will kill the stock. Transfer annually. It is advisable to purify the strain from time to time by spreading a conidial suspension on medium M plus any required nutrient supplements and selecting and testing a single colony in order to maintain the stock.

### 2. *NEUROSPORA CRASSA*

Keep as conidiating cultures in small (10 mm diameter) test tubes on medium VM plus any required nutrient supplements. Use only a small quantity of medium so that the conidia remain well away from the cotton-wool bung, thus reducing the risk of releasing spores when the tube is opened. After incubation seal the bunged tube with the wax sealing tissue known commercially as 'Parafilm'. Store at about 5°C. Transfer annually. Purify the strain from time to time by spreading a conidial suspension on medium VM with sorbose as the main carbon source, together with any required nutrient supplements, and selecting and testing a single colony.

### 3. *SCHIZOPHYLLUM COMMUNE*

Keep monokaryotic mycelia on slopes of medium SCM in screw cap bijou bottles. Store at about 5°C. Transfer annually.

## 4. *SACCHAROMYCES CEREVISIAE*

Keep haploid strains as streaks on slopes of medium YPG in screw cap bijou bottles. Store at about 5°C. Transfer annually. Purify by spreading a cell suspension on medium YPD and selecting and testing a single 'grande' colony (see experiment 8), thus reducing the accumulation of gene mutations and of 'petite' respiratory deficient mutants.

## 5. *PHYTOPHTHORA INFESTANS*

This fungus is more difficult to keep in culture. However, it may be kept at about 15°C on slopes of medium R in large test tubes (20 mm diameter) by transferring to new medium every four to eight weeks. Under these conditions vigorous cultures may be stored for at least one year.

# V. EXPERIMENTS

# 1. LIFE CYCLES AND BREEDING SYSTEMS

Fungi display a very wide variety of life cycles and breeding systems (see Whitehouse 1949; Raper 1966 for reviews). Very many fungi, the imperfect fungi, reproduce only asexually. A very simple asexual life cycle can be demonstrated in the imperfect species, *Aspergillus versicolor*. The asexual spore, or conidium, of this species is normally uninucleate and haploid and the nucleus in each is produced as a result of mitotic divisions only. Therefore each spore is expected to have a genotype identical to that of the parent colony. Rare variants do arise as a result of spontaneous mutation (both chromosomal and extrachromosomal) and of ploidy changes (see experiment 14).

The sexual life cycle in fungi consists essentially of the bringing together of a pair of haploid nuclei, the fusion of these nuclei to form a diploid nucleus, and, usually immediately, the meiotic division of this diploid nucleus to give four (or sometimes eight) haploid sexual spores. Although mostly conforming to this plan, the sexual life cycles found in fungi vary greatly, even within a single genus. The main subdivision is into homothallic and heterothallic systems. In the homothallic species the sexual cycle can be completed upon a single homokaryotic mycelium. This involves a true selfing process, the pairs of haploid nuclei which fuse to give the pre-meiotic diploid nucleus being genetically identical, and consequently the sexually produced spores, like the asexually produced spores, display no segregation at all.

In order to carry out a genetic analysis by way of the sexual cycle in homothallic species it is necessary to obtain a mycelium containing the genetically differing nuclei of both parents of the cross to be analysed so that at least some of the pre-meiotic diploid nuclei are heterozygous, being the result of the fusion of genetically different haploid nuclei. Such a mycelium, where genetically different nuclei are contained within a common cytoplasm, is termed a heterokaryon. In some groups very specialized and stable forms of heterokaryon have developed, for example, the dikaryotic stage in Basidiomycetes (see experiment 5). In other cases, for example, *Aspergillus nidulans*, the heterokaryon is unstable and normally transient (see experiment 3). However, the use of auxotrophic mutant strains may stabilize the heterokaryon in these cases (see experiment 2).

In the heterothallic species the interaction of two different mycelia is essential for the completion of the sexual cycle and the problem of separating the selfed fruit bodies from those of hybrid origin does not arise. The two mycelia involved in a compatible mating may show obvious sexual differences in some species, but more often they are morphologically identical and differ only in their ability to form compatible matings with other strains. These sexual compatibility systems are controlled by genetic factors which vary in complexity from the simple one gene, two allele system found in *Neurospora crassa* (experiment 4) to the complex two gene, multiple allele system found in *Schizophyllum commune* (experiment 5). These two experiments, together with experiment 3, also illustrate the basic principles of Mendelian inheritance.

### Experiment 1. Asexual life cycle

Species: *Aspergillus versicolor*
Strain: Wild-type

Prepare a suspension of conidia following the methods described in section II and dilute to give $3 \times 10^2$ spores per ml. Plate 0·1 ml aliquots on each of from ten to 20 plates of medium M and spread. Incubate at 25°C for seven days and examine. Count and classify the colonies according to morphology.

The majority or all of the colonies which grow will be similar to each other and to the parent strain. Rare morphologically variant colonies may be found to occur with a low frequency (about $10^{-3}$ to $10^{-5}$). If such variants are found, subculture them to determine if they breed true.

### Experiment 2. Heterokaryosis

Species: *Aspergillus nidulans*
Strains: $pyro_4$.  Green conidia.  Requires pyridoxine
         $paba_6\ y$.  Yellow conidia.  Requires *p*-aminobenzoic acid

Prepare conidial suspensions ($10^6$ spores per ml) of both strains and mix together in equal volumes. Spread 0·1 ml aliquots of this mixed suspension on each of two or three plates of medium M and incubate at 35°C for from four to six days. After this period a few centres of active growth will be found among the background of very slow growth. Most of these growth centres will produce both green and yellow conidial heads in about equal proportions. From 12 or 15 of these growth centres make mass hyphal transfers (inocula 2 to 3 mm square), three per plate, to fresh plates of medium M. Incubate at 35°C for from four to seven days.

A proportion of these inocula will give rise to ragged looking but vigorous colonies. Close examination will show the presence of (in this particular case) about equal proportions of closely intermingled green and yellow conidial heads and up to 5 per cent of the total heads will be striped green and yellow thus proving their heterokaryotic origin. Occasionally the heterozygous diploid (see experiment 14) will grow away spontaneously from the edge of a heterokaryotic colony as a green spored sector which rapidly surrounds the heterokaryotic growth due to its superior growth rate.

Make mass hyphal transfers from the heterokaryotic colonies on to plates of medium M, medium M + 10 µg per ml. of pyridoxine + 10 µg per ml of p-aminobenzoic acid, and medium M plus each of these two vitamins singly. Incubate at 35°C for four days and examine the resultant growth. The results obtained will be as given in table 5.1 and they illustrate that the *A. nidulans* heterokaryon is unstable and can only be maintained on a selective medium (M). On any medium where either homokaryotic partner can grow alone the heterokaryotic association breaks down.

**Table 5.1**

| Medium | Growth | Conidial colour |
|---|---|---|
| M | Heterokaryotic | Mixed green and yellow |
| M + pyridoxine + p-aminobenzoic acid | Homokaryotic | Green sectors and yellow sectors |
| M + pyridoxine | Homokaryotic | Green |
| M + p-aminobenzoic acid | Homokaryotic | Yellow |

## Experiment 3.   Sexual life cycle.   Homothallic

Species: *Aspergillus nidulans*
Strains: Wild-type.   Green conidia.   Prototrophic
    *y*.          Yellow conidia.   Prototrophic

Prepare conidial suspensions ($10^6$ spores per ml) of both strains and mix together in equal volumes. Spread 0·1 ml aliquots of this mixed suspension on each of two plates of medium M and incubate at 35°C for from ten to 14 days until the cleistothecia, the large spherical sexual fruit bodies, are mature. The young cleistothecia are pigmentless and as they mature they pass through a red-brown colour to black when fully mature. From an area on these plates where green and yellow conidia are seen to be well intermingled remove 30 of the larger mature cleistothecia. Place these on an agar plate and clean them of any adhering conidia and other debris (mycelium and Hülle cells) by rolling on the agar surface with a rounded glass needle. Pick up a cleaned cleistothecium on a microspatula fashioned from nichrome wire and transfer to the wall of a small test tube (10 mm diameter) containing 0·5 ml of 'Tween 80' solution. Crush the cleistothecium against the wall with the microspatula and shake the tube to suspend the red-brown coloured sexually produced ascospores. Repeat this process to suspend the ascospores from the other cleaned cleistothecia thus obtaining, in total, 30 individual ascospore suspensions. Take a loopful of the first ascospore suspension and streak it on a plate of medium M radially from the edge of the plate to the centre. Repeat this with similar samples from each of the remaining ascospore suspensions. Up to ten such radial streaks may be inoculated on to one plate. Store the ascospore suspensions at 5°C and incubate the streaked plates at 35°C for two to three days.

After incubation classify the growth resulting from each ascospore streak according to conidial colour. The following results are typical: green spored streaks—18; yellow spored streaks—nine; mixed green and yellow spored streaks—three. In this case 90 per cent of the cleistothecia resulted from 'selfing' and just 10 per cent were hybrid in origin.

Return now to one of the ascospore suspensions which the above test has shown was hybrid in origin and, using the haemocytometer slide, estimate the concentration of ascospores in this suspension. (Larger cleistothecia each contain in the region of $10^5$ ascospores.) Dilute the suspension to give about $3 \times 10^2$ spores per ml and spread 0·1 ml aliquots on to each of ten plates of medium M. Incubate at 35°C for three or four days. The resultant colonies, each grown from a single ascospore, should then be classified according to conidial colour. The following results are typical: green spored colonies—153; yellow spored colonies—173. This does not differ significantly from a 1 : 1 proportion of green : yellow colonies ($\chi^2_{(1)}=1·227$; $P=0·30$ to $0·20$, see p. 11) from which we can conclude that the spore colour difference is controlled by a single pair of alternative alleles.

### Experiment 4.  Sexual life cycle.  Heterothallic.  Dipolar

Species: *Neurospora crassa*
Strains: *a arg*-1.     Mating-type *a*, requires arginine
        *A tryp*-2.     Mating-type *A*, requires tryptophan

In this experiment the interest is in the mating-type factors and prototrophic strains could equally well be used. But for class use, in order to reduce contamination risks, it is safer to use auxotrophic strains. In this experiment, therefore, add arginine and tryptophan (150 µg per ml) to all media used.

## Method of mating-type determination

Prepare six tubes (17·5 mm diameter) containing slopes of medium W supplemented with arginine and tryptophan. Insert a strip of filter paper into the tube before autoclaving so that, when the medium sets, the paper is embedded in, but protrudes from, the agar slope. Prepare dense conidial suspensions of both strains following the method described in section II. Inoculate one drop of the suspension of strain $a$ into two tubes, one drop of strain $A$ into a further two tubes, and one drop of a mixed suspension of strains $a$ and $A$ into the third pair of tubes. Incubate at 25°C for about seven days.

The resultant cultures should be examined for the presence or absence of the sexual fruit bodies, the perithecia, which will be clearly visible, if present, as large, black, flask-shaped structures. The following results will be obtained: strain $a$ alone—no perithecia; strain $A$ alone—no perithecia; strains $a + A$ mixed—perithecia present. Unlike *Aspergillus nidulans* the sexual cycle is not completed on a homokaryotic mycelium in pure culture. It is required that mycelia of opposite mating-type are grown together.

## Inheritance of the mating-type factors

Continue the incubation of the above $a \times A$ cross for about a further seven days. By this time the perithecia will have matured and a large number of black ascospores will have been shed on to the sides of the test tubes. Collect loopfuls of these spores and wash them off in a saline solution in order to make an ascospore suspension. Dilute this suspension to give about $5 \times 10^2$ ascospores per ml and place the tube of diluted suspension in a water bath at 60°C for 45 min in order to give the heat shock generally required for the germination of *N. crassa* ascospores. (This heat treatment will also have the effect of killing any contaminating conidia.) Spread 0·1 ml aliquots of this heat-treated suspension on to each of a number of plates of medium VM supplemented with arginine and tryptophan. Incubate overnight at 25°C in order to germinate the ascospores and then, with a microspatula, transfer a sample of at least 30 of these developing colonies to tubes containing slopes of medium VM supplemented as before. (When cutting the inocula take care to avoid contamination with hyphae from adjacent colonies. This process is best carried out using transmitted light under the stereoscopic microscope.) Incubate at 25°C for three or four days until each mycelium produces conidia.

Prepare 60 small (10 mm diameter) tubes with slopes of supplemented medium W and containing strips of filter paper as before. Inoculate 30 of these tubes each with one drop of a conidial suspension of parental strain *a* and, similarly, the second 30 tubes with parental strain *A*. Prepare conidial suspensions from each of the 30 mycelia derived from the ascospores and add one drop of the first of these suspensions to one of the tubes inoculated with parent *a* and one drop to a tube inoculated with parent *A*. Repeat for the other 29 of the progeny. Incubate the 60 small tubes thus inoculated at 25°C for about ten days.

Classify the resultant cultures according to the presence or absence of perithecia. Typical results are given in table 5.2. Thus, in this example, 13 of the sexual progeny were of mating-type *A* and 17 of mating-type *a*. This ratio is not significantly different from 1 : 1 ($\chi^2_{(1)}=0.533$; $P=0.50$). Thus it can be concluded that mating-type is controlled by two alternative alleles at a single genetic locus which segregates in a Mendelian fashion at meiosis to give rise to two classes among the sexual progeny, hence the name dipolar.

Table 5.2

| Sexual progeny | Parent tester | |
| --- | --- | --- |
| | *a* | *A* |
| 13 Mycelia | + | — |
| 17 Mycelia | — | + |

+ = Perithecia present.
— = Perithecia absent.

## Experiment 5.   Sexual life cycle.   Heterothallic.   Tetrapolar

Species: *Schizophyllum commune*
Strains: Wild-type monokaryon. Mating-type A7 B8
        Wild-type monokaryon. Mating-type A8 B7
        Four monokaryotic testers of mating-types A7 B7, A7 B8, A8 B7 and A8 B8

### Method of mating-type determination

Incoculate each of two plates of medium SCM with mass hyphal transfers of the four tester strains in the positions shown in fig. 5.2. Adjacent (about 2 mm distant) to each of these inocula on one of these plates inoculate the A7 B8 monokaryon. For comparative purposes also inoculate this strain in the centre of the plate. Inoculate the A8 B7 monokaryon on the second plate in a similar way.

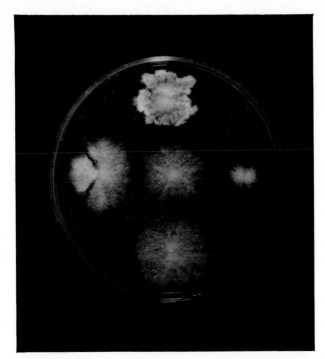

**Fig. 5.1.** Mating reactions in *Schizophyllum commune* resulting from the joint inoculation of the tester strains and of the unknown strain being tested (in this case A8B8) in the positions shown in fig. 5.2.

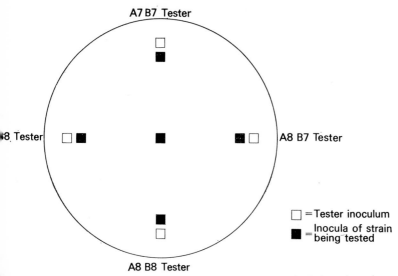

**Fig. 5.2.** Positions of inoculation of the four tester strains and of the unknown strain being tested.

[*Facing p.* 189]

Incubate these two plates at 30°C for three or four days and examine the mating reactions (see fig. 5.1). The following four specific reactions will be found:

(a) Both factors different, e.g. A7 B8 + A8 B7. Compatible reaction or +. This gives the dikaryotic mycelium which has a characteristic morphology. Microscopic examination of the hyphae shows the presence of clamp connections.

(b) The A factor in common, e.g. A7 B8 + A7 B7. 'Flat' reaction or Fl. This combination results in the common A heterokaryon which has a very characteristic morphology in which few aerial hyphae are formed.

(c) The B factor in common, e.g. A7 B8 + A8 B8. 'Barrage' reaction or Bge. In this case a gap or barrage is found along the junction between the two mycelia.

(d) Both factors in common, e.g. A7 B8 + A7 B8. No particular reaction or —. The two mycelia usually grow into each other without any special effect. On classifying the two test plates according to this scheme the results given in table 5.3 will be obtained.

**Table 5.3**

| Tested strain | Testers | | | |
| --- | --- | --- | --- | --- |
| | A7 B7 | A7 B8 | A8 B7 | A8 B8 |
| A7 B8 | Fl. | — | + | Bge. |
| A8 B7 | Bge. | + | — | Fl. |

## Inheritance of mating-type factors

Inoculate the A7 B8 and A8 B7 monokaryons adjacent to each other on a plate of medium SCM and incubate at 30°C for three days in order to establish the growth of the resultant dikaryon. Place this plate at 18°C under continuous illumination for about ten days until the sporophores have developed and matured. (Fruiting will occur, though possibly more slowly, if the plate is placed near the window at room temperature providing this does not exceed about 22°C. The use of medium SF may also assist if fruiting is difficult.) Invert the plate with the mature sporophores over an empty Petri dish for about 2 hours until a white spore print is formed. Add about 3 ml of saline solution to the dish and agitate in order to suspend the basidiospores. Dilute the suspension to give 2–3 × 10² spores per ml and spread 0·1 ml aliquots on each of 30 plates of medium SCM.

Incubate at 30°C for two days when small (about 5 mm diameter) colonies should have developed. Take a sample of these sexually produced monokaryons at random and inoculate them individually on to slopes of medium

SCM in small tubes for storage. In order to take a random sample take
the first plate and transfer an inoculum from every colony which is clearly
well separate from all others. Then start on a second plate and continue
systematically in this way until a large enough sample has been taken. Care
must be taken to ensure that each colony is well separate from the others (both
on the surface and under the medium) so that the mating process could not
have started between neighbouring colonies before the sample is taken. Also,
all suitable colonies must be taken whatever their size. There will be considerable variation in rate of growth among the sexual progeny (see experiment 7 and chapter 3). Incubate the tubes at 30°C for three days until each
mycelium is well established. With experience it should be possible to reject
any dikaryon which may inadvertently be present in this sample on the basis
of morphology, confirmed by the presence of clamp connections.

Determine the mating-type of each of these sexually produced monokaryons following the method already described. Reject any which appear to
have given dikaryons with all four testers. An examination of the 'control'
colony in the centre of the plate in such cases will show that the mycelium
being tested is itself a dikaryon which has escaped the first screening. Also, set
aside any mycelia which gives two + reactions, the other two being either
both Fl. or both Bge.

In a sample of 80 sexually produced monokaryons nine gave two + and
two Bge. reactions. The remainder gave all four different reactions with the
tester strains (table 5.4). The $\chi^2$ analysis of these data given in table 5·5 shows
that the pairs of alleles at both A and B loci segregate in ratios not significantly different from 1 : 1 and that the A and B loci are not linked. The

**Table 5.4**

| Unknown monokaryons | Testers | | | |
|---|---|---|---|---|
| | A7 B7 | A7 B8 | A8 B7 | A8 B8 |
| 20 strains (A7 B7) | — | Fl. | Bge. | + |
| 18 strains (A7 B8) | Fl. | — | + | Bge. |
| 17 strains (A8 B7) | Bge. | + | — | Fl. |
| 16 strains (A8 B8) | + | Bge. | Fl. | — |

**Table 5 5**

| | | |
|---|---|---|
| Total $\chi^2_{(3)}$ | 0·4930 | $P = 0·95$ to $0·90$ |
| Segregation at A locus, $\chi^2_{(1)}$ | 0·3521 | $P = 0·70$ to $0·50$ |
| Segregation at B locus, $\chi^2_{(1)}$ | 0·1268 | $P = 0·80$ to $0·70$ |
| Linkage between A and B loci, $\chi^2_{(1)}$ | 0·0141 | $P = 0·95$ to $0·90$ |

segregation of alleles at the unlinked A and B loci gives rise to four equal classes among the monokaryotic sexual progeny, hence the term tetrapolar. (See chapter 1, p. 15, for a discussion of the analysis of linkage data.)

## Genetic structure of the A locus

The nine monokaryons which gave two + reactions with the tester mycelia can be divided into two groups according to their reactions with the tester strains (table 5.6). A group of four of these strains contain the B7 allele and the remaining five strains contain the B8 allele. However, the A factor in all nine cannot be classified as A7 or A8, but as a new kind of A factor, A*. When these nine strains are crossed together in all combinations the A* factor can be subdivided into two groups, A*1 and A*2 as is shown in table 5.7.

### Table 5.6

| Unknown monokaryons | Testers | | | |
| --- | --- | --- | --- | --- |
| | A7 B7 | A7 B8 | A8 B7 | A8 B8 |
| Four strains (A* B7) | Bge. | + | Bge. | + |
| Five strains (A* B8) | + | Bge. | + | Bge. |

### Table 5.7

| | Group 1 A*1 B7 | Group 2 A*2 B7 | Group 3 A*1 B8 | Group 4 A*2 B8 |
| --- | --- | --- | --- | --- |
| One strain. Group 1, A*1 B7 | — | Bge. | Fl. | + |
| Three strains. Group 2, A*2 B7 | Bge. | — | + | Fl. |
| Three strains. Group 3, A*1 B8 | Fl. | + | — | Bge. |
| Two strains. Group 3, A*2 B8 | + | Fl. | Bge. | — |

The explanation of this behaviour is that the A factor is not a single genetic locus, but consists of two distinct linked loci. This can be represented for our parental strains as follows:

A7 factor:
$$A\alpha7 \qquad\qquad A\beta7$$

A8 factor:
$$A\alpha8 \qquad\qquad A\beta8$$

(The situation is complicated further by the availability of multiple alleles at each of these sub-factors, though this is not illustrated in the present experiment.) In an A7 × A8 cross genetic recombination occurs between the α and β sub-factors in a proportion of meioses dependent upon the genetic distance between them:

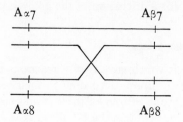

This gives rise to the tetrad:

| $A\alpha7$ | $A\beta7$ | Parental A7 |
|---|---|---|
| $A\alpha8$ | $A\beta8$ | Parental A8 |
| $A\alpha7$ | $A\beta8$ | Recombinant A*1 |
| $A\alpha8$ | $A\beta7$ | Recombinant A*2 |

Each A* recombinant has one sub-factor in common with both parental factors but the difference at only one sub-factor is sufficient to make both A* factors compatible with both parental A factors. In the example given here nine out of 80 monokaryons were recombinants, that is there was 10·125 per cent recombination. The B factor is also constructed in a complex fashion, but the B7 and B8 factors used here have one sub-factor in common so that the recombinants cannot be distinguished from the parental classes.

## 2. VARIATION AND MUTATION

Heritable differences between individuals are a prerequisite for any genetical experiment. Two sources of such differences are available. Naturally occurring variation is encountered among independent isolates from wild populations and similarly spontaneously produced variation is often found among laboratory cultures. For many genetical studies the variation required is deliberately induced as the result of a specific treatment with a mutagen. Fungi provide ideal material for demonstrating naturally occurring and induced variation both at the chromosomal and extrachromosomal level.

In the following two experiments methods of obtaining and studying natural and induced chromosomally located variation are described. For the naturally occurring variation a continuously varying character has been chosen because such variation is an important component in wild populations (see chapter 3). However, it should be kept in mind that the mating-type factors described in experiments 4 and 5 and the pathogenicity differences described in experiment 18 are also examples of naturally occurring variation, but of a discontinuous kind. In experiment 8 the spontaneous occurrence and specific induction of the respiratory deficient cytoplasmic mutant known as 'petite' in yeast is described.

## Experiment 6. Induction of mutation by ultraviolet irradiation

Species: *Aspergillus nidulans*
Strain: Wild-type. Green conidia. Prototrophic

### Production of ultraviolet survival curve

Prepare a suspension of conidia of this strain and adjust to give $10^7$ spores per ml. Pipette equal volumes (about 4 or 5 ml) of this suspension into small (40 or 50 mm diameter) open Petri dishes. Irradiate these suspensions keeping the ultraviolet lamp at a standard distance but varying the dose time. For the first test use 0, 1, 2, 4, 8 and 16 min of irradiation. After irradiation dilute each suspension to give two dilutions containing $5 \times 10^2$ and $5 \times 10^4$ spores per ml respectively. Plate 0·1 ml aliquots of each dilution of each suspension on plates of medium CM and spread, allowing at least two plates of each dilution of each suspension.

Incubate these plates for four days at 30°C and count the number of colonies which develop on each plate. (Some plates may contain too many colonies to count.) The number of colonies on the plates spread with un-irradiated spores will give an estimate of the total number of spores plated. At the higher ultraviolet doses the best estimate of the number of survivors will be obtained from the plates inoculated with the least dilute suspension. Plot the log number of survivors per ml of suspension irradiated as a function of dose time and estimate the dose which gave 1 per cent survival. It may be necessary to carry out a second experiment using a more narrow time scale in order to estimate the 1 per cent survival dose with greater accuracy.

### Mutagenesis experiment

Prepare a conidial suspension of the wild-type strain containing $10^7$ spores per ml. Keep part of this suspension for control platings and irradiate another aliquot to 1 per cent survival. Dilute the irradiated and control suspensions for plating, using a $10^{-2}$ dilution for the control suspension in place of the irradiation treatment. (Account must be made of this dilution difference between the treated and control samples when comparisons of mutant frequencies are being made.) It will now be possible to screen for various classes of mutant.

194

CHAPTER 5

(1) *Spore colour mutants*

Plate dilutions giving about $10^3$ survivors per plate on medium MT, which encourages conidiation. Incubate at 30°C for five days and examine the resultant growth for white, yellow or light green conidial heads among the background of dark green wild type conidia. Up to about 20 such colour mutants will probably be found per plate grown from irradiated conidia. By contrast, only very rare colour mutants will be found on the control plates.

(2) *Morphological mutants*

Plate dilutions giving about 50 survivors per plate on medium M. Incubate at 25°C for from seven to ten days and examine the resultant colonies. On the control plates only very rare ($< 10^{-3}$) morphological variants will be found, whereas the irradiated spores will give rise to many colonies displaying differences in very many aspects of colonial morphology and pigmentation. In experiments of this sort often more than 60 per cent of the colonies grown from irradiated conidia show some mutant morphology. By contrast with the uniformity of the colonies on the control plates, this experiment provides a very striking visual demonstration of the mutagenic effects of ultraviolet irradiation.

(3) *Auxotrophic mutants*

Plate dilutions giving about 150 survivors per plate on medium CM containing sodium deoxycholate (see section II, p. 176). Incubate at 30°C for six or seven days when the small colonies which result from the addition of the deoxycholate will be producing mature conidia. Using a damp pad replicator make two replica platings from each master plate on to medium M. Keep the master plates in the refrigerator and incubate the replica plates at 30°C for three or four days. Carefully compare these plates with the respective master plates, noting the positions where any colonies are absent on the replica plates, but present on the master plates and subculture them on medium CM and retest them also by subculturing on to medium M. Those which now grow on medium CM but not on M will represent auxotrophic mutants. It remains now only to characterize the mutants as fully as possible. This will not be fully described here, but the use of the needle replicator and of the screening techniques described in chapter 3 of Fincham & Day (1971) will facilitate this process.

### Experiment 7. Continuous variation

Species: *Schizophyllum commune*
Strains: Any pair of monokaryons with compatible mating-types

Cross the two monokaryons, fruit the resultant dikaryon and take a random sample of 50 monokaryons following the methods described in experiment 5. Prepare two growth-tubes for each of these monokaryons plus two for each parental strain (see section II). Number the tubes consecutively. Give each of the 50 sexually produced monokaryons and the two parental monokaryons two numbers between 1 and 104, at random. (This randomization process can be carried out by drawing numbers written on folded pieces of paper or from tables of random numbers.) Inoculate the monokaryons into the tubes in duplicate following this randomized order. For example, the first monokaryon may be given the random numbers 36 and 79. Thus this monokaryon will be inoculated into tubes 36 and 79. Incubate at 25°C and measure the growth over ten days.

Table 5.8 Growth per ten days (to nearest mm) of the sexually produced monokaryons

| Progeny No. | Growth | | Progeny No. | Growth | | Progeny No. | Growth | |
|---|---|---|---|---|---|---|---|---|
| 1 | 67 | 61 | 19 | 57 | 57 | 37 | 61 | 63 |
| 2 | 59 | 60 | 20 | 60 | 63 | 38 | 58 | 65 |
| 3 | 67 | 61 | 21 | 55 | 59 | 39 | 58 | 63 |
| 4 | 59 | 60 | 22 | 61 | 66 | 40 | 60 | 61 |
| 5 | 55 | 60 | 23 | 58 | 61 | 41 | 60 | 58 |
| 6 | 61 | 65 | 24 | 64 | 57 | 42 | 49 | 50 |
| 7 | 66 | 67 | 25 | 57 | 58 | 43 | 68 | 65 |
| 8 | 55 | 56 | 26 | 52 | 53 | 44 | 58 | 59 |
| 9 | 64 | 62 | 27 | 60 | 60 | 45 | 61 | 61 |
| 10 | 58 | 64 | 28 | 62 | 58 | 46 | 60 | 58 |
| 11 | 58 | 56 | 29 | 60 | 59 | 47 | 64 | 65 |
| 12 | 61 | 64 | 30 | 57 | 60 | 48 | 57 | 55 |
| 13 | 68 | 70 | 31 | 56 | 59 | 49 | 59 | 62 |
| 14 | 57 | 56 | 32 | 54 | 54 | 50 | 61 | 64 |
| 15 | 59 | 60 | 33 | 58 | 58 | | | |
| 16 | 58 | 59 | 34 | 63 | 67 | | | |
| 17 | 57 | 55 | 35 | 60 | 69 | Parent 1 | 58 | 59 |
| 18 | 58 | 58 | 36 | 63 | 61 | Parent 2 | 57 | 59 |

Decode these results from the random order to the true order (table 5.8) and plot the mean amount of growth as a frequency distribution histogram (fig. 5.3). A simple analysis of variance (see chapter 3) can be carried out on these data. In the case of the results given in table 5.8 this analysis is given in table 5.9. This analysis shows that there is a highly significant genetic component involved in the variation between the sexually produced mono-karyons grown in this experiment. The values of the genetic and environmental components which are estimated from the variance analysis are $\sigma^2{}_G = 10 \cdot 566$ and $\sigma^2{}_E = 5 \cdot 480$. Thus the genetic component accounts for $65 \cdot 85$

Table 5.9

|  | df | SS | MS | Components | F | P |
|---|---|---|---|---|---|---|
| Between monokaryons | 49 | 1304 | 26·612 | $\sigma^2_E + 2\sigma^2_G$ | 4·856 | < 0·001 |
| Duplicate error | 50 | 274 | 5·480 | $\sigma^2_E$ |  |  |

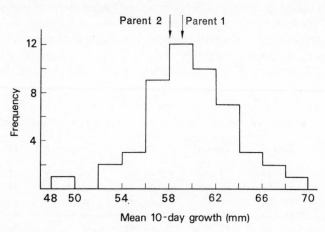

**Fig. 5.3.** Frequency distribution of mean growth rate of sexually produced monokaryons.

per cent and the environmental component for 34·15 per cent of the total variation in this experiment.

It can be seen from fig. 5·3 that the parental growth rates are similar to each other and to the progeny mean. That is, the genes responsible for fast and slow rates of growth are equally dispersed between the two parents. A similar experiment could also be carried out using parental monokaryons in which the genes responsible for fast growth rate are predominantly associated in one parent and those for slow growth rate in the other. The value of fungi for studying the inheritance of quantitative characters, of which growth rate is an example, is further illustrated and discussed in chapter 3.

### Experiment 8. Induction of the cytoplasmic mutant, 'petite'

Species: *Saccharomyces cerevisiae*
Strain: Any wild type respiratory competent strain, i.e. a 'grande' strain.

Inoculate a loopful of the strain into 10 ml of liquid medium YPG in a 100 ml conical flask and incubate, with agitation, for one day at 28°C. Centrifuge 5 ml of this culture and resuspend the cells in liquid YPG to give a suspension containing $10^6$ cells per ml. The mutagenesis is carried out by

growing the culture in 50 ml flasks containing buffered liquid YPG medium
(0·1M phosphate buffer, pH 6·5) with a range of concentrations of acriflavine.
Make up the flasks as shown in table 5.10, using the $10^6$ cells per ml suspension as the inoculum. (To make the 1M phosphate buffer of pH 6·5,
prepare 1M solutions of $Na_2HPO_4$ and $NaH_2PO_4$ and mix together in the
proportions 0·32 of $Na_2HPO_4$ to 0·68 of $NaH_2PO_4$.) Immediately cover the

### Table 5.10

| Flask no. | 1 | 2 | 3 | 4 | 5 | 6 | 7 |
|---|---|---|---|---|---|---|---|
| Acriflavine conc. (µg/ml) | 0 | 0·25 | 0·50 | 0·75 | 1·00 | 1·25 | 1·50 |
| Liquid YPG medium (ml) | 3·25 | 3·250 | 3·25 | 3·250 | 3·25 | 3·250 | 3·25 |
| Phosphate buffer (1M) | 0·50 | 0·500 | 0·50 | 0·500 | 0·50 | 0·500 | 0·50 |
| Acriflavine (10µg/ml) | 0 | 0·125 | 0·25 | 0·375 | 0·50 | 0·625 | 0·75 |
| Water | 0·75 | 0·625 | 0·50 | 0·375 | 0·25 | 0·125 | 0 |
| Inoculum | 0·50 | 0·500 | 0·50 | 0·500 | 0·50 | 0·500 | 0·50 |

flasks with aluminium foil (the action of acriflavine may be influenced by
light) and incubate, with agitation, at 28°C for about 20 hours.

At the start of the experiment each flask should contain about $10^5$ cells
per ml. In order to estimate this figure more accurately dilute a sample of
the inoculum 100 times at the time of the inoculation of the experiment to
give a suspension containing about $10^3$ cells per ml and spread 0·1 ml aliquots
on each of four plates of medium YPD. Incubate for four days at 28°C and
count the numbers of 'petite' colonies and 'grande' colonies which grow.
('Petites' are distinguishable from 'grandes' by their small colony diameter
and very flat thin growth.) In one experiment the following results (summed
over the four replicate plates) were obtained: 'grandes', 534 colonies and
'petites', eight colonies. Thus the best estimation of the number of cells in the
control flasks at the start of the experiment is $542 \times 10^3/4$, or $1·355 \times 10^5$ cells
per ml.

After the incubation of the cultures with the range of acriflavine concentrations centrifuge 3 ml of each and resuspend in the same volume of saline
solution. Dilute to give about $10^3$ cells per ml. (In the experiments whose
results are quoted here, a dilution factor of $10^{-5}$ was required.) Spread 0·1 ml
aliquots of each suspension on plates of medium YPD. Again spread four
replicate plates of each suspension. Incubate at 28°C for four days and count
the number of 'grande' and 'petite' colonies on each plate. In the experiment
illustrated here the results obtained, again summed over the four replicate
plates, are given in table 5.11.

Percentage 'petite' and total colonies are represented graphically as a
function of acriflavine concentration in fig. 5.4. The concentrations of
acriflavine used are clearly not toxic. Averaging over all 28 plates the number

**Table 5.11**

|        | Grande | Petite | Total colonies | % petite |
|--------|--------|--------|----------------|----------|
| Flask 1 | 350 | 6 | 356 | 1·69 |
| 2 | 255 | 74 | 329 | 22·49 |
| 3 | 91 | 286 | 377 | 75·86 |
| 4 | 31 | 338 | 369 | 91·60 |
| 5 | 8 | 306 | 314 | 97·45 |
| 6 | 3 | 361 | 364 | 99·18 |
| 7 | 3 | 386 | 389 | 99·23 |

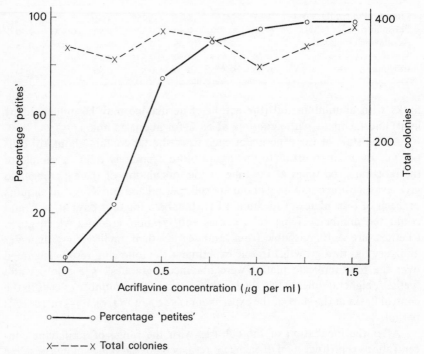

o———o Percentage 'petites'

x– – – –x Total colonies

**Fig. 5.4.** Percentage 'petites' and total colonies as a function of acriflavine concentration.

of cells plated per 0·1 ml aliquot was 89·214. The average number of cells per ml in the cultures after the 20 hours incubation was thus $89·214 \times 10^6$. At the start of the experiment the flasks contained $1·355 \times 10^5$ cells per ml. Thus it can be estimated that during the 20 hours of incubation the cultures have undergone something over nine cell generations, while over 99 per cent of the cells have been converted to 'petites' at the higher concentrations of acriflavine.

# 3. GENETIC ANALYSIS BY WAY OF MEIOTIC SYSTEMS

In higher plants and animals an analysis of the products of meiosis is usually restricted to random samples of the products of large numbers of meioses. In fungi, on the other hand, the products of any one meiosis can often be recovered together and it is thus possible to analyse samples of individual meioses by examination of the meiotic products in their original tetrads. Equally, the isolation of the meiotic products can be carried out after they have been released from the tetrads and mixed in a random fashion. Each of these forms of analysis, namely random spore analysis and ordered and unordered tetrad analysis, are illustrated in experiments 9, 10 and 11.

The first three experiments in this section treat the gene as a single indivisible unit. However, the sites of independently derived mutations at a particular genetic locus can be shown to be separable by recombination, thus demonstrating that independent sites of mutation exist within the gene. The frequency of recombination between two such mutant sites is very low, but, by the use of selective techniques and assisted by the presence of outside markers, such recombinational events can be studied. In experiment 12 an example of a fine structure analysis in *Aspergillus nidulans* is described.

Finally, the extrachromosomal location of a particular mutant can often be demonstrated by an analysis of the sexually produced progeny of a cross. One of the best methods is found in species where it is possible to carry out reciprocal crosses, thus allowing the detection of the uniparental inheritance of a character. The maternal inheritance of the slow growing respiratory deficient mutant, 'poky', in *Neurospora crassa* is described in experiment 13.

### Experiment 9. Random spore analysis of a three-point cross

Species: *Aspergillus nidulans*
Strains: $pro_1\ paba_6\ y$. Yellow spores. Requires proline and $p$-aminobenzoic
acid.
$pyro_5$.　　　　　Green spores. Requires pyridoxine

Synthesize a heterokaryon from the two haploid strains as described in experiment 2. Grow the heterokaryon for about 14 days until the cleistothecia are mature. Sample the ascospores from the individual cleistothecia and obtain a suspension of ascospores derived from a hybrid fruit body as described in experiment 3. Dilute this suspension to about $3 \times 10^2$ spores per ml and spread 0·1 ml aliquots on each of ten plates of medium M supplemented with 100 μg per ml of proline and 10 μg per ml each of $p$-aminobenzoic acid and pyridoxine.

Incubate at 35°C for four days and classify each resultant colony according to conidial colour. Typical results are 273 green and 262 yellow colonies. Since this difference is controlled by a single gene equal numbers of green and yellow colonies are expected. The results obtained do not differ significantly from this expectation ($\chi^2_{(1)}$=0·226; $P$=0·70 to 0·50). Occasionally cleistothecia are found which give aberrant ratios. These should be avoided since they may be the result of 'twin' cleistothecia.

Take a small mass conidial inoculum from each colony of a random sample of colonies from these plates following the sampling method described in experiment 5. Inoculate these in accurate 5 × 5 arrangements upon plates of medium M supplemented with proline, $p$-aminobenzoic acid and pyridoxine. Incubate these plates for three days at 35°C to provide master plates for use with the needle replicator (see section II). A sample size of about 100 (four master plates) is adequate, though a larger sample was taken in the example quoted here. Replicate each master plate onto the following media: M + proline + $p$-aminobenzoic acid + pyridoxine; M + proline + $p$-amino-benzoic acid; M + proline + pyridoxine; and M + $p$-aminobenzoic acid + pyridoxine. Incubate these plates for two to three days at 35°C and score the conidial colour of the colonies on the fully supplemented plates and score presence or absence of growth on the plates where only two of the three supplements had been added.

From the growth responses obtained on this differential range of media it is possible to deduce the genotype of each individual in this sample of the sexually produced progeny. The genotypes of a sample of 321 such individuals is given in table 5.12. For example, 63 individuals had green spores and grew on all media and hence they carry the wild-type or + allele at all loci. Similarly 58 individuals had green spores and grew on all media except that lacking pyridoxine, hence they carry the wild-type allele at all loci except that responsible for the production of pyridoxine where they carry the mutant allele.

Examination of table 5.12 shows that the parental genotypes, $y\ pro_1\ paba_6$ and $pyro_5$, are among the most frequent classes recovered. However, the wild-type and $y\ pro_1\ paba_6\ pyro_5$ classes are equally frequent suggesting free segregation of the $pyro_5$ marker from the other three. But before pursuing this further it must first be tested that there are no viability differences, that is that the mutant and wild-type alleles at each locus are equally frequent. The $\chi^2$ values, each for 1 degree of freedom, for testing this expectation at each locus are given in table 5.13 and in no case is there any significant deviation from the expectation. The examination of the linkage relationships can thus be continued without having to consider possible upset from differential viability.

In table 5.14 the joint segregation of each of the possible pairs of loci is examined. For example, in the case of the $y$ to $pro_1$ comparison the parental strains had the genotypes $y^+pro^+$ and $y\ pro$. The two possible recombinant classes are thus $y^+\ pro$ and $y\ pro^+$. From the results given in table 5.12 it is

Table 5.12

| Genotype | | | | No. of individuals |
|---|---|---|---|---|
| + | + | + | + | 63 |
| + | + | + | pyro | 58 |
| + | + | paba | + | 3 |
| + | + | paba | pyro | 2 |
| + | pro | + | + | 3 |
| + | pro | + | pyro | 7 |
| + | pro | paba | + | 11 |
| + | pro | paba | pyro | 15 |
| y | + | + | + | 17 |
| y | + | + | pyro | 14 |
| y | + | paba | + | 9 |
| y | + | paba | pyro | 3 |
| y | pro | + | + | 2 |
| y | pro | + | pyro | 0 |
| y | pro | paba | + | 51 |
| y | pro | paba | pyro | 63 |

Table 5.13

| Locus | Mutant allele | Wild-type allele | $\chi^2_{(1)}$ | $P$ |
|---|---|---|---|---|
| y | 159 | 162 | 0·0280 | 0·90 to 0·80 |
| pro | 152 | 169 | 0·9003 | 0·50 to 0·30 |
| paba | 157 | 164 | 0·1526 | 0·70 to 0·50 |
| pyro | 162 | 159 | 0·0280 | 0·90 to 0·80 |

found that these four genotypes were recovered with the following frequencies: $y^+ pro^+$ 126; $y\ pro$ 116; $y^+ pro$ 36; and $y\ pro^+$ 43. Thus there are 242 parental combinations and 79 non-parental combinations of these two markers. In the absence of linkage the parental and non-parental classes are expected to be equally frequent. Testing for a deviation from this expectation we have a $\chi^2_{(1)} = 82\cdot7695$ ($P < 0\cdot001$) showing there to be a significant excess of parental classes and giving good evidence for the linkage of the $y$ and $pro_1$ loci. The percentage of recombinants is taken as the estimate of map distance and in this case is 24·61. The other pairwise comparisons given in table 5.14 show that the $pyro_5$ locus is not linked to the other loci, all three of which are linked.

The relative order of the three linked loci can be deduced from the recombination frequencies. The $y$ to $pro_1$ recombination frequency is the largest and it can thus be assumed that the $paba_6$ locus is situated between

Table 5.14

| Comparison | Parentals | Recombinants | $\chi^2_{(1)}$ | P | % recombinants |
|---|---|---|---|---|---|
| *y–pro* | 242 | 79 | 82·7695 | <0·001 | 24·61 |
| *y–paba* | 257 | 64 | 116·0405 | <0·001 | 19·94 |
| *y–pyro* | 161 | 160 | 0·0031 | 0·98 to 0·95 | |
| *pro–paba* | 292 | 29 | 215·4798 | <0·001 | 9·03 |
| *pro–pyro* | 144 | 177 | 3·3925 | 0·10 to 0·05 | |
| *paba–pyro* | 153 | 168 | 0·7009 | 0·50 to 0·30 | |

them, giving the following linkage map:

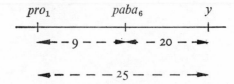

The original cross with respect to these three linked loci was thus as follows:

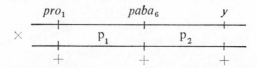

From table 5.14 our estimates of the recombination frequencies $p_1$ and $p_2$ are 0·0903 and 0·1994 respectively. Thus the proportion of individuals in the progeny sample which have resulted from simultaneous crossover events both between the $pro_1$ and $paba_6$, and the $paba_6$ and $y$ loci is expected to be $p_1 p_2$, or 0·0903 × 0·1994 = 0·0180. In our sample the number of individuals which have arisen from such a double crossover can be seen from table 5.12 to be seven. (The double crossover classes are $pro^+$ $paba$ $y^+$ $pyro^+$, $pro^+$ $paba$ $y^+$ $pyro$, $pro$ $paba^+$ $y$ $pyro^+$ and $pro$ $paba^+$ $y$ $pyro$.) The observed frequency of double recombinants is thus 7/321 = 0·0218. The ratio of the observed frequency of double recombinants over the expected frequency is termed the coefficient of coincidence. In this present example this is 0·0218/0·0180 = 1·2. This slight excess of observed double recombinants also known as negative interference, is usually observed in *A. nidulans* with the markers used in this experiment.

### Experiment 10.  Unordered tetrad analysis

Species: *Saccharomyces cerevisiae*
Strains: A number of haploid strains of opposite mating-type with suitably
         linked complementary auxotrophic markers are available

In the description of this experiment a detailed discussion of the culture techniques only will be given. The technique of tetrad dissection requires the availability of a microdissector and of the skill required to operate this machine. As a result it is felt that this experiment will be much more restricted in its application for teaching purposes than the other experiments. However, in order to carry out a cross between two haploid strains of yeast of opposite mating-types the following procedure should be used.

Inoculate two 50 ml conical flasks, each containing 5 ml of liquid medium YPG, one with the $a$ strain and the other with the $\alpha$ strain. Incubate, with agitation, for one day at 28°C. Add 0·5 ml of each culture to 4 ml of liquid medium YPG in a single 50 ml conical flask and incubate with agitation at 28°C for 5 hours. A microscopic examination of this culture will show the presence of a proportion of dumb-bell shaped zygotes. Centrifuge the culture and resuspend in saline. Spread 0·1 ml aliquots of this suspension on each of two or three plates of medium YM and similarly plate $10^{-1}$, $10^{-2}$ and $10^{-3}$ dilutions of this suspension. Incubate at 28°C for from five to seven days. The colonies which result will have developed from the prototrophic diploid zygotes and one of the four dilutions should give plates each with a reasonable number (50 to 100) of well-separated colonies.

With a loop remove a sample of cells from one of these diploid colonies and inoculate into 5 ml of liquid medium PSP. Incubate, with agitation, at 28°C for 24 hours and inoculate 0·1 ml of this culture into 5 ml of liquid medium SP and incubate, with agitation, for three days. Microscopic examination of this culture will now show the presence of a proportion of asci inside which the tetrad of ascospores can be seen.

In order to dissect the tetrad it is first necessary to remove the ascus wall by enzyme digestion. Dissolve 15 mg of cysteine in the contents (1 ml) of one ampoule of snail enzyme (Suc d'Helix Pomatia available from Industrie Biologique Française SA, 35 à 49 Quai du Moulin de Cage, 92-Gennevilliers, France). Add three drops of the sporulating culture to three drops of the enzyme solution in a small tube and incubate at 28°C for 30 to 45 min until the ascus wall is digested and place in a refrigerator until dissection. The actual design of the dissection chamber and technique of dissection will vary but in general it is best carried out on a plaquette of agar with dimensions of about $40 \times 20 \times 2$ mm. Streak a loopful of the digested ascus suspension on one edge of the agar plaquette and, by use of the microdissector, accurately dissect out the individual spores of each tetrad towards the opposite edge. Dissect about ten tetrads on each plaquette. After dissection place the plaquette on the surface of a plate of medium YPG and incubate at 28°C for two or three days until the spores have germinated and developed into small colonies. The appearance of such a plaquette is shown in fig. 5.5. In this example six out of ten tetrads have fully germinated. This is a good result. With a small loop transfer each colony of a complete tetrad to a plate of medium YPG.

Transfer four tetrads in this way to one plate. After incubation these plates can then be used as master plates for pad replication on to medium YM, and on to the differential series of supplemented medium YM in order to classify each spore for the presence of the wild-type or mutant alleles of the auxotrophic markers used. To do this follow the procedures described in experiment 9.

In respect of the genes segregating at any pair of loci, *a* and *b*, three types of tetrad may be recognized: 1. Parental ditype (PD). These are tetrads where two spores have the genotype of one parent, and two that of the other. So, in the cross $a\ b^+ \times a^+\ b$, this class of tetrad has two spores $a\ b^+$ and two $a^+\ b$. 2. Non-parental ditype (NPD). In these tetrads two spores have one genotype

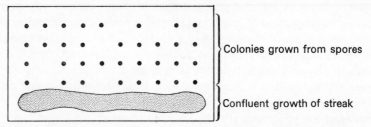

**Fig. 5.5.** Agar plaquette showing colonies developed from spores after dissection and incubation. Ten tetrads were dissected vertically away from the streak at the base of the plaquette.

and two the complementary genotype, neither of which are like the parents. Thus in this cross two spores will be $a^+\ b^+$ and two $a\ b$. 3. Tetratype (T). The tetrad contains one of each possible product, two of which are parental and two non-parental in genotype, i.e. one each of $a\ b^+$, $a^+\ b$, $a^+b^+$ and $a\ b$. The tetrads can therefore be classified into these three groups in respect of any pair of loci segregating in the cross. In the absence of linkage a 1 : 1 ratio of PD and NPD tetrads is expected and this can be tested as a $\chi^2$ for 1 degree of freedom. If linkage is indicated by this test, there being a significant excess of PD tetrads, the recombination frequency can be estimated as in experiment 9, that is as the total number of recombinant spores divided by the total number of spores scored.

For establishing linkage between genes unordered tetrad analysis is technically more difficult and statistically less informative than random spore analysis. But for a general discussion of the special uses of tetrad analysis see Fincham & Day (1971) chapters 4 and 8.

### Experiment 11.  Ordered tetrad analysis

Species: *Neurospora crassa*
Strains: *STA* 4. Wild-type. Mating-type A
        *a asco*. Mating-type *a*. Colourless ascospores

In this experiment it would be better to use an auxotrophic derivative of the *STA* 4 strain in order to reduce contamination risks. Also, the *asco* strain requires lysine for growth and lysine will be added to all media used at a concentration of 500 μg per ml.

Prepare conidial suspensions of both strains and mix in equal proportions. Inoculate one drop of this mixed suspension on to slopes of medium W supplemented with lysine following the general methods for the production of perithecia in this species outlined in experiment 4. Set up 12 tubes in this way and incubate at 25°C.

In order to obtain good results from this experiment it is essential that the asci are examined and scored at exactly the right stage of development. The ascospores should be mature enough to ensure that all those carrying the wild-type allele at the *asco* locus develop their full black pigmentation. Equally, it is essential that the asci remain intact and if left for too long the asci eject the ascospores. In order to ensure the availability of cultures at the correct stage of development it is suggested that the following method is used. Continue the incubation of the cultures until the perithecia approach their full mature size and the perithecial walls begin to accumulate a black pigmentation. Then make squash preparations of one or two perithecia each day until it can be seen that the majority of asci in any one perithecium are just approaching maturity. At this stage remove two tubes from the incubator and place them in the refrigerator at 5°C. Transfer a further two tubes to the refrigerator each day over a period of six days in total. A squash preparation of a perithecium from the culture removed on the last day should show that the asci are very mature and are ejecting the ascospores, and preparations made from the cultures removed at intermediate times should reveal one which is at the correct stage of maturity. The cultures may be left in the refrigerator for up to three or four weeks before scoring if required.

A sample of asci each containing eight ascospores should be classified according to the positions of the dark (wild-type) and colourless (*asco*) spores which they contain. Since the eight spores in each ascus consist of four pairs, the members of each pair being identical mitotic products of one nucleus of the original tetrad, six types of asci can be recognized (table 5.15) and these can be grouped according to whether segregation occurred at the first or second division of meiosis (see Fincham & Day 1971, chapter 4).

In this experiment no attempt is being made to detect linkage between two loci since this would proceed along the lines already indicated in experiments 9 and 10. Instead the ease with which linkage of a genetic locus, the *asco* locus in this case, to its centromere can be detected by the analysis of ordered tetrads may be demonstrated. The value, one half of the frequency of second division segregation, is a measure of the frequency of recombination between a locus and its centromere (Fincham & Day 1971, chapter 4). Thus in the sample of 168 asci, 32 or a proportion of 0·1905, were found to show

**Table 5.15**

| Ascus type | Spore pairs | | | | Segregation | Number obtained |
|---|---|---|---|---|---|---|
| | 1·2 | 3·4 | 5·6 | 7·8 | | |
| i | + | + | asco | asco | } 1st division | 66 |
| ii | asco | asco | + | + | | 70 |
| iii | + | asco | + | asco | ⎫ | 8 |
| iv | + | asco | asco | + | ⎬ 2nd division | 9 |
| v | asco | + | asco | + | ⎭ | 6 |
| vi | asco | + | + | asco | | 9 |

second division segregation. The frequency of recombination between *asco* and its centromere was therefore 0·1905/2 or 9·52 per cent recombination. Where, as in this case, the locus and the centromere are fairly closely linked then this value is a good estimate of the map distance.

Invariably, in experiments of this kind, aberrant asci are found. In the experiment referred to here 12 aberrant asci were found and these are listed in table 5.16. The first of these is probably the result of a rare overlap of spindles during the post-meiotic mitosis giving rise to an exchange of position between spores four and five. It is otherwise a normal example of first division segregation and was included in table 5.15 as such. The remainder all involve deviations from a 4 : 4 segregation of *asco* and its wild-type allele. In those showing 5 : 3 segregations a pair of spores which should be the identical products of one of the final mitoses in fact display segregation. The 6 : 2 and 2 : 6 asci on the other hand do not show segregation during the final mitotic division, but have resulted from abnormal 3 : 1 or 1 : 3 segregations during meiosis. For a discussion of the possible explanations and importance of these aberrant tetrads see Fincham & Day (1971, chapter 8).

**Table 5.16**

| Spore position in ascus | | | | | | | | + : asco | Frequency |
|---|---|---|---|---|---|---|---|---|---|
| 1 | 2 | 3 | 4 | 5 | 6 | 7 | 8 | | |
| asco | asco | asco | + | asco | + | + | + | 4 : 4 | 1 |
| asco | asco | + | asco | + | + | + | + | 5 : 3 | 2 |
| + | + | + | + | asco | asco | + | asco | 5 : 3 | 1 |
| + | + | + | + | asco | asco | asco | + | 5 : 3 | 2 |
| + | + | + | + | asco | asco | + | + | 6 : 2 | 3 |
| + | + | asco | asco | + | + | + | + | 6 : 2 | 1 |
| + | + | + | + | + | + | asco | asco | 6 : 2 | 1 |
| asco | asco | asco | asco | + | + | asco | asco | 2 : 6 | 1 |

## Experiment 12.   Analysis of genetic fine structure

Species: *Aspergillus nidulans*

Strains: Alleles at the *paba$_1$* locus in various combinations with the outside
markers *ad$_9$*, *y* and *bi$_1$*. The mutant alleles available are indicated
on the following map of the *paba$_1$* gene:

```
ad₉      paba₅     paba₁ paba₁₂    paba₆     paba₁₃    paba₁₈    y      bi₁
 +── ── ──+────────+──+────────────+─────────+─────────+── ── ──+───+
```

In the following experiment one of the many possible crosses is described.
The strains used are:

> *paba$_5$ bi$_1$*    Green conidia. Requires *p*-aminobenzoic acid and
> biotin.
>
> *ad$_9$ paba$_{13}$ y*    Yellow conidia. Requires *p*-aminobenzoic acid and
> adenine.

Prepare suspensions of conidia ($10^7$ per ml) of both strains and mix them
in equal proportions. Spread 0·1 ml aliquots of this mixed suspension on each
of two or three plates of medium M supplemented with *p*-aminobenzoic acid
(10 μg per ml) and incubate at 30°C. After seven days heterokaryotic growth
will be present and after 14 days many mature cleistothecia will have developed
on this growth. Remove and clean about 100 cleistothecia as described in
experiment 3. Place all of the cleaned cleistothecia in a tube with about 0·5 ml
of saline solution and crush them with a glass rod. Make up the volume to
about 2 ml and allow the cleistothecial wall debris to settle so that the
ascospore suspension can be removed with a Pasteur pipette. Dilute the
suspension to give 20 ml containing $3 \times 10^5$ spores per ml and a further
10 ml containing $6 \times 10^2$ spores per ml.

Pour 5 ml of medium M supplemented with adenine (50 μg per ml) and
biotin (1 μg per ml) and containing 1 per cent agar instead of the normal 1·5
per cent into each of 20 test tubes. Cool to 45°C and keep in a water-bath at
this temperature. Add 1 ml of the $3 \times 10^5$ per ml suspension to one of these
tubes and immediately mix well and pour over the surface of a plate of the
same medium (M + adenine + biotin) and allow to set. Repeat this procedure for the other 19 ml of this suspension. Plate the 10 ml of the more dilute
suspension in a similar way but in medium M + *p*-aminobenzoic acid (10 μg
per ml) + biotin. Incubate all 30 plates at 35°C for three days. After this
period there will be much background heterokaryotic growth but the
homokaryotic recombinants which are selected by these two media will
be well established on the surface of the media and easily distinguishable.
Count the number of the homokaryotic colonies which grow on each plate
and classify them according to conidial colour. Typical results, summed over
the 20 plates in the case of the medium supplemented with adenine and biotin

and over the ten plates in the case of the medium supplemented with $p$-amino-benzoic acid and biotin, are given in table 5.17.

**Table 5.17**

| Plating medium | Total colonies | | Total spores plated |
|---|---|---|---|
| | Yellow | Green | |
| Adenine–biotin medium | 16 | 25 | $6 \times 10^6$ |
| $p$-Aminobenzoic acid–biotin medium | 223 | 1176 | $6 \times 10^3$ |

Isolate all colonies which grow on the medium supplemented with adenine and biotin and inoculate them on to master plates of the same medium for use with the needle replicator. After incubation replicate these colonies on to medium M, medium M supplemented with adenine and medium M supplemented with biotin, and from the resultant growth responses determine the genotypes of each colony. The genotypes of the 41 colonies listed in table 5.17 and which were selected on the adenine biotin medium are given in table 5.18. The cross used in this experiment, together with the various detectable regions of crossing over, can be represented as follows:

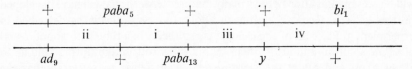

All of the genotypes listed in table 5.18 are of course $p$-aminobenzoic acid independent, that is they result from a crossover in region i. The positions of the minimal number of crossovers required to explain the origin of the 41 $p$-aminobenzoic acid independent colonies are also indicated in table 5.18. The most frequent genotype recovered is $ad_9 \; y^+ \; bi_1$. This is consistent with the order of the sites of mutation in this cross being as indicated above. If the relative order of the two *paba* alleles were the reverse then the largest group of

**Table 5.18**

| Genotype | | | Frequency | Crossover positions | | |
|---|---|---|---|---|---|---|
| $ad_9$ | + | $bi_1$ | 18 | i | | |
| + | + | $bi_1$ | 3 | i | ii | |
| $ad_9$ | $y$ | + | 8 | i | iii | |
| $ad_9$ | + | + | 3 | i | iv | |
| + | $y$ | + | 5 | i | ii | iii |
| + | + | + | 1 | i | ii | iv |
| $ad_9$ | $y$ | $bi_1$ | 3 | i | iii | iv |

$p$-aminobenzoic acid independent progeny would be expected to have the genotype $ad^+\ y\ bi^+$.

In this experiment the selection procedure used (namely the selection only of $p$-aminobenzoic acid independent recombinants) has resulted in the recovery of half of the recombinants. The reciprocal recombinant class carrying both mutant *paba* alleles remains undetected. Thus, if all of the ascospores plated had resulted from cleistothecia of hybrid origin the frequency of recombination between the two mutant sites at the $paba_1$ locus would be equal to $2b/m$, where $b$ is the number of selected $p$-aminobenzoic acid independent colonies and $m$ is the total number of spores plated on the adenine–biotin medium.

But as was seen in experiments 3 and 9, not all of the spores plated in this experiment would have been derived from cleistothecia of hybrid origin. In order to estimate the proportion which were, the known recombination frequency between the $ad_9$ and $y$ markers of $0 \cdot 16$ can be used. Thus if all the ascospores had been derived from hybrid cleistothecia this frequency of $0 \cdot 16$ is equal to $2a/n$ where $a$ is the number of yellow colonies selected on the $p$-aminobenzoic acid biotin medium and $n$ is the number of spores plated on that medium. Therefore the estimate of the proportion of ascospores actually derived from hybrid cleistothecia is $2a/(n \times 0 \cdot 16)$ and the recombination frequency between the pair of mutant sites at the $paba_1$ locus, $q$, is equal to $(b \times n \times 0 \cdot 16)/(m \times a)$. In the present example $n = 6 \times 10^3$, $m = 6 \times 10^6$, $a = 223$ and $b = 41$, and the frequency of recombination $q = 2 \cdot 94 \times 10^{-5}$. Thus, in this experiment the relative order of two mutant sites within a single gene has been determined together with an estimate of the frequency of recombination between them.

### Experiment 13. Uniparental inheritance of a cytoplasmic character

Species: *Neurospora crassa*
Strains: *OR A*. Wild-type. Mating-type $A$
  *mi-1 nic-2 a*. Slow growth. Requires nicotinic acid. Mating-type $a$

Prepare conidial suspensions of both strains and inoculate them separately on to slopes of medium W supplemented with nicotinic acid (10 µg per ml) in large tubes (17·5 mm diameter) and also containing strips of filter paper as described in experiment 4. Incubate at 25°C for seven days when protoperithecia will be seen as aggregates of hyphae on the mycelia of both strains. Fertilize each of these mycelia by pouring a suspension of conidia of the second strain in each case into the tube and out again. Continue incubation at 25°C. After one day the perithecia will be seen to be developing rapidly and after ten to 14 days they will be mature and will have shed many ascospores on to the sides of the tubes.

Collect and germinate a sample of ascospores and then collect a random sample of the resultant developing colonies all as described in experiment 4. Take a sample of at least 30 from both the ♀ *OR A* × ♂ *mi*-1 *nic*-2 *a* cross where *OR A* is the protoperithecial parent and from the reciprocal ♀ *mi*-1 *nic*-2 *a* × ♂ *OR A* cross. Transfer each of these sexually produced strains to a slope of medium VM supplemented with nicotinic acid in a small (10 mm diameter) tube. Incubate at 28°C for five days and classify each of the resultant mycelia as normal or 'poky'. The mutant form is readily distinguished by its small amount of growth and sparse conidiation as compared to the normal.

Finally, test each of these mycelia for the segregation of the *nic*-2 and mating-type alleles. Inoculate conidial suspensions of each mycelium on to slopes of medium VM to test for the presence of the *nic*-2 allele, and on to slopes of medium W, supplemented with nicotinic acid and already inoculated with a tester strain of either mating-type in order to determine mating-type (see experiment 4). In an experiment where 32 of the progeny from each reciprocal cross were tested the results given in table 5.19 were obtained.

**Table 5.19**

| *OR A* × *mi*-1 *nic*-2 *a* | | *mi*-1 *nic*-2 *a* × *OR A* | |
|---|---|---|---|
| 'Poky' progeny | 0 | Normal progeny | 0 |
| Normal *nic*+ *A* | 10 | 'Poky' *nic*+ *A* | 8 |
| Normal *nic*-2 *A* | 5 | 'Poky' *nic*-2 *A* | 7 |
| Normal *nic*+ *a* | 7 | 'Poky' *nic*+ *a* | 10 |
| Normal *nic*-2 *a* | 10 | 'Poky' *nic*-2 *a* | 7 |

Thus, in this experiment the 'poky' mutant was inherited entirely in a maternal manner. At the same time alleles at the two chromosomal loci, *nic*-2 and mating-type, segregated in a Mendelian fashion. These two loci are in fact located on opposite arms of linkage group I but this loose linkage cannot be detected here.

## 4. GENETIC ANALYSIS BY WAY OF THE PARASEXUAL CYCLE

Following the association of unlike haploid nuclei in a heterokaryon (see experiment 2) a sequence of events, which has been termed the 'parasexual cycle', may occur. This sequence is initiated by the fusion of a pair of unlike nuclei to give a heterozygous diploid nucleus. During their subsequent mitotic divisions the diploid nuclei undergo recombination and re-assortment and ultimately haploid nuclei emerge carrying new combinations of the genes for which the original haploid nuclei differed. The outcome is therefore similar to that of the sexual cycle in that it leads to genetic recombination and segregation. This is particularly convenient in the imperfect fungi (see experiment 1) where the parasexual cycle provides a method for the genetic analysis of such species. Each stage in the parasexual cycle occurs with a low frequency and selective methods are required to facilitate this method of analysis. For a more detailed discussion of parasexuality see Fincham & Day (1971, Chapter 5).

## Experiment 14. The parasexual cycle

Species: *Aspergillus nidulans*
Strains: $pyro_5$. Green conidia. Requires pyridoxine
$an_1$ $pro_1$ $paba_6$ $y$ $bi_1$. Yellow conidia. Requires thiamin, proline, p-aminobenzoic acid and biotin

The linkage relationships between these markers as found by meiotic analysis are as follows:

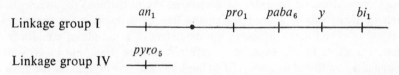

Linkage group I

Linkage group IV

Synthesize the heterokaryon between these two strains on medium M as described in experiment 2. Prepare a dense suspension of conidia from the heterokaryotic growth and adjust to $10^6$ spores per ml. Plate 1 ml aliquots of this suspension, using the technique described in experiment 12, in a surface layer of soft agar on plates of medium M. Incubate at 35°C for three or four days when there will be a background of resynthesized heterokaryotic growth and, in addition, a number of green spored rapidly growing colonies. These will be the diploid colonies which have grown from the rare heterozygous diploid spores which have formed on the heterokaryon. In the experiment from which results will be quoted here 23 diploid colonies were produced per $10^7$ conidia plated. The genotype of this diploid can be expressed as follows:

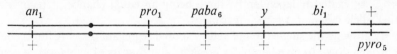

Purify one of these diploids by spreading a low density conidial suspension on medium M and isolating one of the resultant colonies and keep as the stock diploid. Measure the relative conidial diameter of a sample of conidia from the diploid and from both parent haploid strains by the use of a micrometer eyepiece. The mean conidial diameter of the diploid strain will be found to be significantly larger than that of the haploid strains.

Prepare a suspension of conidia of the diploid strain and dilute to $3 \times 10^2$ spores per ml. Spread 0·1 ml aliquots per plate of this suspension on two series of media: (a) medium M supplemented with thiamin, proline, p-amino-benzoic acid, pyridoxine (all at 10 µg per ml) and biotin (1 µg per ml) and p-fluorophenylalanine or FPA (60 µg per ml); (b) this same fully supplemented medium without the FPA. Prepare 20 plates of the first medium and 40 of the second in this way and incubate both sets of plates at 35°C for seven days.

### (1) *Growth on medium with FPA*

The colonies which grow on the medium with FPA will be inhibited, growing slowly and with much brown pigmentation. Most colonies will be producing fairly vigorous sectors. Isolate a sample of yellow spored sectors, taking only one sample from any one colony, and inoculate on plates for use as master plates for the needle replicator. Replicate on to the differential range of supplemented media in order to determine the nutritional requirements of each of these yellow spored segregants (see experiment 9 for the general method involved here). Measure the diameters of a sample of conidia from each segregant. In the experiment referred to here a sample of 60 yellow segregants was found to consist of 59 haploids and one diploid on the basis of conidial diameter.

The 59 haploid segregants were found to fall into only two genotypic classes:

$an_1$   $pro_1$   $paba_6$   $y$   $bi_1$   $pyro^+$     28 segregants;
$an_1$   $pro_1$   $paba_6$   $y$   $bi_1$   $pyro_5$     31 segregants

These results demonstrate that the result of treatment with FPA is to break down the diploid colony to give haploid segregants. By selecting yellow haploids only all of the linked markers are also recovered. There was no segregation for any of the linkage group I markers. The unlinked marker, $pyro_5$, segregated independently, there being an equal chance of recovering either allele ($\chi^2_{(1)} = 0·153$; $P = 0·70$).

By growing the single diploid yellow spored segregant on FPA it too could be broken down to give haploid segregants and thus its exact genotype could be determined. By this method the genotype of this diploid was found to be as follows:

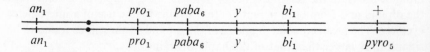

This diploid segregant is what is known as a non-disjunctional diploid, and it probably arose in the following manner. At some stage in the growth of the diploid colony an aneuploid nucleus arose in which chromosome I was represented three times as follows:

$$
\begin{array}{lllll}
an_1 & pro_1 & paba_6 & y & bi_1 \\
an_1 & pro_1 & paba_6 & y & bi_1 \\
+ & + & + & + & +
\end{array}
$$

Aneuploids in *A. nidulans* are very unstable and the subsequent loss of the chromosome containing the wild-type alleles left a diploid homozygous for all the markers on linkage group I.

## (2) *Growth on medium without FPA*

The diploids which grow on the series of plates without the FPA will be normal in appearance. They will be found to be almost entirely green spored except for single conidial heads, or patches of conidial heads, in which the spore colour is yellow. From each colony take one sample of yellow spores from one of these areas using a fine glass needle. Stab these samples into plates of medium M supplemented with thiamin, proline, *p*-aminobenzoic acid, biotin and pyridoxine. After incubation most of these inocula will give rise to mixed yellow and green colonies because of the difficulty of isolating single heads without contamination by the surrounding conidia. However, with a second isolation at this stage it should be possible to obtain purified yellow segregants. Use the needle replicator to inoculate this sample of yellow segregants on to the differential range of supplemented media to determine their nutritional requirements, and measure a sample of conidia from each to determine the ploidy. (With very little experience a separation into probable haploids and diploids can be made on morphological grounds.)

   In the experiment referred to a sample of 90 yellow segregants was divided into 19 haploids and 71 diploids. The genotypes of the haploids and the probable genotypes of the diploids are given in table 5.20. To be certain of the genotypes of the diploids each should be haploidized by the use of FPA and the segregation among these haploids examined. However, the labour involved would be too great for a class, and the genotypes of the diploids as given in table 5.20 are the most probable based on the simplest explanation possible assuming mitotic recombination. The position of the crossover is indicated in each case. The presence of the heterozygous $an_1$ gene in the case of the two mycelia with the whole of the right arm of linkage group I homozygous shows that they originate from a crossover event and are

H

not non-disjunction diploids. This also illustrates how the parasexual cycle can be used to locate centromeres. The single mycelium homozygous for the $pyro_5$ gene either had had a second crossover event in that linkage group between the allele and the centromere, or was a non-disjunction diploid for that linkage group.

**Table 5.20.**

| Genotype | Frequency |
|---|---|
| | |

In this experiment the main features of the parasexual cycle have been illustrated. The use of FPA to induce the breakdown of the diploid to haploids in which, in *A. nidulans*, linkages are rarely broken illustrates the basis of the method which is widely used in order to establish linkage group relationships in this species. The recovery of diploid mitotic recombinants shows how the parasexual cycle can also be used to establish the order of linked genes. From table 5.20 it can be seen that, of the crossover events which were selected, 35·2 per cent occurred between the $paba_6$ and $y$ genes, 62 per cent between $pro_1$ and $paba_6$ and 2·8 per cent between the centromere and $pro_1$. The following mitotic map giving the order of the linked genes used and the relative mitotic map distances where known can thus be drawn:

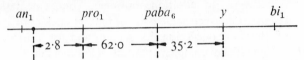

This map should be compared with the meiotic map obtained for this region in experiment 9. The order of the genes is always found to be the same by both methods but the relative frequencies of recombination between them may vary.

## 5. GENETIC CONTROL OF METABOLISM

Although it was in the fungi that the genetic control of biochemical reactions was first demonstrated, it is the bacteria which now provide the most convenient organisms for these studies (see chapter 6). However, fungi, and particularly *Neurospora crassa*, can be used to provide clear demonstrations of some of the principles of biochemical genetics. If we have a biosynthetic pathway A→C and two mutants, $M_1$ and $M_2$, which block this pathway as indicated,

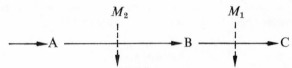

then mutant $M_1$ will grow if substance C is added to the culture medium. That is, the block between B and C has been bypassed. However, during this growth the first part of the pathway, →A→B, functions and the substance B accumulates in the culture medium. Similarly mutant $M_2$ will grow if substances B or C are added to the medium, and substance A accumulates. This accumulation may be detected by chemical or physical means, or by a bioassay method. Experiments 15 and 16 illustrate examples of some of the effects of mutations in biosynthetic pathways.

In experiment 12 a method of demonstrating that there are many sites of mutation within a gene was described. The use of the phenomenon of complementation between mutants of entirely different functional systems, such as between *p*-aminobenzoic acid and pyridoxine requiring auxotrophs, has been made extensively in these experiments. For example, these auxotrophs were used in experiments 2 and 14 for the selection of heterokaryons and diploids in *Aspergillus nidulans*. In experiment 17 complementation between strains of *Saccharomyces cerevisiae* carrying mutant alleles at two different genes both of which are concerned with the biosynthesis of adenine is demonstrated, thus showing the independence of function of these genes in this synthetic pathway. In addition, however, complementation is also demonstrated to occur between some pairs of mutant alleles of the same gene. For further information concerning this section and for a discussion of the significance of interallelic complementation see Fincham & Day (1971), chapters 7, 9 and 10.

### Experiment 15.   Genetic control of a biosynthetic pathway; (a) Arginine

Species: *Neurospora crassa*

Strains: *arg-1 a*

   *arg-3 a*

   *arg-5 a*. Three mutant strains, each carrying a mutation resulting in a requirement for arginine, but in a different gene in each case. All of mating-type *a*

Prepare conidial suspensions of all three strains and inoculate each on to slopes of medium VM, and on to medium VM supplemented separately with arginine, ornithine and citrulline (all at 150 μg per ml). Ornithine and citrulline are intermediates in the arginine biosynthetic pathway. After incubation at 25° for two days classify each tube for the presence or absence of growth. The results which will be obtained are given in table 5.21.

### Table 5.21

| Mutant | Medium | | | |
| | VM | VM + ornithine | VM + citrulline | VM + arginine |
| --- | --- | --- | --- | --- |
| *arg–5* | − | + | + | + |
| *arg–3* | − | − | + | + |
| *arg–1* | − | − | − | + |

From these results it is clear that these three mutants are each responsible for a different step in the biosynthesis of arginine. The three mutants block the synthetic pathway as follows:

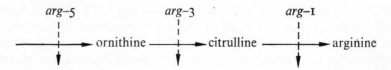

## Experiment 16.  Genetic control of a biosynthetic pathway; (b) Tryptophan

Species: *Neurospora crassa*

Strains: The following eight tryptophan requiring mutant strains are available:

*tryp*-1 A9,  *tryp*-1 A10,  *tryp*-1 A20
*tryp*-2 A80
*tryp*-3 A78,  *tryp*-3 td71,  *tryp*-3 A72
*tryp*-4 A35

Grow the required strains on slopes of medium VM supplemented with 150 µg per ml of tryptophan to produce conidiating cultures and prepare and wash conidial suspensions of each strain. (In the description which follows the properties of all eight strains will be discussed.)

### Growth on intermediary metabolites

This test is carried out in 10 ml of liquid medium in 50 ml conical flasks. Prepare one flask of each of the following media for each strain: medium VM, medium VM supplemented with anthranilic acid (10 µg per ml), medium VM supplemented with indole (10 µg per ml) and medium VM supplemented with tryptophan (20 µg per ml). Inoculate each flask with one drop of the appropriate conidial suspension and incubate at 25°C for three days. After this period examine each flask for the presence or absence of growth. The results obtained should be as indicated in table 5.22.

**Table 5.22**

| Mutant | Medium | | | |
|---|---|---|---|---|
| | VM | VM + anthranilate | VM + indole | VM + tryptophan |
| *tryp*-1 A9 | — | — | + | + |
| A10 | — | + | + | + |
| A20 | —* | —* | + | + |
| *tryp*-2 A80 | — | + | + | + |
| *tryp*-3 A78 | — | — | + | + |
| td71 | —* | —* | —* | + |
| A72 | — | — | — | + |
| *tryp*-4 A35 | — | — | + | + |

+ = Growth.   — = No growth.   * = Slight growth due to leaky mutant.

### Accumulation of intermediary metabolites

Continue the incubation of the flasks for a total period of seven days. The limited quantities of the various supplements which had been added will have been completely used up during this period and growth will have ceased.

*Accumulation of a substance fluorescent in ultraviolet light*

Examine all the flasks for the presence or absence of a bluish fluorescence when placed under an ultraviolet lamp. At the same time examine a set of uninoculated flasks which had been incubated for the full duration of the experiment as controls. The strength of the fluorescence can be estimated on an arbitrary scale and the results obtained should be as given in table 5.23.

Table 5.23

| Mutant | Medium | | | |
|---|---|---|---|---|
|  | VM | VM + anthranilate | VM + indole | VM + tryptophan |
| *tryp*–1 A9 | o | 3 | o | o |
| A10 | o | o | o | o |
| A20 | 2 | 3 | 4 | 4 |
| *tryp*–2 A80 | o | o | o | o |
| *tryp*–3 A78 | 1 | 3 | 4 | 4 |
| td71 | 2 | 3 | 2 | 3 |
| A72 | 1 | 3 | 1 | 4 |
| *tryp*–4 A35 | 1 | 3 | 4 | 4 |
| Uninoculated | o | 3 | o | o |

o = No fluorescence.  1 = Very weak.  2 = Weak.  3 = Strong.  4 = Very strong.

These results should be compared to the growth responses given in table 5.22. The pattern of accumulation of the fluorescent substance is clearly seen in the flasks which had originally been inoculated with tryptophan and indole and this is repeated in the VM flasks, even in those cases where visible germination of the conidia had not occurred. The fluorescence in the uninoculated flask to which anthranilic acid had been added shows that this substance is itself fluorescent and the pattern of fluorescence in the other flasks to which anthranilic acid had been added reflects those strains which were able, and those which were not able, to utilise this substance.

The main fluorescent substance which is being accumulated in these cultures is, in fact, anthranilic acid. Proof of this identity of the fluorescent substance would require its extraction from the medium and purification and this will not be attempted in the present experiment. However a comparison of the paper-chromatographic behaviour of the fluorescent accumulate with pure anthranilic acid and other fluorescent substances would be instructive. (Compare the $R_F$ values obtained using a 2 : 1 solution of *n*-propyl alcohol and 1 per cent ammonia, and a 2 : 1 solution of *n*-propyl alcohol and 1 per cent acetic acid as solvents.)

## Accumulation of indole

Carry out the following test on the residual medium contained in each flask which had originally been supplemented with tryptophan. Take 0·5 ml of the residual medium and place in a tube. Add 1 ml of Ehrlich's reagent (5 per cent p-dimethylaminobenzaldehyde in 95 per cent ethanol) and 8·5 ml of acid alcohol (8 ml conc. HCl in 92 ml of 95 per cent ethanol). Mix and leave for 15 min. The development of a purple colour indicates the presence of indole.

In the case of the present eight mutant strains a fairly strong yellow colour will be present in the medium of all those which accumulate anthranilic acid and the purple colour, indicating the accumulation of indole, will be present only in the case of mutant tryp-3 td71. (It will also be noticed that this flask has a strong characteristic smell of indole.)

## Growth of the mutant strains on culture accumulates

Remove the mycelium from the eight flasks which had originally contained tryptophan and sterilize the remaining medium by passing it through a 'Millipore' filter or by autoclaving at 120°C for 10 min. Prepare 64 tubes (17·5 mm diameter) each containing 4 ml of liquid medium VM. To each of eight of these add 1 ml of the sterile extract from the tryp-1 A9 culture. Similarly, to each of a further eight tubes add 1 ml of the extract from the tryp-1 A10 culture, and so on. Finally, add one drop of freshly prepared conidial suspension of the various tryp mutants to each tube so that each of the eight mutants is inoculated into medium containing extracts from all eight mutants. Incubate at 25°C for three days and examine for the presence or absence of growth. The results which will be obtained are given in table 5.24.

**Table 5.24**

| Inoculum | Extract | | | | | | | |
|---|---|---|---|---|---|---|---|---|
| | A9 | A10 | A20 | A80 | A78 | td71 | A72 | A35 |
| tryp-1 A9 | — | — | — | — | — | (+) | — | — |
| A10 | — | — | + | — | + | + | + | + |
| A20 | —* | —* | —* | —* | —* | (+) | —* | —* |
| tryp-2 A80 | — | — | + | — | + | + | + | + |
| tryp-3 A78 | — | — | — | — | — | (+) | — | — |
| td71 | —* | —* | —* | —* | —* | —* | —* | —* |
| A72 | — | — | — | — | — | — | — | — |
| tryp-4 A35 | — | — | — | — | — | (+) | — | — |

+ = Growth. (+) = Reduced amount of growth. — = No growth.
—* = Slight growth due to leaky mutant.

This particular pathway is complicated and for a full explanation refer to Fincham & Day (1971), and to DeMoss, Jackson and Chalmers (1967) for references to the original work. However, a summary of the enzymes affected by the various mutants and the steps affected by these enzymes is given in fig. 5·6. A list of the steps affected by each of the mutant strains used in this experiment and the substances shown to be accumulated by each are given in table 5·25. The complex effects of the various *tryp-1* mutants are due to three

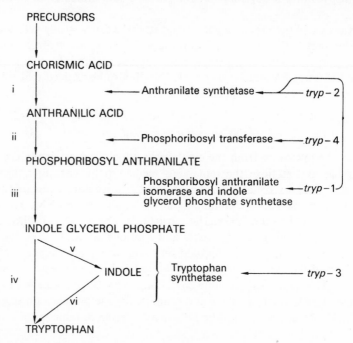

**Fig. 5.6.** Pathway of tryptophan synthesis showing the various enzyme activities affected by the four *tryp* genes in *Neurospora crassa*.

**Table 5.25**

| Mutant | Affected steps (see fig. 5.6) | Substances shown to be accumulated |
|---|---|---|
| *tryp-1* A9 | i and iii | |
| A10 | i | |
| A20 | iii | Anthranilic acid |
| *tryp-2* A80 | i | |
| *tryp-3* A78 | iv and v | Anthranilic acid |
| td71 | iv and vi | Indole and anthranilic acid |
| A72 | iv, v and vi | Anthranilic acid |
| *tryp-4* A35 | ii | Anthranilic acid |

enzyme activities being located on an aggregate of the products of both the *tryp*-1 and *tryp*-2 genes. Thus all *tryp*-2 mutants and some *tryp*-1 mutants are deficient in anthranilate synthetase activity. Note that anthranilic acid is accumulated by *tryp*-3 and some *tryp*-1 mutants as well as by the *tryp*-4 mutant. This is possibly due to the occurrence of the reverse reaction until equilibrium is reached.

### Experiment 17. Interallelic complementation

Species: *Saccharomyces cerevisiae*

Strains: *ad*-1. Requires adenine. Accumulates red pigment.

*ad*-2. Six independent mutants at this locus. All require adenine and accumulate a red pigment. The mutants available are *ad*-2.0, *ad*-2.1, *ad*-2.6, *ad*-2.7, *ad*-2.R4 and *ad*-2.R8. All seven mutant alleles are required in strains with both the *a* and α mating-types

Grow all 14 strains on plates of medium CM. (This medium is used in this experiment because it promotes a strong red pigmentation.) Using a small loop streak the seven strains with the *a* mating-type in parallel straight lines about 7·5 mm apart on one plate of medium CM. Streak the seven strains of mating-type α in a similar way across a second plate. Incubate at 28°C for two or three days until all streaks have produced good confluent growth. Then, using a pad, replicate one of these streaked plates on to two plates of medium CM. Finally, using a fresh pad each time, replicate the second streaked plate containing the other mating-type on to the same two plates of medium CM, but with the streaks at right angles to the first set. In this way the 49 possible crosses (including 'selfs' and 'reciprocals') are made at the intersections. Incubate at 28°C for three or four days.

After incubation the lattice of replicated streaks will have grown and will have a red pigmentation. At each intersection haploid strains of *a* and α mating-types are mixed together and diploids are formed. If the two mutant alleles in any one of these combinations do not complement each other then the diploid cells formed are still adenine requiring and pigmented. However, in each combination where complementation occurs the diploid has a wild phenotype and is white in colour. The results which will be obtained are given in table 5·26. In order to demonstrate that the diploids produced at the white intersections are in fact prototrophic replicate one of these plates on to medium YM. Incubate at 28°C for about four days when it will be seen that the only growth on this medium has resulted from the white cells on the CM plates. (The *ad*-2.R8 mutant is leaky and will grow slightly in this test.) From the results given in table 5.26 a complementation map of the *ad*-2 gene can be

**Table 5.26**

| a strains | α strains | | | | | | |
|---|---|---|---|---|---|---|---|
| | ad–1 | ad–2·0 | ad–2·1 | ad–2·6 | ad–2·7 | ad–2.R4 | ad–2.R8 |
| ad–1 | − | + | + | + | + | + | + |
| ad–2·0 | + | − | − | − | − | − | − |
| ad–2·1 | + | − | − | − | + | + | + |
| ad–2·6 | + | − | − | − | − | − | + |
| ad–2·7 | + | − | + | − | − | − | − |
| ad–2.R4 | + | − | + | − | − | − | + |
| ad–2.R8 | + | − | + | + | − | + | − |

− = No complementation.  + = Complementation.

constructed. In this map non-complementing mutants are represented by overlapping lines.

ad-2.0
ad-2.6
ad-2.7
ad-2.1
ad-2.R4
ad-2.R8

# 6. GENETICS OF HOST–PARASITE RELATIONSHIPS

One of the most obvious economic importances of fungal genetics concerns host–parasite relationships and the problems of disease resistance and sensitivity. This is a subject which is relatively difficult to demonstrate under class conditions because most pathogenic fungi are difficult to culture under artificial conditions. However, it is possible to maintain the potato blight fungus, *Phytophthora infestans*, in culture for a considerable period and also to grow it easily on detached potato leaves. In the following experiment the part played by single gene mutations, both in the host and parasite, in the control of disease resistance is demonstrated. For a discussion of variation in pathogenicity see Fincham & Day (1965), chapter 11.

## Experiment 18. Genetic control of virulence

Species: *Phytophthora infestans*
Strains: *Phytophthora* races $p_1$, $p_4$ and $p_{1.4}$
    Potato varieties $R_1$, $R_4$ and $R_{1.4}$

Grow cultures of the *Phytophthora* strains on plates of medium R at 18°C until they produce many sporangia. Add sterile water to these cultures and agitate to make dense suspensions of sporangia and transfer these to tubes. These sporangial suspensions may be used as the inocula in this experiment or, if preferred, a zoospore suspension could be used. To obtain zoospores place the suspension of sporangia at 11°C for about 6 to 10 hours. After this period most of the sporangia will have liberated the biflagellate motile zoospores.

Place clean dry detached leaves of the various potato varieties on wet cotton-wool covered with gauze in flat dishes (for example in photographic developing dishes). Then, with a Pasteur pipette, place drops of the suspensions of sporangia or zoospores on the leaves so that each race of fungus is inoculated on to leaves of each variety of potato. The whole experiment should be carried out at least in duplicate. Cover the dishes with sheets of glass and incubate at 18°C with continuous illumination for up to ten days. During this period the cotton-wool upon which the leaves are lying should be kept moist by the daily addition of sterile water. The leaves should be inspected each day and the development of an active growth of the fungus on each leaf should be recorded as it occurs.

The growth of *Phytophthora* on these leaves will reproduce the typical field symptoms of potato blight. The fungus will first grow into the leaf causing a black colouration. Later, the mycelium will produce many sporangiophores which grow mainly from the lower side of the leaf. Secondary bacterial infection will rapidly follow and rot the infected leaf. Where there is no attack and the leaves used were undamaged they will remain healthy until the end of the experiment. The results obtained in the present experiment with these races and varieties will be as given in table 5.27.

#### Table 5.27

| Potato | *Phytophthora* | | |
|--------|:---:|:---:|:---:|
| | $p_1$ | $p_4$ | $p_{1.4}$ |
| $R_1$ | + | − | + |
| $R_4$ | − | + | + |
| $R_{1.4}$ | − | − | + |

+ = Attack.   − = No attack.

## VI. REFERENCES

ALEXOPOULOS C.J. (1962) *Introductory Mycology*, 2nd Edn. Wiley, New York.

DeMoss J.A., Jackson R.W. & Chalmers J.H. (1967) Genetic control of the structure and activity of an enzyme aggregate in the tryptophan pathway of *Neurospora crassa*. *Genetics*, **56**, 413–424.

FINCHAM J.R.S. & Day P.R. (1965) *Fungal Genetics*, 2nd Edn. Blackwell, Oxford.
FINCHAM J.R.S. & DAY P.R. (1971) *Fungal Genetics*, 3rd Edn. Blackwell, Oxford.
RAPER J.R. (1966) Life cycles, basic patterns of sexuality, and sexual mechanisms. In
    Ainsworth G.C. & Sussman A.S. (eds), *The Fungi*, Vol. 2, pp. 473–511. Academic
    Press, New York.
WHITEHOUSE H.L.K. (1949) Heterothallism and sex in the fungi. *Biol. Rev. Cambridge
    Phil. Soc.*, **24**, 411–447.

# CHAPTER 6

# BACTERIAL AND BACTERIOPHAGE GENETICS

R. C. CLOWES

## I. INTRODUCTION

A series of experiments to illustrate those basic genetic concepts which are most suitably demonstrated with bacteria and their phages should cover the following topics.

1.  General methods and techniques, maintenance and growth of pure cultures (single colony plating, viable cell and total cell counts), preparation of phage lysates and assay by overlay method.
2.  Isolation of auxotrophs by penicillin selection.
3.  *Luria-Delbrück* fluctuation test.
4.  Transformation in *B. subtilis* with linked markers.
5.  Induced enzyme synthesis and genetic regulation in *lac* genes of *E. coli* following transfer by conjugation.
6.  Conjugation in *E. coli* K12
    (a) F factor transfer
    (b) Hfr crosses and zygotic induction
    (c) Mapping by interrupted mating.
7.  Physiology and genetics of T-phages
    (a) Adsorption and assay of phage
    (b) One step growth curve
    (c) Phage crosses (intercistronic and intracistronic (*rII*))
    (d) Complementation of *rII* and 'amber' mutants (spot tests).

This chapter attempts to cover the basic techniques required for such experiments.

## II. GENERAL METHODS AND TECHNIQUES

### 1. STERILITY

One essential feature of working with micro-organisms is to develop a sterile technique, i.e. to ensure that the various operations are carried out avoiding accidental contamination by extraneous micro-organisms which float about in the air and which are present on all non-sterilized surfaces. This is achieved by the use of growth media which are sterilized before use (see appendix 6.1 and section III, 1, a), by working with sterile apparatus (see III, 1, b) and by keeping such media and apparatus from subsequent contamination with unwanted micro-organisms. In developing this technique, it is useful to assume that all exposed surfaces are contaminated and the contaminating micro-organisms are falling as a light rain from above.*

All media and sterile material is therefore uncovered and exposed for minimum periods. Pipettes used for transfer and dilution are sterilized, usually in batches in cans (see III, 5, b). They are handled only at the mouth end. Care is taken that the tip does not touch any non-sterile surface. Sterile liquid media is dispensed into sterile capped bottles or tubes. To transfer samples, the container is usually held in the left hand and the pipette in the right between thumb and first three fingers.† The cap is then removed by, and retained between the fourth (small) finger and the palm, the sample is taken into the pipette, the cap is immediately replaced and the sample in the pipette is then transferred. The pipette is finally discarded into a jar filled with water. A similar procedure of uncapping and retaining the cap is followed if an inoculum is taken by a wire loop or needle.

For transfer of an arbitrary number of bacteria from one medium to another, a wire set in a holder is often used. This wire is usually 25 to 50 mm long and is made of metal resistant to oxidation.‡ It may either be straight, when it is used to 'stab' into the depth of media, or turned at the end into a small loop and used either to inoculate solid media by drawing it gently across the surface or to transfer small volumes of culture. The wire is heated to red heat in a bunsen burner before each operation. On removal it will cool in a matter of seconds and can then be used. After use, it is again reflamed.

To avoid repetition, the words 'sterile' and 'sterilization' have been omitted from most of the text. It should be assumed that all solutions, media and glassware used in each experiment are sterile.

## 2. MEDIA PREPARATION AND PLATE POURING

The composition of the various media are shown in appendix 6.1 (pp. 280–286). Solid constituents of media are usually dissolved and made up to

* All operations requiring sterile precautions can with advantage be performed in the vicinity of a bunsen flame where convection currents reduce the 'fallout' of micro-organisms.
†During the period when the container is without the cap, it should be held in an inclined position so that any 'fallout' micro-organisms come to rest in the neck. The neck is then passed through a bunsen flame before recapping.
‡ 'Nichrome' or platinum wire (24 SWG) is most suitable.

volume in bulk quantities, dispensed in appropriate-sized screw-capped (s/c) bottles (see appendix 6.5) and sterilized by steam under pressure.* (IMPORTANT: Screw-caps should be loosened before pressure is applied. They are tightened on cooling, and the sterile media can then be stored in this condition for indefinite periods at room temperature.)

To pour 'plates', agar medium, if taken from store, is melted by a short exposure to steam at 121°† and, where necessary, the other pre-warmed ingredients are added and mixed.‡ The medium (at about 60–70°) is then dispensed into Petri dishes removing the lids only at the time of pouring. Pour about 25 ml per plate for growth of micro-organisms as separate colonies, or about 40 ml for phage and other experiments when confluent growth of micro-organisms as a 'lawn' is required. Bubbles may be removed by bunsen-burner flame. The dish lids are then replaced leaving a small vent until the agar has set. Excess condensation is removed (the plates are 'dried') by incubation in an inverted position with the lids removed in a warm room for about an hour at 37° or, in an oven, set about 60° for about an hour with oven doors open to permit air circulation. Alternatively, plates can be dried by incubation overnight in the inverted position with the lids on. Many problems arising from contamination may be avoided if agar plates (particularly those used in 'spreading' techniques) are used the same day, or no more than one day after pouring, storing for short periods if necessary in a refrigerator or cold room.

## 3. PRESERVATION OF BACTERIAL CULTURES

Stock cultures of bacterial strains are preserved as *stabs* in 'stab' medium (appendix 6.1 (22)) in small, tightly capped tubes which can be hermetically sealed, as, for example, by a plastic stopper,§ kept at room temperature in the dark.

A sterile straight wire, touched to a well-characterised bacterial colony of the required strain is then 'stabbed' to the bottom of the tube, which is incubated overnight before storage. From these master 'stab' cultures, *nutrient agar slopes* (or slants) (appendix 6.1 (21)) are inoculated (usually by

* Media can alternatively be made up in large Erlenmayer flasks and used immediately after sterilization (e.g. agar medium can be dispensed into plates before it gels so as to avoid subsequent 'melting'), or capped with foil and stored for a limited period.
† Temperatures are expressed in degrees centigrade throughout.
‡ WARNING: Care should be taken to allow molten agar medium heated under pressure to cool to at least 90° before addition of other components. Agar remains superheated for some time and if liquids are added at this time, or if the contents are agitated, a spray of very hot agar may be produced which can cause severe burns, particularly to the face.
§ e.g. ½ dram (1·8 ml) vials Camlab (Glass) Cambridge are suitable.

drawing a wire loop, touched to the surface or depth of the 'stab' growth, several times across the surface of the 'slope') and similarly incubated overnight before storage. Single colonies of the strains are then isolated from these slopes by streaking on a plate of nutrient agar (*stock plate*) (see section II, 6), each colony originating from a single bacterial cell.

## 4. PRESERVATION OF PHAGE PREPARATIONS

Phages are usually preserved as suspensions in liquid, either as lysates in broth or more preferably in a non-nutrient medium such as phage buffer (appendix 6·1 (17)). They are conveniently stored in small capped tubes (such as those used for bacterial 'stabs' above) at 4° to restrict growth of microorganisms. Some workers prefer to add a drop of chloroform to each tube, as a further precaution against contamination, although of course this precludes the use of plastic tubes. After storage for long periods it is advisable to assay such preparations before use.

## 5. PREPARATION OF LIQUID-GROWN BACTERIAL CULTURES

### (a) 'Overnight' cultures

For many experiments, the starting material is an overnight (o/n) bacterial culture. An inoculum from a single colony on the stock plate is taken by a sterile loop into medium* and incubated overnight (for about 18 hours at 37°) without aeration, when the concentration of cells will usually have reached about $10^9$ per ml.

### (b) Exponential or logarithmically growing (log) cultures

From an o/n culture, a dilution is made into fresh medium which is then aerated at 37° to achieve a cell density of approximately $2 \times 10^8$/ml. Two techniques are commonly employed. The culture may be diluted into a small

---

*For most experiments, an inoculum is conveniently made into 5 ml broth (appendix 6.1 (1)) in a 15 ml screw-capped bottle. This method has the advantage that the culture can then be centrifuged at low speeds (3000 rev/min) in the same bottle if the cells require 'washing' (see III, 4), and may then be resuspended in a different medium. However, some workers prefer to take their inoculum direct from a slope (slant) without recourse to the intermediate stock plate. The stock-plate method, however, has the advantage that the purity of the original culture may be seen from the uniformity in the morphology of the colonies and, moreover, any contamination of this plate is more readily noticed.

s/c bottle and incubated on a 33 rev/min rotor (see III, 3, b), a 1 in 10 dilution in broth requiring about 1½ hours and 1 in 50 dilution about 2½ hours. Alternatively, the culture can be diluted into a bubbler tube and aerated by bubbling air from an aeration pump (see III, 3, a), this method being a little more efficient. Here again, the use of s/c bottles permits subsequent centrifugation without transferring to a centrifuge tube.

## 6. STREAKING FOR SINGLE COLONIES

A sterile wire loop dipped into a bacterial culture, or touched on a bacterial colony, is rubbed several times across the surface of part of a plate of solid medium (fig. 6.1A).* The wire is sterilized by flaming and when cool, it is drawn half a dozen times at right angles across the original streaks (B). It is again sterilized and the cross-streaks repeated at (C) and again at (D). After appropriate incubation of the plate (with the agar layer uppermost, to avoid condensation) (18 hours at 37° for most strains on nutrient agar and 24 hours on minimal agar), growth of the individual cells will give rise to independent colonies of several mm diameter (usually in D, if the original inoculum was dense). If the original culture was pure, all the colonies will be similar in appearance although colonies that are well separated from others tend to be larger.

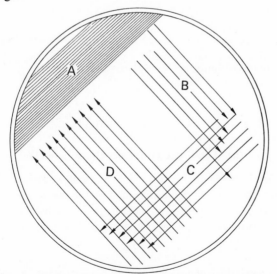

Fig. 6.1. Streaking a plate for single colonies of bacteria.

*If the colony is a recent one, do not remove too much growth, bearing in mind that each colony contains more than $10^8$ cells. With colonies on older plates, many cells will be dead and a larger inoculum may be needed.

## 7. DILUTIONS

Most bacterial cultures and phage suspensions are grown and used at high concentrations (bacteria $10^8$ to $10^9$ cells/ml; phages, $10^7$ to $10^{11}$ particles/ml). To count the individual cells (by their ability to form colonies—see II, 8), or the individual phage particles (by their ability to form plaques—see II, 9), considerable dilution is usually necessary. This is conveniently achieved by a series of small dilutions (either tenfold or hundredfold); for example, a 1 in a million ($10^{-6}$) dilution is obtained by three successive 1 in 100 ($10^{-2}$) dilutions. Each dilution may be conveniently performed by transferring 0·1 ml by sterile pipette into 10 ml of sterile liquid in a sterile tube; after swirling to mix, a *fresh*, sterile pipette is used for the next dilution.

Standard dilutions may be achieved as follows:

$$\left( \frac{0·1}{0·9} \text{ indicates } 0·1 \text{ ml into } 0·9 \text{ ml} \right)$$

$$10^{-1} = \frac{0·1}{0·9}$$

$$10^{-4} = \frac{0·1}{10} \times \frac{0·1}{10}$$

$$10^{-7} = \frac{0·1}{10} \times \frac{0·1}{10} \times \frac{0·1}{10} \times \frac{1}{9}$$

$$10^{-2} = \frac{0·1}{10}$$

$$10^{-5} = \frac{0·1}{10} \times \frac{0·1}{10} \times \frac{1}{9}$$

$$10^{-8} = \frac{0·1}{10} \times \frac{0·1}{10} \times \frac{0·1}{10} \times \frac{0·1}{10}$$

$$10^{-3} = \frac{0·1}{10} \times \frac{1}{9}$$

$$10^{-6} = \frac{0·1}{10} \times \frac{0·1}{10} \times \frac{0·1}{10}$$

$$10^{-9} = \frac{0·1}{10} \times \frac{0·1}{10} \times \frac{0·1}{10} \times \frac{0·1}{10} \times \frac{1}{9}$$

Dilutions are made in buffer unless otherwise specified.

## 8. ASSAY OF BACTERIAL CELLS*

### (a) 'Total' bacterial cells ('total' count)

The number of bacteria (living or dead) present in suspensions with cell densities greater than $10^7$/ml can be counted by the use of a counting chamber (Helber or Petroff–Hauser) (see appendix 6.5) viewed by dark-field or phase-contrast microscopy. All the cells in the culture, whether viable or not, are counted by this technique.

*A very rough approximation of dense suspensions of bacteria may be made by estimating the turbidity of the culture. The smallest concentration of bacteria such as *Escherichia coli* showing just perceptible turbidity is about $5 \times 10^6$ cells per ml. Overnight cultures will grow to about $5 \times 10^8$ cells per ml and a culture vigorously aerated for about 24 hours reaches its maximum turbidity of about $2 \times 10^9$ cells per ml.

## (b) Colony-forming bacteria ('viable' count)

A liquid culture is diluted so that its cell density is about 1 to 5 × 10³/ml. Samples of 0·1 ml may then be assayed either

(i) *By spreading:* a 0·1 ml drop is placed by pipette on an agar plate and is spread over the surface using a bent glass rod (a 25 mm right-angled bend on a 3 mm diameter glass rod) which has been pre-sterilized by flaming after immersion in alcohol. After 18 hours incubation, each viable cell will produce a colony. (With the organisms used in these experiments, the size will be 2–3 mm on nutrient media or about 1 mm on minimal media.) or

(ii) *By overlay:* the 0·1 ml sample is added to 2·5 ml of soft agar (appendix 6.1 (11)) in a small tube, held molten at 46° in a waterbath. The whole contents of the tube are then poured on the surface of an agar plate, which is gently rocked to distribute it as a thin layer over the entire surface. Thin disc-like colonies (1–2 mm diameter in nutrient media) are produced within the soft-agar overlay which may be counted after 24 hours incubation.


## 9. STANDARD ASSAY FOR BACTERIOPHAGE

The most widely used method for assaying phage stocks is the agar-layer method. A small volume (0·05–0·5 ml) of a phage suspension (about 200 particles) is pipetted into a small tube, containing 2·5 ml of molten 'soft' agar and about 2 × 10⁸ bacteria which are sensitive to the phage (*indicator* bacteria). (These indicator bacteria are conveniently added as 3–4 drops from an o/n broth culture using a 0·1 ml pipette. The agar is maintained in the molten state by holding the tube in a 46° waterbath.) The mixture of phage, indicator bacteria and soft agar is then poured over the surface of a nutrient agar plate (i.e. 'plated'), and the plate is gently rocked to distribute the mixture evenly over the surface as a thin layer (overlay). When the top layer has solidified (10–15 min), the plate is inverted and placed in an incubator at 37° and incubated overnight. The next day, a continuous lawn of bacterial growth will be seen over the surface of the agar, except where a clone of phage particles has lysed the bacteria, and produced a visible clear area or 'plaque'. In the case of the phages and host bacteria used in these experiments, it has been shown that a single phage particle will produce a plaque. The number of plaques, therefore, gives a direct assay of the number of phage particles put on the plate.

The titre of a phage preparation may then be expressed as a concentration of 'plaque-forming units' (pfu), equivalent in the case of the 'T' phages to the actual concentration of particles.

## 10. PREPARATION OF LYSATES OF VIRULENT ('T') PHAGES*

About 20 ml of phage-broth (appendix 6.1 (15)) in a bubbler tube is inoculated with 0·5 ml of an o/n broth culture of the bacterial host strain (*E. coli* B (EMG 31) for T phages) and aerated by bubbling at 37° in a waterbath for about 2½ hours, until a concentration of about $1 \times 10^8$ cells/ml is reached.† Phage is then added at a concentration of $10^9$ pfu/ml, so that all bacteria are infected. Aeration is continued for 2–3 hours when the turbid culture will often (but not always) 'clear' and a final titre of $10^{10}$ to $10^{11}$ pfu/ml will usually be achieved. Several drops of chloroform are added, shaken to disperse and the lysate is then spun at low speed in a small, angle centrifuge to sediment bacterial debris. The clear supernatant is decanted and several more drops of chloroform are added to it to kill residual bacteria. After incubation at 37° for about ½ hour, the chloroform is allowed to settle and the top, aqueous layer decanted off. For further concentration, the phage may be sedimented as a pellet, after high-speed centrifugation (12000 *g* for 1 hour) which can then be taken up in a small volume of phage buffer.

## III. APPARATUS AND FACILITIES REQUIRED

It is realised that facilities will vary extensively, but the following list of apparatus is compiled with a view to general economy and convenience.

## 1. STERILIZATION

(a) *Steam sterilization* under pressure at from 109° to 126° is required for sterilizing media and other labile materials. For large classes, an autoclave is an obvious advantage, but for most purposes a large domestic pressure cooker will serve equally well.

---

* These details apply to the production of large amounts of high-titre phage suspensions. Where the amount of phage available as a starting inoculum is less than required here, it may be necessary for a pre-preparation of the inoculum. This may be achieved by inoculating a small volume (0·5 ml) of log host bacteria with a lower concentration of phage particles and continuing aeration for several hours before centrifuging and assaying. The initial phage inoculum should always be pre-assayed by the agar overlay method. If its titre is very low, a new stock may be prepared from a single phage plaque. The plaque is stabbed with a straight wire which is immediately rinsed in a vigorously aerated 0·5 ml log culture of the host bacteria in phage-broth. After 2–3 h further incubation, a fairly high titre stock should be produced, which after assay can be used in the preparation of high-titre lysates as above.

† Measured by total count.

(b) *Dry heat sterilization* is needed for pipettes, tubes and other glass and metal apparatus, and is achieved in a thermostatic oven operating up to 180° (preferably with time-switch). Exposure to 160° for 2 hours is satisfactory for most purposes. A similar oven may be used for drying plates.

## 2. INCUBATION

(a) *Thermostatic waterbaths*—a minimum of two per class, operating at 37° and 46° are required.

(b) *Thermostatic incubators*—generally operating at 37° are required for incubation. A thermostatic hot-room is an advantage for large classes, and is particularly useful for drying plates.

## 3. AERATION

(a) *By bubbling* is achieved by small aquarium aerators. The piston type is more robust than the moving diaphragm type (see appendix 6.5).

(b) *By rotation* of capped bottles is sometimes more convenient. A rotor inclined at 45° and circulating at 33 rev/min is suitable for this purpose. (A suitable machine is commercially available (see appendix 6.5) or may be constructed from a gramophone turntable with attached 'Terry' clips.) Screw-capped bottles should be filled only to a third of capacity to contain sufficient air for suitable growth of the culture.

## 4. CENTRIFUGATION

Some experiments require that the cells in a culture medium be separated from the liquid, and then resuspended in fresh medium. This is most efficiently achieved by sedimenting the cells by spinning the culture in a small bench centrifuge, centrifugation for 10–15 min at 3000–5000 rev/min being usually sufficient. The centrifuge is then allowed to come gently to rest (to avoid resuspending some of the pelleted bacteria), the container is carefully removed, uncapped and the supernatant liquid gently decanted off, the container being retained in the inverted state for a short time to allow proper draining. A fresh volume of sterile liquid may then be added and the cells resuspended. If this first addition is buffer, and the resulting suspension is recentrifuged and the buffer decanted off—the cells are said to have been 'washed'. The cells may then be resuspended in the final medium after one or more such washes. Cells may be concentrated by resuspending in a smaller final volume than the original.

For concentrating bacteriophage, an ultracentrifuge capable of reaching 10000–12000 rev/min is necessary but this facility is not essential.

## 5. GLASSWARE

(a) *Petri dishes* of glass can with advantage be replaced with plastic, pre-sterilized disposable dishes. A 90 mm dish is suitable for most experiments.

(b) *Pipettes.* A selection of 0·1 ml, 1 ml and 10 ml pipettes are required for most experiments, a maximum number of about 30 × 0·1 ml, 50 × 1 ml and 50 × 10 ml per group being sufficient for any single experiment. After use, they are well rinsed, dried and sterilized in cans by dry heat. The micro-organisms used in the experiments are all of non-pathogenic groups, e.g. *Escherichia coli.* Accordingly, the use of unplugged pipettes for dispensing all cultures (saving much time in cleaning and preparing pipettes for re-use) is now common practice, and is considered a safe procedure.

(c) *Test-tubes.* Two sizes only are required for most purposes, a 150 × 16 mm ('large') and a 75 × 13 mm ('small'). They are conveniently 'capped' with aluminium slip-on caps (see appendix 6.5) and can then be sterilized in racks by dry heat. (For some phage preparations a larger (150 × 22 mm) test-tube is useful. This can be converted into a 'bubbler' by the use of a 5 mm internal diameter (i.d.) delivery tube,* plugged lightly with cotton-wool at one end passing through 10 mm hole bored in a slip-on cap, which is lined with a rubber grommet. The bubblers should be steam sterilized, condensation in the cotton-wool being avoided by pre-wrapping the plugged end in aluminium foil.)

(d) *Bottles.* A wide range of screw-capped (s/c) bottles with rubber liners in 500 ml, 100 ml, 25 ml and 10 ml sizes is very convenient for making up media and growing cultures. The smaller sizes can also be used for centri-fuging at low speed. They are sterilized by steam under pressure.

## 6. OTHER APPARATUS

(a) *Colony or plaque counts.* Many experiments require the counting of many hundreds of bacterial colonies or plaques. A hand-tally counter is the most economical, but a more sophisticated electronic device which may be pur-chased (see appendix 6.5) or constructed (see appendix 6.4), has obvious advantages.

(b) *Agitator or blender* is required for experiment 10—see note 1b (p. 272).

---

* A suitable tube with a constriction which prevents the sterile plug from slipping is a commercially available pasteur pipette (appendix 6.5).

(c) *Spectrophotometer*. A standard model nephelometer or other device capable of measuring differences in turbidity and colour, either by transmission or reflection is required.

## 7. GROUP APPARATUS

In addition to the specialised requirements set out at the beginning of each experiment, each group of workers requires several small items of equipment for most experiments. These include:
(i)   two inoculating wires, one straight and one with an end loop,
(ii)  bunsen burner (a type with a side lever, permitting flame to be turned down to a pilot flame is most suitable,
(iii) small container of alcohol (95 per cent) and a glass spreader,
(iv)  sterile pipettes in 0·1 ml, 1·0 ml and 10 ml sizes,
(v)   glass-marking pencil or pen.
A list of supply houses for many specialized items of equipment is provided in appendix 6.5.

## IV. THE EXPERIMENTS

The experiments are set out in several sections.

The first section states the purpose and gives some background to the experiment and often includes relevent reading references.

The second section, headed REQUIREMENTS, indicates the day-to-day materials required for each group of workers (most conveniently one or two persons). In addition to these materials, each group requires several small items of equipment (as shown in III, 7). The composition of the various media is shown in detail in appendix 6.1.

The next section, headed METHODS, sets out the day-to-day operations for the experiment in detail.

Finally in some experiments, a series of CONCLUSIONS or NOTES underline the essential features of the data obtained. All temperatures are expressed in degrees Centigrade throughout. A standard temperature of 37° is used unless otherwise stated.

All appropriate apparatus and media for the experiments should be provided sterile. Details of items of equipment are given in section III and of media in appendix 6.1.

### Bacterial strains

Standard genetic symbols are used to define the bacterial strains used (see Demerec *et al.* (1966) *Genetics*, **54**, 61). This system uses triletter italic symbols for genetic loci; *ind, leu, met, pro, thi, thr* and *tyr* indicating genes controlling the ability of a bacterial strain to synthesize indole, leucine, methionine, proline, thiamin (vitamin $B_1$), threonine or tyrosine, respectively. (A minus superscript indicates a mutant gene with a requirement for one of these compounds, and a plus superscript indicates the active 'wild-type' gene with the ability to dispense with the requirement.) *lac* indicates a gene controlling the ability of the cell to ferment lactose (a plus superscript referring to the active 'wild-type' state and a minus superscript the mutant, non-fermenting state).

*str* stands for a gene controlling resistance to streptomycin, *str-r* being the resistant and *str-s* the sensitive state.

(In all cases similar roman triletters with initial capitals are used as abbrevations for the compound when added to media.)

T1, T2, T4 and T6 are 'T' coliphages. T2-*s* indicates a bacterial strain sensitive to this phage and *T2-r* a resistant mutant (sometimes shown as /2).

λ is a temperate phage. λ-*s* indicates a bacterial strain that is sensitive to this phage, λ-*r* a resistant strain. (λ)$^+$ indicates a strain of *E. coli* in which the bacteriophage is established in the latent, prophage state (a lysogenic strain). (λ)$^-$ shows a similar strain where the prophage is not present.

Lack of appropriate symbols implies the wild (unmutated) state. In *E. coli*, the wild strain is independent of organic growth requirements (amino acids, vitamins, purines, or pyrimidines), it has the ability to ferment many carbohydrates ((e.g. lactose (*lac*), maltose (*mal*), galactose (*gal*), arabinose (*ara*), xylose (*xyl*),) it is sensitive to all common T phages and many drugs (e.g. streptomycin (*str*), azide (*azi*) ) and carries the λ prophage (λ)$^+$.

The bacterial strains used are given numbers with an 'EMG' prefix throughout the text. These numbers refer to the strain list in *Experiments in Microbial Genetics* (eds Clowes, R.C. & Hayes, W. (1968) Blackwell: Oxford) where fuller details of the origin and derivation of these strains can be found.

## EXPERIMENT 1. PURE CULTURES OF BACTERIA: THEIR MAINTENANCE, ASSAY AND CHARACTERIZATION

The purpose of this experiment is to provide experience in maintaining a pure culture of a single bacterial strain, to assay and characterize it.

### (a) Requirements

*Part I*

NA plate (stock plate) streaked for single colonies of the
*E. coli* K12 strain 58–161 (*met⁻lac⁺*) (EMG 1)

Overnight broth culture of strain EMG 1
Seven Nutrient Agar (NA) plates
One Minimal Medium (MM) plate
One MM plate supplemented with methionine (MM + Met)
One EMB + lactose (EMB Lac) plate
100 ml buffer
Six large tubes

## (b) Method

### Part 1 ($\frac{1}{2}$ h)

(1)  Prepare six dilution (buffer) tubes (2 × 9 ml, 4 × 10 ml).
Mark the backs of the 2 MM plates and the EMB plate in quadrants and number each 1–4. Number six of the nutrient plates 1–6.
(2)  Dilute the broth culture provided to $10^{-5}$ using three of the dilution tubes (see II, 7).
(3)  Repeat a similar dilution with the three remaining tubes.
(4)  From one of the final dilution tubes, take three 0·1 ml samples and spread each over the surface of a NA plate (see II, 8, b (i)) (plates 1–3).
(5)  From the duplicate final dilution tube, spread 0·1 ml samples, each over one of the three remaining plates (plates 4–6).
(6)  With a wire loop, touch one of the colonies on the stock plate and streak the remaining NA plate (plate 7) to obtain separate colonies (see II, 6).
(7)  With the wire loop, touch another colony on the stock plate and streak several times across quadrant one of the unsupplemented MM plate. Without sterilizing the wire, streak in turn across quadrant one of both the MM + Met and the EMB Lac plates.
(8)  Sterilize the wire and repeat for three further colonies from the stock plate, streaking in turn on quadrants 2, 3 and 4 on the same three plates.
(9)  Incubate all plates overnight.

### Part 2.   Day following Part 1 ($\frac{1}{2}$ h)

(1)  Examine the MM, MM + Met and EMB Lac plates. No growth should be seen on the unsupplemented MM plate, but lines of bacterial growth following the inocula should be seen on the two other plates on all four quadrants.

Note:
(a)  Growth on MM + Met is translucent while on EMB Lac it is blue-black.
(b)  Growth on MM + Met but not on unsupplemented MM shows that the strain requires methionine for growth.
(c)  The coloured growth on EMB shows that the organism can ferment the carbohydrate in the plate, in this instance lactose (a non-fermenting strain would produce a pinkish-coloured growth).

(d) MM and MM + Met are *selective media* and permit growth only of bacteria with specific growth requirements (MM permits growth only of organisms able to grow without an amino acid, vitamin, purine or pyrimidine or other complex organic supplement; MM + Met permits growth only of these same organisms plus those that specifically require a methoinine supplement).

(e) EMB Lac is a *differentiating medium*. All *E. coli* strains will grow on this medium, those fermenting lactose producing blue-black coloured growth, those unable to ferment lactose producing a pinkish coloured growth.

(f) Growth on all quadrants of each plate is similar, showing each colony picked has the same properties as the others (the stock plate contains a 'pure' culture).

(2) Examine the NA plates 1–6 spread with the $10^{-5}$ dilutions of the broth culture. Note that each contains isolated colonies of similar size and appearance.

(3) Count the number of colonies on each of the plates and thus assay independently from each of the two dilution series the number of bacterial cells in the original broth culture.

Note that plates 1–3 should show approximately the same numbers of colonies. However, certain small variations due to chance errors in sampling are unavoidable. If the dilutions were carried out correctly, plates 4–6 should also show similar numbers of colonies again with similar variations between plates 4–6 as between plates 1–3 due to statistical variation between samples.

(4) The remaining NA plate (7) should show separated colonies resembling the stock plate.

**Note to instructors**

Several different types of K12 strains may be distributed, one to each group, providing of course that the appropriately supplemented MM plates are made available. It would be advantageous to use several different auxotrophs of K12 (e.g. EMG 8, 9, 10 and 28) including some lactose-fermenting and some lactose non-fermenting strains.

## EXPERIMENT 2. THE LURIA AND DELBRÜCK FLUCTUATION TEST

When bacteriophage is added to a turbid culture of bacterial cells which are sensitive to the phage, after a few hours the culture becomes clear due to multiplication of the virus inside the cells which ultimately burst (or lyse) to release the particles of phage progeny. After further incubation, the culture may again become turbid due to the growth of bacteria which are resistant to the bacteriophage.

Two hypotheses have been advanced to account for the origin of these resistant variants:

(1) *The adaptation hypothesis*, according to which every cell has a small probability of being induced by the phage to adapt itself so that it can survive and grow in the presence of the phage, this adaptation then being passed on to its descendants.

(2) *The spontaneous mutation hypothesis*, which states that every cell has a small probability of mutating during its life-time from phage-sensitivity to phage-resistance, whether phage is present or not. The progeny of such a resistant cell will also be resistant unless back-mutation occurs.

Luria & Delbrück (1943) devised *the fluctuation test* to decide between these two hypotheses. An important difference between the two alternative ideas is that, according to the adaptation hypothesis, the bacterial population is homogeneous before the phage is added. Whilst, according to the mutation hypothesis, the population is not homogeneous, since mutation to resistance may occur at any time during the growth of the culture before the phage is added. The number of bacteria resistant to the phage will thus depend upon whether the first mutation to phage-resistance occurred early or late in the growth of the culture.

Thus, according to the adaptation hypothesis the probability of any bacterium becoming resistant after contact with the phage should be the same for all the bacteria in the culture. The adaptation hypothesis therefore predicts that there will be no large fluctuations in the numbers of resistant bacteria from culture to culture in a parallel series to which phage is added. The fluctuations should in fact be no greater than those encountered in a series of samples all taken from the same culture.

On the other hand, according to the hypothesis of spontaneous mutation, the time of occurrence of a mutation in a series of parallel cultures will be subject to random variation. Cultures in which a mutation occurs early will contain large numbers of resistant cells, while cultures in which mutation occurs late will contain very few resistant cells. The mutation hypothesis thus leads to the prediction that there will be larger fluctuations in the numbers of resistant mutants from culture to culture in a parallel series, than from a series of samples taken from the same culture.

The experimental test therefore consists in determining the numbers of resistant bacteria present in two series of samples, one series from parallel cultures and the other series taken from the same bulk culture.

The experiment carried out by Luria and Delbrück measured resistance to phage T1 in *E. coli* B, and the variance calculated from the number of resistant bacteria in a series of ten samples from the same culture was found to be 54, while the variance for the samples taken from ten parallel cultures was 3498. Clearly, the variance between parallel and independent cultures was greater than the variance between samples taken from the same culture, indicating that the spontaneous mutation hypothesis is correct.

The fluctuation test has been applied to investigate the origin of a large number of bacterial variants in several bacterial species. The characters investigated include resistance to bacteriophages of various types, resistance to various antibiotic drugs (e.g. streptomycin, penicillin, sulphonamide) and to radiation, independence of growth factor requirements, and the ability to ferment various carbohydrates. In each case the variants have shown a 'clonal' distribution indicative of mutation.

**References**

LURIA S.E. & DELBRÜCK M. (1943) *Genetics*, **28**, 491.

### (a)  Intention of experiment

The experiment is designed:
(1)   to apply the fluctuation test to determine the origin of resistant variants to phage T6 in *E. coli* K12 and,
(2)   to measure the mutation rate from T6–sensitivity to T6–resistance.

N.B. Some cultures of *E. coli* contain variants which produce large mucoid colonies with a slimy, glistening appearance due to the production of quantities of polysaccharide material. These bacteria are genetically sensitive to phage T6 but they are able to form colonies in the presence of T6 because the polysaccharide material prevents the phage from adsorbing to the cell wall. Thus, only those colonies which have the normal (i.e. non-mucoid) *E. coli* morphology should be scored when counting the numbers of T6–resistant clones.

### (b)  Requirements

*Part 1*

1 ml of an o/n broth culture of wild-type *E. coli* K12 (EMG 2) sensitive to phage T6 (*T6–s*).
100 ml broth
100 ml buffer
Four large tubes
Ten small tubes

*Part 2*

5 ml of phage T6 at a concentration of $5 \times 10^{10}$ particles/ml (see II, 10)
24 NA plates
100 ml buffer
Six large tubes

## (c) Method

*Part 1 ($\frac{1}{2}$ h)*

(1) Prepare one large tube of 10 ml broth
Ten small tubes of 1 ml broth
Three dilution tubes (2 × 10 ml, 1 × 9 ml)
(2) Dilute the bacterial culture $10^{-5}$
(3) Add 0·1 ml of this dilution to each of the ten small broth tubes. (These tubes comprise the parallel series of cultures.)
(4) Add 1 ml of this dilution to the 10 ml of broth in a large tube (this tube is the large volume culture).
(5) Incubate all the tubes overnight.

*Part 2. Day following Part 1 ($\frac{1}{2}$ h)*

(1) Prepare 6 × 10 ml dilution tubes
(2) Spread 0·2 ml of the T6 phage suspension on each of 20 NA plates. Spread the suspension evenly over the whole surface taking special care to ensure that it is spread out to the edges of the plate. Allow a few minutes for the plates to dry.
(3) Spread 0·1 ml from each of the ten small tubes in the parallel series, each on a separate plate seeded with the phage (separate pipette for each). Spread the sample evenly over the surface taking special care to *avoid* spreading it to the outer edges of the plate.
(4) Similarly spread 0·1 ml aliquots from the large tube on to each of the remaining ten plates previously seeded with phage T6 (separate pipette for each to avoid contaminating culture with T6).
(5) Pool the small cultures. Assay the number of viable bacteria by diluting $10^{-6}$ and spreading 0·1 ml on to each of two NA plates. Similarly assay the large culture.
(6) Incubate all the plates overnight.

*Part 3. Day following Part 2 ($1\frac{1}{2}$ h)*

(1) Count the number of T6–resistant colonies on each of the phage-spread plates. Calculate the variance for each series of samples separately.

$$\text{Variance} = \frac{\Sigma (x - \bar{x})^2}{n - 1}$$

$\bar{x}$ = mean of the observed numbers of T6–resistant colonies.
$x$ = observed number of T6–resistant colonies in each sample.
$n$ = number of observations, i.e. ten.
$\Sigma$ = the sum of all these values.
(2) Compare the two values obtained. Is the origin of phage T6–resistant variants best explained by the adaptation hypothesis or by the mutation hypothesis?*

* The significance of the difference can be determined by a variance ratio ($F$) test which is explained in most statistical textbooks.

(3) Count the number of colonies on the four assay plates (total viable bacteria). Calculate the number of bacteria per sample taken from the two sets of cultures. Calculate the mutation rate T6-s——→T6-r from the formula:

$r = aN_t \log_e (aN_t C)$ (Luria & Delbrück 1943).

$r$ = average number of mutants per sample.

$C$ = number of cultures or samples, i.e. ten.

$a$ = mutation rate.

$N_t$ = number of bacteria per sample.

Example: $r = 30$

$\qquad C = 10$

$\qquad N_t = 2 \times 10^8$

From the graph of $r$ plotted against $aN_t$ for various values of $C$ (see appendix 6.2), we see that when $r = 30$, $aN_t = 7.3$.

Therefore, $30 = a \times 2 \times 10^8 \log_e 73$

Therefore, $a = \dfrac{30}{2 \times 10^8 \times 4.295} = 3.5 \times 10^{-8}$

## EXPERIMENT 3. TRANSFORMATION

A few species of bacteria can undergo a genetic alteration simply by the uptake from the medium of molecules of DNA extracted from another strain of the same species that has a hereditable distinguishing feature. This effect, termed *transformation*, was discovered by Griffith in 1928 who observed the acquisition of virulence by an avirulent strain of a pneumococcus after exposure to the heat-killed virulent form. Avery, Macleod & McCarty (1944) discovered that the active agent in transformation, when chemically pure, was DNA. Transformation still provides the most convincing evidence that genetic information is carried by DNA rather than by protein. There has been a recent resurgence of interest in the complex processes involved in transformation because it offers a relatively direct way of studying the processes of recombination.

The events between the addition of the DNA to a bacterial culture and the expression of the newly acquired genetic trait can be divided into four stages:

1. The DNA is absorbed by the cell. At this time it can still be removed by washing and can be inactivated by the addition of the enzyme, deoxyribonuclease (DNase) to the culture.

2. The absorbed DNA becomes irreversibly bound. It is no longer susceptible to DNase in the medium.

3. The DNA begins the process of integration—the so-called 'eclipse phase' because little active transforming DNA can be isolated with the DNA of the recipient. (This is not true of *Haemophilus*).

4. Part of the donor DNA, possibly only one strand, is integrated, probably by substitution, into the genome of the recipient.

The three species in which transformation is well established are *Streptococcus pneumoniae* (in U.S., *Diplococcus pneumoniae*), *Haemophilus influenzae* and *Bacillus subtilis*. Although under the best conditions a rather larger fraction of the recipient cells of the first two species can be transformed, *B. subtilis* is the most useful for student demonstration. This species, unlike the others, is non-exacting, and will grow well in a simple, defined medium. It is, therefore, possible to use a wide range of mutants with simple auxotrophic markers. In contrast, the first two transformation systems have rather complicated growth requirements, and drug resistances are the only convenient markers.

### References

ANAGNOSTOPOULOS C. & SPIZIZEN J. (1961) *J. Bact.*, **81**, 741.
AVERY O.T., MACLEOD C.M. & McCARTY M. (1944) *J. exp. Med.*, **79**, 137.
GRIFFITH F. (1928) *J. Hyg.*, **27**, 113.
RAVIN A.W. (1961) *Adv. Genet.*, **10**, 61.

### (a) Intention of experiment

The experiment utilizes an *ind⁻ tyr⁻* doubly auxotrophic strain requiring indole and tyrosine, in which the two mutational sites are closely linked. Transformation of either, or of both, of these genes can be observed by selection for independence to either or both nutritional requirements. The experiment is designed to demonstrate three points:

(1) *Efficiency of transformation*. Approximately 1 in 1000 recipient cells can be transformed to independence of a given growth requirement.

(2) *Susceptibility to DNase*. The addition of DNase to the culture destroys the transforming activity of the DNA if added early, but not if added late in the experiment.

(3) *Genetic linkage*. The separation and purification of transforming DNA from bacteria leads to the breakdown of the bacterial genome (chromonema) into fragments. These are, however, large compared to the size of a gene, and the more closely two markers are linked genetically, the greater is the chance that they will remain on the same fragment of DNA. Over a wide range of DNA concentrations, the number of cells transformed for a single marker is a simple linear function of the concentration. To obtain double transformation of two *unlinked* markers, two separate molecules must be taken up and the chance of double transformation is therefore proportional to the square of the concentration. At a fixed, low concentration of DNA, the frequency of double transformants will therefore be only a small fraction of the cells transformed for a single marker. However, if the markers are sufficiently near

one another to be retained frequently on a single fragment of DNA, then the frequency of co-transformation will be close to that of transformation for a single marker. They are then referred to as linked markers.

In this experiment, the indole and tyrosine markers are sufficiently close for some 70 per cent of the *ind⁻tyr⁻* cells transformed for one marker to be transformed for the other by DNA from an *ind⁺tyr⁺* strain. (If DNA from an *ind⁺tyr⁻* strain is mixed with DNA from an *ind⁻tyr⁺* strain, to give the same final concentration, the frequency of double transformation to *ind⁺tyr⁺* is about 5 per cent of the frequency of transformation for a single marker of *ind⁺* or *tyr⁺*.)

The experiment requires two preliminary operations, the isolation of transforming DNA and the preparation of 'competent' cultures. The DNA isolated from two or three grams of cells provides sufficient material for many transformation experiments and retains its activity for long periods (if sterile). It need not, therefore, be prepared for each experiment and can be provided as a reagent from a previously prepared stock.

*B. subtilis* is only transformed at high frequency if subjected to a special regime which produces cells said to be 'competent'. Competent cultures are not so easy to prepare and some trial and error may be necessary in each laboratory before the best method is achieved.

### DNA extraction

Grow 1 litre of the appropriate strain of *B. subtilis* in nutrient broth to about $5 \times 10^8$ cells/ml, harvest by centrifugation and wash in cold ($4°$) saline (0·14M). Resuspend the pellet in 40 ml of cold saline (containing 0·001M EDTA (ethylene diamine tetra-acetic acid) per gram wet weight of packed cells)* and ensure that the pellet is well dispersed. Add crystalline lysozyme to a concentration of 0·2 mg/ml and incubate with gentle shaking at $37°$. After 5–10 min the cells lyse and the suspension clears but becomes very viscous. Add an equal volume of 90 per cent phenol. (It is preferable to use freshly distilled phenol and it is imperative that immediately before use the phenol is shaken twice with buffer (0·1M borate) to remove traces of acid.) Shake gently for 20 min so that the phases remain well dispersed. Spin at 5000 rev/min for 5 min. Remove as much of the aqueous (top) layer as possible without disturbing the interface, at which the denatured protein accumulates. Remove the dissolved phenol from the aqueous layer either by shaking twice for 1 min with 4 ml of ether in a separating funnel, or if possible, by dialysis overnight against a large volume (2 litres or more) of saline-EDTA. A second centrifugation (20 min at 10000 rev/min) is usually necessary to remove residual denatured protein. The DNA is then precipitated by adding an equal volume of 2-ethoxyethanol slowly down the side of the tube and

* *B. subtilis* cells from 1 litre at $5 \times 10^8$/ml have a wet weight of about 1 g.

swirling the vessel; the fibres of DNA are removed by winding them on a glass rod. Wash the fibres once with alcohol, drain on filter paper and redissolve in one-third the original volume of 0·01M NaCl and 0·001M EDTA and leave overnight.

Determine the DNA concentration. The optical density of native DNA is approximately 20 (at 260 nm (nanometre)) for 1 mg/ml. Adjust the concentration to between 1·5 and 2 mg/ml and then add one-tenth volume of 1M NaCl. If the solution is at all turbid, precipitate the denatured protein by spinning at 10 000 rev/min for 20 min. This preparation should be stored at 4°.

**Competent cultures**

Inoculate the growth medium (Subtilis Minimal Salts (SMS—appendix 6.1 (18)) + 0·5 per cent glucose + 0·02 per cent casein hydrolysate) supplemented with 20 μg/ml indole and 10 μg/ml tyrosine, with a single colony of the recipient strain and aerate gently overnight. Dilute into fresh medium to give an optical density, at 450 nm, of 0·08 O.D. units (in 10 mm cells). Aerate *very vigorously* (by bubbling). Follow the O.D. at half-hourly intervals. It should increase at about 0·4 O.D. units per hour. After about 3 hours as the optical density reaches 1·7–1·8, the rate of increase should decline and when it reaches 0·2 O.D. units per hour, dilute with an equal volume of pre-warmed 'starvation medium' (SMS + 0·5 per cent glucose without other supplements). Continue the vigorous aeration for 90 min. The culture should now be maximally competent and remain so for about another hour. At the expense of perhaps 80 per cent of the competent cells, these cultures can be kept at −20° for a week in 15 per cent glycerol. Add 1·5 ml of glycerol per 9 ml of culture and dispense in 0·9 ml aliquots and freeze quickly.

### (b) Requirements

*Per class of 20 groups*
Three solutions of DNA at a concentration of 100 μg/ml.
3 ml DNA 'A': DNA extracted from the prototrophic strain of *B. subtilis* EMG50 (*ind+tyr+*)
2 ml DNA 'B': DNA extracted from an indole-requiring *B. subtilis* strain EMG51 (*ind−tyr+*)
2 ml DNA 'C': DNA extracted from a tyrosine-requiring *B. subtilis* strain EMG52 (*ind+tyr−*)
5 ml DNase soln; containing 400 μg/ml dissolved in 0·2M MgSO$_4$

*Per group*
5 ml of competent culture of *B. subtilis* strain EMG53 (*ind−tyr−*) containing about 10$^7$ cells/ml
Six plates of Subtilis Minimal Agar (SMA—appendix 6.1 (20))

I

Eight plates of SMA + 10 μg/ml tyrosine (SMA + Tyr)
Eight plates of SMA + 20 μg/ml indole (SMA + Ind)
Two plates of SMA + 10 μg/ml of tyrosine + 20 μg/ml of indole (SMA + Tyr + Ind)
2 × 100 ml buffer
Four small tubes
Ten large tubes

### (c) Method

#### Part 1 ($1\frac{1}{2}$ h)

(1) Prepare 20 dilution tubes (2 × 10 ml; 18 × 9 ml)
   Label the small tubes A to D and place in 37° waterbath
   Label SMA plates 3, 4, 15, 16, 17 and 18
   Label SMA + Tyr plates 1, 5, 6, 7, 11, 19, 20 and 21
   Label SMA + Ind plates, 2, 8, 9, 10, 12, 22, 23 and 24
   Label SMA + Tyr + Ind plates, 13 and 14

(2) Follow protocol below

| Time (min) | Tube A | B | C | D |
|---|---|---|---|---|
| 0 | Add 0·05 ml DNA 'A' + 0·05 ml DNase | Add 0·05 ml DNA 'A' | Add 0·1 ml buffer | Add 0·05 ml of a 1 : 1 mixture, of DNAs 'B'+'C' |
| 15 | Add 0·9 ml competent culture | Add 0·9 ml competent culture | Add 0·9 ml competent culture | Add 0·9 ml competent culture |
| 35 | — | Add 0·05 ml DNase | — | Add 0·05 ml DNase |
| 45 / Plate and dilute | Spread 0·2 ml on: (i) SMA + Tyr at 10° diln. Plate 1 | Spread 0·2 ml on: (i) SMA at 10$^{-1}$ and 10$^{-2}$ dilns. Plates 3, 4 | Spread 0·2 ml on: (i) SMA + Tyr at 10° diln. Plate 11 | Spread 0·2 ml on: (i) SMA at 10°, 10$^{-1}$, 10$^{-2}$ and 10$^{-3}$ dilns. Plates 15–18 |
| | (ii) SMA + Ind at 10° diln. Plate 2 | (ii) SMA + Tyr at 10$^{-1}$, 10$^{-2}$ and 10$^{-3}$ dilns. Plates 5, 6, 7 | (ii) SMA + Ind at 10° diln. Plate 12 | (ii) SMA + Tyr at 10$^{-1}$, 10$^{-2}$ and 10$^{-3}$ dilns. Plates 19, 20, 21 |
| | — | (iii) SMA + Ind at 10$^{-1}$, 10$^{-2}$ and 10$^{-3}$ dilns. Plates 8, 9, 10 | (iii) SMA + Ind + Tyr at 10$^{-5}$ and 10$^{-6}$ dilns. Plates 13, 14 | (iii) SMA + Ind at 10$^{-1}$, 10$^{-2}$ and 10$^{-3}$ dilns. Plates 22, 23, 24 |

(3) Incubate all plates for 48 hours.

*Part 2.   Two days following Part 1 (1 h)*

Count colonies on all the plates

Note:

(a) Tube C is the control. Plates 11 and 12 should therefore show no colonies. Plates 13 and 14 are supplemented with both requirements of the competent culture and these plates will therefore give an assay of the numbers of cells in the competent culture.

(b) Tube B shows *linked transformation* of both *ind* and *tyr* markers. Colonies on plates 3 and 4 are transformants to *ind*+ *tyr*+; colonies on plates 5, 6 and 7 are *ind*+ *tyr*− in addition to *ind*+ *tyr*+ transformants; and on plates 8, 9 and 10, *ind*− *tyr*+ together with *ind*+ *tyr*+ transformants.

(c) Tube A shows that the active transforming agent is destroyed by DNase and no colonies should be seen on plates 1 and 2.

(d) Tube D shows independent transformation of *ind* and *tyr* markers. Colonies should be seen on plates 19 to 21 (largely due to *ind*+*tyr*− transformants) and also on plates 22 to 24 (largely *ind*−*tyr*+ transformants).

(e) Calculate the independent transformation frequency to *ind*+*tyr*− and to *ind*−*tyr*+ found from tubes B and D. Calculate joint (linked) transformation frequency to *ind*+*tyr*+ from tubes B and D.

(f) Why is the joint transformation frequency much higher from tube B?

# EXPERIMENTS 4–6. EXPERIMENTS WITH 'VIRULENT' BACTERIOPHAGES

The following experiments have been designed to acquaint the student with some of the main techniques used in research with virulent phage. Most experiments will be performed with the two closely related phages T2 and T4, which have as their host bacterium *Escherichia coli* strain B.

## (a) Standard assay for bacteriophage

The method used in these experiments is the agar-layer method detailed on p. 231.

## (b) Synchronous adsorption of bacteriophage

In many phage experiments it is essential to have a population of infected cells in each of which phage development has proceeded for the same length of time. The simplest method of achieving this is to adsorb the phage particles to bacteria which are temporarily non-metabolizing due to some imposed inhibition, and then to start phage development synchronously in all infected cells by suddenly releasing the inhibition. A convenient inhibitor is cyanide—the presence of concentrations of M/500 (conveniently by the addition of 0·04M solution of KCN) completely arrests phage development without interfering with adsorption, and a subsequent dilution at $10^{-4}$ removes all effects of the drug.

## Overnight cultures

Unless specifically stated otherwise, all overnight cultures in this section are prepared by inoculating into 10 ml of phage broth contained in a large 150 × 19 mm tube and incubating overnight at 37° without aeration.

## References

ADAMS M.H. (1959) *Bacteriophages*. Interscience, New York.
HAYES W. (1968) *The Genetics of Bacteria and their Viruses*, 2nd Edn. Blackwell Scientific Publications, Oxford.
STENT G.S. (1963) *Molecular Biology of Bacterial Viruses*. W.H. Freeman, London.

# EXPERIMENT 4
# PHAGE ASSAY AND PLAQUE MORPHOLOGY

The various stages involved in the attack of bacteria by bacterial viruses have been separated into several steps. The first step involves the attachment of the virus by its 'tail' to specific receptor sites on the bacterial cell wall. The DNA core of the virus is then injected into the cell, the protein coat remaining on the outside. The injected phage DNA then 'takes over' the metabolism of the cell and diverts it to the manufacture of many copies of the phage DNA genome, and also of the various components of its protein structure which are finally assembled together to form many hundreds of phage particles. A phage enzyme then lyses the bacterium from the inside allowing the phages to be liberated. In a liquid culture, these progeny phages may then attach to un-infected bacteria where the process is repeated for many cycles, leading in some instances to a killing of all the bacterial cells in the culture.

If a small volume of a liquid culture of host bacteria containing a hundred million cells or so is infected with a few hundred phage particles and im-mediately mixed with soft agar and then poured over a nutrient plate, the agar will set and immobilize the bacteria in the overlay. On incubation, each of the initially phage-infected cells will lyse and the progeny particles will infect the neighbouring cells. After many such cycles of infection, the cells within a small area will be killed. However, the majority of plated cells will grow into closely spaced microcolonies forming a confluent 'lawn' over the plate except for those areas in the vicinity of an originally infected cell where a zone devoid of bacterial growth will be seen; this area is termed a 'plaque'.

If all phage particles can infect and give rise to lysis, then the numbers of plaques is a direct measure of the number of phage particles originally added and thus provides a simple and accurate assay for counting phage particles.

In this experiment three different 'T' phages are used, together with their normal host strain *E. coli* B. The phages can most readily be distinguished by the morphology of the plaques that they produce. In addition, two other bacterial strains are provided, a mutant of *E. coli* B which has lost the receptors for T2 phage and a mutant of *E. coli* K12 (as *E. coli* B, normally sensitive to all T phages in the wild state) which, like the mutants investigated in experiment 2, has lost receptors specific for T6 phage and is thus T6-resistant.

It will be obvious that if phage infection is prevented, as for example by the use of a host strain which has mutated to lose the phage receptors, no plaques will be seen. By the use of these three strains as indicators, therefore, the three phages can more precisely be distinguished by the specificity of their cell-wall receptor sites.

## (a) Requirements

1 ml o/n phage broth cultures of the following strains:
EMG 31—*E. coli* B ('wild')
EMG 33—*E. coli* B *T2-r* (sometimes termed *E. coli* B/2)
EMG 1—*E. coli* K12 *T6-r*
0·5 ml of a preparation of ONE* of the phages-
   T1 (at $2 \times 10^8$ pfu/ml) T2 or T6 (both at $10^9$ pfu/ml)
Three TNA plates (see appendix 6.1 (16))
50 ml buffer
50 ml soft agar (molten at 46°)
Three large tubes
Nine small tubes

## (b) Method

### Part 1 ($\frac{1}{2}$ h)

(1)   Prepare $3 \times 10$ ml dilution tubes, $9 \times 2.5$ ml tubes of soft agar (maintain at 46°).
(2)   Dilute the phage preparation provided to $10^{-6}$.
(3)   Add to each of three soft agar tubes, 0·1 ml of EMG 31.
(4)   Add to each of three further soft agar tubes, 0·1 ml of EMG 33.
(5)   Add to each of the three remaining soft agar tubes, 0·1 ml of EMG 1.
(6)   Add to each soft agar tube, 0·2 ml of the diluted phage suspension and pour each carefully over the surface of a TNA plate, rock to spread as a layer and leave to set. Label plates using EMG 31 as an indicator 1–3. Label

* Different groups should be provided with different phages which, for added interest, can be presented unidentified.

plates using EMG 33 as an indicator 4–6. Label plates using EMG 1 as an indicator 7–9.

(7)   When overlays have set (10 min), incubate all plates overnight.

## Part 2.   Day following Part 1 ($\frac{1}{2}$ h)

(1)   Note whether there are plaques on all plates, or only on some, and thus identify the phage provided.

(2)   Compare the morphology of plaques produced by different phages used by other groups on the same indicator strains.

(3)   Count the plaques on all plates, which should give similar numbers irrespective of the indicator, within the error of sampling.

(4)   Estimate the titre of the phage provided.

## EXPERIMENT 5. PHAGE CROSSES

Several types of mutants having distinct phenotypic properties have been isolated in the phages T2 and T4. The mutant types used in these following experiments are:

(1) *'Host-range' mutants of phage T2*. As we have seen in experiment 4, bacterial mutant strains can be isolated which are resistant to phages and *E. coli* B/2 is such a strain resistant to phage T2. It is, however, possible to isolate from the 'wild-type' T2 phage, mutants which can infect and lyse the bacterial strain B/2. Such phage mutants have a more extensive host-range than the T2 wild-type, and are designated by the term T2*h*.

(2) *'Rapid-lysis' mutants of phages T2 and T4*. The plaques produced by wild-type T2 (and T4) have a clear centre surrounded by a turbid halo. Mutants have been isolated which produce larger and uniformly clear plaques. These are called 'rapid-lysis' or *r* mutants. Independent isolates of these *r* mutations are genetically distinct. The *r* mutants map in three distinct regions and are thereby classified as *rI*, *rII* and *rIII* mutants. Mutants of the *rII* group can be further distinguished from other *r* mutants by the property that they are unable to produce plaques on the lysogenic strain of *E. coli* K12 (carrying the λ prophage, i.e. $(\lambda)^+$ strain), here called $K(\lambda)$.

In a phage cross, sensitive bacteria are infected with two phage mutants which are genetically distinct, under conditions such that each cell is infected with at least one of *both* types of phages (multiple, mixed-infection). The phage progeny arising from these infected bacteria are formed of phage particles of the parental types, and also of recombinant phage particles possessing characters derived from each of the parents. In crosses where only two phage genetic markers are used, the percentage of recombinants in the progeny serves as an index of the proximity of the markers on the phage chromosome.

## References

Benzer S. (1955) *Proc. Nat. Acad. Sci. (Wash.)*, **41**, 344.
Hershey A.D. & Rotman R. (1949) *Genetics*, **34**, 44.

Two crosses will be performed in this experiment:

(1) *T2hrI* × *T2 'wild'*. The four possible genotypes arising in this cross can be distinguished phenotypically by plating with a 'mixed-indicator' which is a mixture of cultures of the bacterial strains *E. coli* B and *E. coli* B/2. A phage with the host-range mutation (T2*h*) can lyse both B and B/2, and its plaques are thus clear. However, the wild-type phage (T2*h*+) can lyse only the B cells in the mixed-indicator layer. The B/2 cells grow within a zone of lysis of B and thus produce a 'turbid' plaque. The plaque types that are seen are therefore as follows:

|  | Genotype | Plaque type |
|---|---|---|
| Parents | h rI | Clear, large |
|  | + + | Turbid, small |
| Recombinants | + rI | Turbid, large |
|  | h + | Clear, small |

(2) *T4rII(147)* × *T4rII(271)*. This cross is between two independently isolated *rII* mutants, which form 'r' plaques on *E. coli* B but neither of which can form plaques on *E. coli* K(λ). In this cross, as can be seen below, only the wild-type recombinant can be distinguished phenotypically from the parent phage. The number of such recombinants is easily determined by assaying the progeny directly on K(λ).

|  | Genotype | Plaque type on *E. coli* | |
|---|---|---|---|
|  | rII | B | K(λ) |
| Parents |  |  |  |
|  | 147 + | r (large) | –* |
|  | + 271 | r (large) | –* |
| Recombinants | 147 271 | r (large) | –* |
|  | + + | r+ (small) | r+ (small) |

* No plaques formed.

In addition to the strains B, B/2 and K12(λ) used in previous experiments, a streptomycin-resistant strain of B, B/S (sometimes referred to as *E. coli* B *str-r*) is used as an indicator. If streptomycin is added to a mixture of phage infected and non-infected streptomycin-sensitive *E. coli* B, both the infected and uninfected cells are killed. If a large number of streptomycin-resistant *E. coli* B (*B/S*) are now added, any phage particles that remained unadsorbed to the original *E. coli* B will now adsorb to *B/S* and can be measured if this mixture is now plated. This technique allows measurement of the actual numbers of particles adsorbed to the host *E. coli* B rather than the numbers added.

As a preamble to the experiment, strain B is subcultured in phage-broth in the usual way to produce a log culture at $10^8$ cells/ml. It is then centrifuged and resuspended in fresh, warm phage-broth of $\frac{1}{3}$ original volume containing M/250 KCN, so that cell concentration of B is now $3 \times 10^8$ cells/ml.

### (a) Requirements

*Host bacteria*—2 × 0·5 ml of a log culture of *E. coli* B ($3 \times 10^8$ cells/ml) (EMG 31) in phage-broth containing M/250 of KCN (see above).

*Indicator bacteria*—5 ml of overnight cultures of *E. coli* B (EMG 31), *E. coli* B/S (EMG 32), *E. coli* B/2 (EMG 33) and *E. coli* K12(λ) (EMG 2) in phage-broth. Take 2 ml B and mix with 1 ml B/2. This is the 'mixed indicator'— label this indicator 'M'.

*Phage*—a mixture of the two parental phages is prepared for each cross containing $2 \times 10^9$ particles per ml of each parental type

Cross A—T2*hrI* × T2 'wild'
Cross B—T4*rII(147)* × T4*rII(271)*
16 TNA plates
Two Strep TNA plates
2 ml (3 per cent) streptomycin solution
50 ml phage broth
100 ml phage buffer
100 ml soft agar (molten at 46°)
18 large tubes
18 small tubes

### (b) Method

#### Part 1 ( 2 h)

(1) Prepare: 8 × 10 ml; 6 × 9 ml buffer tubes
    4 × 10 ml broth tubes. Place in 37° bath
    18 tubes molten soft agar. Add 0·3 ml streptomycin solution to 2 of these tubes (label 17 and 18). Place in 46° bath
(2) Label TNA plates 1–16, Strep TNA plates 17–18
(3) Put the two tubes containing 0·5 ml host bacteria in 37° waterbath and label 'A' and 'B'. Leave for about 10 min
(4) During this time, assay the parental phage mixtures by diluting and plating, *viz.*

Cross A—$10^{-7}$, 2 × 0·4 ml with M (plates 1 and 2)
Cross B—$10^{-4}$, 0·1 ml with K(λ) (plate 3)
↓
        $10^{-7}$, 0·4 ml with B (plate 4)

(5) Then proceed as follows:
*Time* (min)
0 Add 0·5 ml of phage mixture A to tube 'A'
3 Add 0·5 ml of phage mixture B to tube 'B'
8 Dilute $10^{-4}$ from tube 'A' into warm broth; label final dilution tube 'A1' and leave at 37°
9 Dilute $10^{-1}$ from tube 'A1' and plate 0·05 ml with B (5)
                                      0·02 ml with B (6)
                                        0·02 ml with B/S and streptomycin (17)
11 Dilute $10^{-4}$ from tube 'B' into warm broth; label final dilution tube 'B1' and leave at 37°
12 Dilute $10^{-1}$ from tube 'B1' and plate 0·05 ml with B (7)
                                        0·2 ml with B (8)
                                        0·2 ml with B/S and streptomycin (18)
90 Dilute from tube 'A1' $10^{-2}$ plate 0·1 ml with M (9)
                      ↓     plate 0·4 ml with M (10)
                  $10^{-3}$ plate 0·1 ml with M (11)
                         plate 0·4 ml with M (12)
93 Dilute from tube 'B1' $10^{-1}$ plate 0·1 ml with K(λ) (13)
                      ↓     plate 0·2 ml with K(λ) (14)
                  $10^{-3}$ plate 0·1 ml with B (15)
                         plate 0·2 ml with B (16)
(6) Incubate all plates overnight.

## Part 2. *Day following Part 1 (1 h)*
Cross A—T2*hrI* × T2 wild
(1) Calculate the titre of both phage types in the parental mixture (1, 2).
(2) From the numbers of plaques on plate (17), calculate the percentage of phages that adsorbed. Assuming the titre of *E. coli* B was $3 \times 10^8$/ml, and that both types of phages were adsorbed to the same extent, what was the multiplicity of infection of each type of phage?
(3) Score all four plaque-types observable in the progeny of the cross (9, 10, 11, 12). Calculate the fraction of recombinants in the progeny.
Cross B—T4*rII(147)* × T4*rII(271)*
(1) Calculate the titre of the phage in the parental mixture (4). How could the number of each parental type in the parental mixture be found experimentally? Why are there no plaques on (3)?
(2) Calculate the fraction of wild-type recombinants in progeny of the cross (13, 14, 15, 16).
(3) With similar assumptions as in Cross A, calculate the multiplicity of infection and the average burst size in the cross.

## EXPERIMENT 6
## SPOT TESTS WITH *rII* MUTANTS OF T4 PHAGE
(Can be performed concurrently with experiments 5 and 7)

Benzer has shown that the *rII* region of the T4 chromosome can be divided into two functional parts, called the *A* and *B* cistrons. Phage having a mutation in either cistron are unable to produce plaques on plates seeded with *E. coli* K($\lambda$). However, if cells of K($\lambda$) are infected with both an *rIIA* and an *rIIB* mutant, then phage production does occur in these cells as a result of 'complementation', phage *rIIA* being non-mutant in the function that is defective in an *rIIB* mutant (and *vice versa*). Thus both phages together have a full, normal complement of functions of the wild ($r^+$) phage.

This can be simply demonstrated as a spot test by plating $10^8$ particles of an *rIIA* mutant with K($\lambda$) and adding to the plate separate drops containing $10^6$ particles either of an *rIIA* or an *rIIB* mutant. In the spot where both *rIIA* and *rIIB* phage are present, extensive lysis occurs, but in the other spot there is no lysis. This test forms the basis of a quick method for locating any newly-isolated *rII* mutant in either the *A* or *B* cistron.

**Reference**
BENZER S. (1957) *Chemical Basis of Heredity*, eds. McElroy W.D. & Glass B. Johns Hopkins Press, Baltimore.

### (a) Requirements

*Indicator bacteria*—2 ml of o/n cultures in phage-broth of *E. coli* B (EMG 31) and *E. coli* K($\lambda$) (EMG 2)
*Phages*—0·1 ml of T4 *rIIA164* and T4*rIIB196* at $10^9$ particles/ml (these are representative *rII* mutants of T4 in *A* and *B* cistrons) 0·1 ml of T4 wild type, T4*rIIA147*, T4*rIIB114*, and T4*rIIABH23* (a deletion across the *rIIA* and *rIIB* cistrons), all at $10^7$ particles/ml*
Four TNA plates
20 ml soft agar (molten at 46°)
Four small tubes

### (b) Method
#### *Part 1* ($\frac{1}{2}$ h)
(1) Prepare four tubes of molten soft agar. Place in 46° bath.
(2) Label plates 1–4 and mark the backs into four quadrants.
(3) Overlay the plates with the following indicator bacteria and phage:
  (1) B
  (2) K($\lambda$)
  (3) K($\lambda$) + 0·1 ml *rIIA164*

* To add interest in the experiment, these phages should be labelled in an arbitrary way by the instructor and presented to the students as four unknowns.

(4) K($\lambda$) + 0·1 ml *rIIB196*
(4) Leave plates on bench for 15 min for overlay to set.
(5) With a 0·1 ml pipette, spot one drop of each of the four phage preparations provided at $10^7$ particles/ml on each plate, one in each quadrant.
(6) Allow spots to soak in.
(7) Incubate plates overnight.

*Part 2.    Day following Part 1 ($\frac{1}{4}$ h)*
From the pattern of phage lysis, what can be said about the genetic structure of the phages added as spots?

## EXPERIMENT 7
### SPOT TESTS WITH 'CONDITIONAL-LETHAL' PHAGE MUTANTS
(Can be carried out in the same period as experiments 5 and 6)

Normally the wild-type phage T4 plates equally well either on *E. coli* B or on a derivative strain of *E. coli* K12 designated CR63. However, a class of phage mutations called '*amber*' (*am*) (which may occur in any cistron in phage T4) has been isolated by Epstein, which have lost their ability to plate on B, but retain their ability to plate with high efficiency on CR63. For these 'amber' mutants, CR63 is termed a 'permissive' strain and strain B is 'non-permissive'.

Phage T4 normally plates equally well both at 25° and 42°. Again mutants can be found in most cistrons which still plate at the low temperature but have lost the ability to plate at the higher one. These 'temperature-sensitive' (*ts*) mutants respond similarly in both strains B and CR63.

Complementation tests similar to that described above for the *rII* mutants, can be used to determine whether two 'conditional-lethal' mutants of either *am* or *ts* type are in the same, or in different functional units or cistrons.

### Reference
EPSTEIN *et al.* (1963) *Cold Spr. Harb. Symp. quant. Biol.*, **28**, 375.

### (a) Requirements
**Phages**
(i)    0·2 ml of an *amber* mutant of T4 (*am*A) in gene *56* at $10^9$ pfu/ml.
(ii)   0·2 ml of each of four T4 conditional-lethal mutants and of wild-type T4, each at $10^7$ pfu/ml.*

* Suitable phages for use would be two *am* mutants (one in gene *56* and the other in another gene, e.g. gene *1*) and two *ts* mutants (one in gene *56* and the other in another gene, e.g. gene *19*).
These phage preparations, together with the wild-type T4 control, should be labelled in an arbitrary manner by the instructor.

256 CHAPTER 6

*Indicator bacteria*—2 ml o/n culture of *E. coli* B and *E. coli* CR63 (EMG13) in phage broth.
Six TNA plates
Ten ml soft agar (molten at 46°)
Six small tubes

### (b) Method
#### Part 1 ($\frac{1}{2}$ h)
(1)  Prepare six tubes of molten soft agar. Place in 46° bath.
(2)  Label TNA plates 1–6 and mark the backs into five segments.
(3)  Overlay the plates with the following indicator bacteria and phage:
   (1) and (2) CR63
   (3) and (4) B
   (5) and (6) B + 0·1 ml T4*am*A
(4)  Leave the plates on the bench for 15 min.
(5)  With a 0·1 ml pipette, spot one drop of each of the five unknown phage preparations provided at $10^7$ pfu/ml on each plate, one spot per segment.
(6)  Allow spots to soak in.
(7)  Incubate plates (1, 3, 5) overnight at 25°.
(8)  Incubate plates (2, 4, 6) overnight at 42°.

#### Part 2.   Day following Part 1 ($\frac{1}{2}$ h)
From the pattern of phage lysis, what can be said about the genetic structure of the phages added as spots?

### EXPERIMENTS 8–10
### CONJUGATION IN *E. coli* K12 MEDIATED BY THE SEX FACTOR F

Conjugation is the only mechanism mediating genetic recombination in bacteria which permits transfer of a large part, and occasionally even the whole, of the chromosome of one parental bacterium to another, so that a complete linkage map can be constructed by recombinant analysis. In addition, it is the only system from which populations of persistent, partially diploid bacteria (intermediate males carrying F prime factors) can be derived and tests of dominance performed.

   On the basis of ability to conjugate and transfer genetic material, strains of *E. coli* K12 can be broadly classified into two sexes, 'male', ($\delta$) and 'female' ($\female$ or F⁻), which are genetically and physiologically determined. Cultures of $\female$ strains are infertile when mixed; maximum fertility is shown by mixtures of $\delta$ and $\female$ strains; mixtures of $\delta$ strains are usually poorly fertile. The character of maleness is conferred by a 'sex factor' called F (for 'fertility') which is composed of DNA and determines two sexual functions of the ♂ bacteria which harbour it:

(1) The ability to form conjugal unions with ♀ bacteria;
(2) The one-way transfer, from ♂ to ♀, of genetic material.

The ♀ bacteria play no active role in either of these processes.

The factor F exists in ♂ bacteria in one or the other of two mutually-exclusive states between which it may alternate with low probability ($c$. $10^{-4}$ per cell generation). In one state ($F^+$♂), the sex factor is unassociated with the bacterial chromosome and replicates independently of it. In the other state (Hfr ♂), the sex factor is inserted into the continuity of the circular bacterial chromosome so that the two form a single unit of replication. All ♂ bacteria, irrespective of the state of the sex factor which they carry, conjugate with ♀ bacteria with high efficiency. However, the state of the sex factor has a profound influence on the nature of the subsequent genetic transfer.

## Properties of $F^+$ ♂ bacteria

(1) On conjugation, the sex factor is transferred with high efficiency to females, converting them to the $F^+$ ♂ state, so that the character of maleness spreads like an epidemic through the ♀ population. When young broth cultures of ♂ and ♀ bacteria are mixed, from about five to over 90 per cent of the ♀ population may be infected with the sex factor in 1 hour, depending on the particular strains used.

(2) Chromosome transfer is very low, as shown by the fact that recombinants for genes located on the bacterial chromosome appear with a frequency of less than about $10^{-4}$. This low frequency is due, in part at least, to the generation of Hfr ♂ bacteria in the $F^+$ population by a spontaneous change in the state of the sex factor.

(3) The sex factor is readily eliminated from $F^+$ ♂ bacteria by growth in the presence of acridine orange or, in the case of thymine-requiring (*thy*⁻) mutants, under conditions of partial thymine deprivation. The ♂ bacteria are thereby converted to females.

## Properties of Hfr ('high frequency of recombinants') ♂ bacteria

(1) On conjugation, the circular bacterial chromosome opens up at a specific point and is transferred, at high frequency, as a linear structure from one particular extremity termed the 'leading locus' or $O$ ('origine'). The speed of chromosome transfer is sensitive to temperature variation but is constant under standard conditions, so that the genes enter the ♀ bacteria in the same sequence as their arrangement on the chromosome and at fixed time intervals proportional to their absolute distances from $O$. At 37°, transfer of the whole

chromosome occupies about 100 min. During transfer, the chromosome tends to break in a random way so that the transfer of genes, and the frequency of their inheritance by recombinants, falls exponentially as their distance from $O$ increases. This gradient of transfer forms the main basis of mapping by genetic analysis in this system. Unlike all other systems, the gradient is not a function of the frequency of recombination but results from the exclusion of genes from the zygotes (see Hayes 1968, p. 659).

(2) Hfr ♂ bacteria may arise, in F⁺ ♂ populations, from insertion of the sex factor into the circular chromosome at any one of a considerable (though probably limited) number of sites (see fig. 6.3). There is good evidence that this insertion is mediated by a recombination event between a circular sex factor and the circular chromosome which leaves the circularity unimpaired. On conjugation, the chromsome opens up to become linear at the site of sex factor insertion, the leading extremity ($O$) being determined by the orientation of the sex factor. Thus although all the bacteria in a culture of any particular Hfr isolate behave homogeneously in transfer, different Hfr isolates from the same F⁺ strain may transfer the chromosome from different starting points and in opposite directions.

(3) Hfr ♂ bacteria do not normally transfer the F factor to females on conjugation. In mixtures, all the non-recombinant ♀ bacteria, as well as all recombinants formed at high frequency, remain ♀. But, if recombinants are selected which inherit the last (terminal) genes to be transferred, and which arise at low frequency due to chromosome breakage, these are frequently found to be Hfr males with the same transfer characteristics as the parental Hfr ♂ parent. Thus, irrespective of the type of Hfr strain, the functional part of the sex factor is always transferred last, at the terminal extremity of the chromosome.

(4) The sex factor of Hfr ♂ bacteria is not eliminated by treatment with acridine orange, nor by thymine deprivation.

The properties and inter-relationships of F⁺ and Hfr males are summarized in fig. 6.2

### F-prime (F′) factors

Hfr ♂ bacteria tend to revert to the F⁺ state, presumably due to reversal of the act of recombination which led to insertion of the sex factor into the chromosome. Occasionally, however, a sex factor is released which has incorporated into its structure a fragment of bacterial chromosome. This fragment is usually, though not necessarily, the terminal extremity of the chromosome which was transferred last by the Hfr strain during conjugation (fig. 6.3). Such substituted sex factors are termed F-prime (F′) factors (see Hayes 1968, pp. 674–679).

When an F′ factor is transferred to ♀ bacteria it gives rise to a new type of

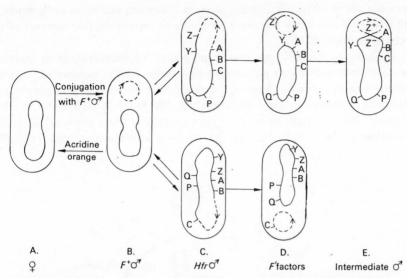

A.              B.              C.              D.              E.
♀              F⁺♂            Hfr♂           F′factors       Intermediate ♂

**Fig. 6.2.** Properties and inter-relationships of F⁺, Hfr and F-prime donors. The interrupted lines represent the bacterial chromosome, and the dotted lines the sex factor, the arrow head indicating the polarity of transfer. The letters A–Z show the locations of various bacterial genes.

♂ which displays some of the properties of both F⁺ and Hfr males, and so is called an 'intermediate' ♂. An *intermediate* ♂ strain not only promotes conjugation and independent transfer of its F′ factor, but also transfer of its chromosome with the same polarity and general sequence of markers as the Hfr strain in which the F′ factor originated, but with rather lower efficiency. However, the first marker to be transferred is now one of the alleles of the gene carried by the F′ factor. For example, if the original Hfr ♂ transferred its genes with the sequence *AB-Z*, an *F-Z⁺* factor derived from it, when transferred to a *Z⁻* ♀, promotes chromosome transfer with the sequence *Z±-AB-Z∓*. This is because the intermediate ♂ is heterozygous $Z^+/Z^-$ so that pairing and recombination between the two alleles occurs frequently, leading to rapid alternation between the inserted Hfr and released F⁺ states (see fig. 6.2D, E, and Hayes 1968, p. 677).

## Zygotic induction

When lysogenic Hfr ♂ bacteria, which carry an inducible prophage (such as λ, fig. 6.3) at a specific chromosome location, are mated with non-lysogenic females, as soon as the prophage penetrates the ♀ bacteria (zygotes) during chromosome transfer, it enters the vegetative state so that the zygotes are lysed and free phage particles liberated. This is called 'zygotic induction' and does not occur when the ♀ strain alone is lysogenic, nor when both parents are lysogenic and carry the same prophage or distinguishable mutants of it (see Hayes 1968, p. 662).

Zygotic induction is an important source of perturbation in genetic analysis since it leads to progressive diminution in the number of recombinants inheriting proximal genes, the closer these are linked to the prophage, while the potential to yield recombinants for genes located distal to the prophage is eliminated, since all zygotes containing them are lysed. Zygotic induction is also a valuable tool for studying chromosome transfer itself,

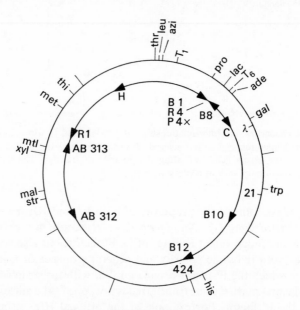

**Fig. 6.3.** The chromosome of *E. coli K*12

1. The figures shows the arrangement and approximate locations on the circular chromosome of a number of commonly used loci, with special reference to those employed in the experiments described. Their designations appear on the outside of the circle. The locations of three prophages (λ, 21 and 424) are shown on the inside of the circle.

2. The radial projections divide the chromosome into nine segments, each equivalent to 10 minutes transfer time in interrupted mating experiments at 37°. The map and distances between loci are derived from the data Taylor & Thoman (1964) and indicate the relative locations of the markers with some accuracy.

3. The arrows on the inner circle indicate the leading extremity and direction of transfer of the chromosome of various Hfr strains. Note that the time taken by an Hfr strain to begin to transfer its most proximal marker may exceed by 3 to 4 minutes the apparent distance between the leading locus and the proximal marker as shown on the map. (This extra time is that needed to build a conjugation tube before chromosome transfer can begin.)

since inducible prophages behave as chromosomal markers which are expressed directly on entering the zygote, without the intervention of recombination. Thus, for a given Hfr ♂ strain, the frequencies of zygotic induction for a series of inducible prophages, located at different chromosomal regions (e.g. λ, 21, 424 in fig. 6.3) indicate directly the frequencies with which these regions are transferred. The fact that the frequency of zygotic induction falls exponentially with the distance of the prophage location from the leading chromosomal extremity, $O$, shows that the chromosome breaks randomly during transfer. Again, by comparing the frequency of zygotic induction with the frequency with which a proximal marker, closely linked to the prophage site, appears among recombinants in comparable crosses between non-lysogenic strains, an estimate can be made of the probability with which a transferred marker is integrated into a recombinant chromosome ('coefficient of integration' = $c$. 0·5) (see Hayes 1968, p. 685).

## References

HAYES W. (1960) *Symp. Soc. gen. Microbiol.*, **10**, 12.
HAYES W. (1966) *Proc. R. Soc. (B)*, **164**, 230.
HAYES W. (1968) *Genetics of Bacteria and their Viruses.* Blackwell, Oxford.
TAYLOR A.L. & THOMAN M.S. (1964) *Genetics*, **50**, 659.

# EXPERIMENT 8. F⁺ × F⁻ CROSSES

The experiments demonstrate the dependence of mating on the presence of the sex factor, $F$, in one of the parents; they also show the low frequency of recombinant formation in comparison with that of transfer of the sex factor.

KEY (fig. 6.3)

| | |
|---|---|
| *ade* = adenine | *mal* = maltose fermentation |
| *azi* = sodium azide, resistance or sensitivity | *met* = methionine |
| | *mtl* = mannitol fermentation |
| *gal* = galactose fermentation | *pro* = proline |
| *his* = histidine | *str* = streptomycin resistance or sensitivity |
| *lac* = lactose fermentation | |
| *leu* = leucine | *thi* = thiamine (vitamin B1). |

*thr* = threonine
*trp* = tryptophan
*T*1, *T*6 = resistance or sensitivity to phages T1 and T6 respectively

Hfr strains: 
AB = Adelberg
B = Broda
C = Cavalli
H = Hayes
R = Reeves
P4x = Jacob's strain (J2)

(1) *Recombinant formation.* Log broth cultures of an F$^+$ and an F$^-$ strain *of the same genotype*, both streptomycin-sensitive (*str-s*), are each mixed with a culture of a streptomycin-resistant (*str-r*) F$^-$ strain of different genotype and incubated. The mixtures are then washed, and plated on minimal or other selective medium on which only recombinant bacteria can grow to yield colonies. The washed mixtures are also diluted and plated on nutrient agar + streptomycin (NA + SM) for colony counts indicating the number of viable F$^-$ bacteria present. F$^+$ × F$^-$ crosses alone yield recombinants, and only at low frequency.

(2) *F factor transfer.* Individual colonies of the F$^-$ *str-r* strain, selectively re-isolated from both F$^+$ × F$^-$ and F$^-$ × F$^-$ mixtures on NA + SM are cultured in separate tubes of broth and then crossed to an F$^-$ strain of different genotype, as above. Only those cultures derived from bacteria to which the sex factor was transferred, by conjugation during the initial cross, can yield recombinants. The proportion of originally F$^-$ bacteria thus converted by F$^+$ males to the F$^+$ state is high.

### (a) Requirements

#### Part 1

Log broth cultures, in 15 ml s/c bottles, of each of the following *E. coli* K12 strains, labelled:

EMG21 F$^+$ (*met$^-$ str-s*), 5 ml
EMG14 F$^-$ (*met$^-$ str-s*), 5 ml
EMG10 F$^-$ (*thr$^-$ leu$^-$ thi$^-$ str-r*), 2 × 5 ml
Two MA + thiamin (vitamin B$_1$) at 5 µg/ml + SM (200 µg/ml) plates (MA + B$_1$ + SM)
Four NA + SM (200 µg/ml) plates
100 ml buffer
Eight large tubes

#### Part 2

10 ml log broth culture strain EMG14
Two MA + B$_1$ plates (*no* streptomycin added)
20 ml buffer
10 ml broth
Six small tubes
Template for micro-plating technique (see appendix 6.3)

### (b) Method

#### Part 1 (2$\frac{1}{2}$ h)

(1) Prepare eight dilution tubes (4 × 10 ml, 4 × 9 ml)
(2) Mix 5 ml each of cultures as follows:

$$21 + 10 \, (= F^+ \times F^-)$$
$$14 + 10 \, (= F^- \times F^-)$$

(3)  Place mixtures in waterbath at $37°$ for $1\frac{1}{2}$ hours. Then centrifuge and resuspend in 5 ml buffer.

(4)  Spread $0.1$ ml of each mixture (undiluted) on $MA + B_1 + SM$ for prototrophic ($met^+$ $thr^+$ $leu^+$) recombinants, plates (1) and (2).

(5)  Dilute both mixtures $10^{-5}$ and $10^{-6}$. Spread $0.1$ ml of each dilution on $NA + SM$ to assess the total number of viable $F^-$ parental cells and to re-isolate these cells in order to test them for inheritance of the sex factor.

Mixture $21 + 10$, plates (3) and (4)
Mixture $14 + 10$, plates (5) and (6)

(6)  Incubate plates (3) to (6) overnight and plates (1) and (2) for 42–48 hours.

### Part 2.  Day following Part 1 (about 6–7 h including incubation time)

### (A) Recombinant formation

(1)  Count, by marking on the back of the plates, the total number of colonies arising on plates (3) or (4), whichever contains the higher countable number of colonies. Do not stab the colonies since some are required for sub-culture.

(2)  Assess the total number of viable $F^-$ cells initially present in $1.0$ ml of the mating mixture and record the result.

### (B) F factor transfer

(1)  Prepare six small tubes containing 1 ml broth.

(2)  From one of plates (3) or (4), pick three well-isolated colonies of the recovered $F^-$ strain, and sub-culture each to a separate tube of broth.

(3)  Similarly pick and sub-culture three colonies from plates (5) or (6).

(4)  Incubate all six broth cultures in a waterbath at $37°$.

(5)  When the cultures are markedly turbid (4–5 h), add to each 1 ml of the log broth culture of strain EMG14 provided. (Considerable time can be saved in incubation period if a shaking waterbath is available.)

(6)  Incubate in waterbath at $37°$ for $1\frac{1}{2}$ hours.

(7)  Centrifuge each mixture, wash once in 2 ml buffer, and resuspend in $0.5$ ml buffer.

(Since, in these crosses, the presumptive $F^+$ donor strain is $str$-$r$ while the $F^-$ recipient bacteria, which will constitute the zygotes, are $str$-$s$, SM cannot be added to the selective medium. Under these conditions cross-feeding and considerable syntrophic growth may obscure or mimic recombinant colonies unless nutrients are thoroughly removed by proper washing.

(8)  Using the template provided (appendix 6.3), spread *loopfuls* of each mixture, in duplicate, over 25 mm diameter areas on $MA + B_1$ plates.

Spread duplicate areas from three colonies derived from (3) or (4). Plate (7). Spread duplicate areas from three colonies derived from (5) or (6). Plate (8). (This is a qualitative test, requiring only a 'yes' or 'no' answer. The technique is economical of material and saves time. The wire loop should have a diameter of about 3 mm. For a loopful of adequate volume to be picked up, the loop should be immersed and removed vertically in a plane parallel to the liquid surface.)

(9) Incubate the plates 42–48 hours.

### Part 3.   Day following Part 2 ($\frac{1}{2}$ h)

#### (A) Recombinant formation

(1) Count the number of prototrophic recombinant colonies on plates (1) and (2) and express in terms of the number of recombinants arising from 1·0 ml of the mating mixture.

(2) From this recombinant count and the count of viable $F^-$ bacteria calculated from the number of colonies on plates (3) and (4) assess the recombination frequency, in terms of the number of $F^-$ bacteria yielding one recombinant, from the ratio:

$$\frac{\text{No. prototrophic recombinants per ml mixture}}{\text{No. viable } F^- \text{ bacteria per ml mixture}}$$

(3) Record your result for comparison with that of the Hfr $\times$ $F^-$ cross in experiment 9, part 3(A)(3) (p. 268).

### Part 4.   Day following Part 3 ($\frac{1}{2}$ h)

#### (A) F factor transfer

(1) Examine the $MA + B_1$ plates (7) and (8), for areas showing prototrophic colonies. Their occurrence shows that the particular isolates of the initially $F^-$ bacterium have become $F^+$. The number of recombinant colonies per area will probably fall in the range 5–20. Only isolates from the $F^+ \times F^-$ cross (plate 7) should yield recombinants.

(2) Correlate the class results: assess the frequency of $F$ transfer in terms of the proportion of $F^-$ cells initially present in the $F^+ \times F^-$ cross which have been converted to the $F^+$ state. Compare this frequency with the frequency with which recombinants are formed.

### (c)  Conclusions

Using approximately equal population densities of the suggested $F^+$ and $F^-$ strains, the frequency of recombinants should be about one per $10^4-10^5$ ♀ bacteria, while about 50 per cent of the parental ♀ bacteria should be converted to the $F^+$ state under the recommended conditions.

# EXPERIMENT 9. (A) Hfr × F⁻ CROSSES
# (B) ZYGOTIC INDUCTION

Three crosses are made, involving the same non-lysogenic F⁻ strain but different Hfr strains; the results are compared with respect both to the genetic constitution of the recombinants and to the occurrence of zygotic induction. Two of the Hfr strains (Hfr*H*) differ only in that one is lysogenic for the inducible prophage, λ, while the other is non-lysogenic. The third Hfr strain (Hfr*C*) transfers its loci in the reverse order to Hfr*H* and, although lysogenic, does not show zygotic induction, since the λ prophage locus is terminal on its chromosome and is, therefore, transferred to the F⁻ bacteria at only a very low frequency (for the sequence of transfer of various loci by these Hfr strains, refer to fig. 6.3, p. 260). Log broth cultures of the Hfr strains, which are *str-s*, are throughly washed to remove free λ phage particles, and diluted 1/10 into fresh broth at 37°. Equal volumes of diluted Hfr suspension and log broth cultures of the F⁻ *str-r* strain are mixed and incubated to allow time for transfer of the proximal part of the chromosome. The mixture is then treated with SM to prevent further multiplication of the Hfr parent and the spontaneous liberation of free λ particles. The treated mixture is finally diluted and plated on minimal or other selective medium for recombinants; it is also plated, together with λ-sensitive, *str-r* indicator bacteria, on NA + SM for 'infectious centres', i.e. zygotes which have received prophage and will lyse to liberate infective phage λ. The recombinants are finally purified and scored for inheritance of unselected Hfr markers.

## (a) Requirements

### Part 1

5 ml log broth culture of *one* of the Hfr strains below, washed twice, re-suspended, and diluted 1/10 in fresh broth at 37°.*

EMG23 = Hfr*H*(λ)⁻ (Hayes strain: prototrophic, *T1-s lac⁺ T6-s str-s*).
EMG24 = Hfr*H*(λ)⁺ (strain 23, lysogenized with λ phage).
EMG25 = Hfr*C*(λ)⁺ (Cavalli strain: *met⁻ T1-s lac⁺ T6-s str-s*).
5 ml log broth culture of strain EMG26 F⁻ (λ)⁻ (P678: *thr⁻ leu⁻ thi⁻ T1-r lac⁻ T6-r* (λ)⁻ *λ-r str-r*). (This strain, although non-lysogenic, lacks λ phage receptors and so is resistant to infection by free phage; it can yield infectious centres only by transfer of prophage to it during conjugation.)
5 ml o/n broth culture of strain EMG10—a *str-r* (λ)⁻, λ-s indicator strain for free phage λ particles (labelled 'I').
Two plates MA + B₁ + SM (methionine added for Hfr*C* cross).
Two plates MA + B₁ (without SM—methionine added for Hfr*C* cross).

* Different groups will use different Hfr strains.

Two NA + SM plates (40 ml agar) *kept at* 37°.

One plate NA + SM.

One large tube containing 4·5 ml buffer + SM (200 μg/ml) in 37° bath

Three large tubes of buffer (1 × 10 ml; 2 × 9 ml) in 37° bath.

Two small tubes containing 3 ml soft agar held molten at 46°.

Waterbath at 46°.

### Part 3

Three plates MA + B1 + SM (methionine added for the HfrC cross only).

### Part 4

60 small test tubes in rack, sterile, capped.

Three plates MA + SM (methionine added for the HfrC cross).

Three plates MA + B1 + SM (*no* methionine; required for HfrC cross only).

Three plates EMB-lactose agar.

1 ml phage T1 suspension at *c.* $10^{10}$ pfu/ml

1 ml phage T6 suspension at *c.* $10^{10}$ pfu/ml

100 ml buffer.

Seven large tubes.

### (b) Method

#### Part 1 ($2\frac{1}{2}$ h)

(1) Mix the Hfr and F⁻ cultures and incubate in waterbath at 37° for 1 hour.

(2) Transfer 0·5 ml mixture to 4·5 ml buffer + SM at 37° (= 1/10 dil.) to kill the Hfr bacteria and prevent their spontaneous liberation of phage λ.

(3) Maintain inoculated buffer-SM solution in waterbath at 37° for 20 min.

(4) Further dilute $10^{-2}$, $10^{-3}$ and $10^{-4}$ in warm buffer (= final $10^{-3}$, $10^{-4}$ and $10^{-5}$ dil's) and maintain at 37°.

### (A) To demonstrate genetic recombination

(1) Spread 0·1 ml of the $10^{-3}$ and $10^{-4}$ final dil's on MA + B1 + SM (for the HfrC cross this medium supplemented with methionine should be used). Label plates (1) and (2).

(2) Spread 0·1 ml of the $10^{-3}$ and $10^{-4}$ dil's on MA + B1 (without SM) for approximate viable count of the viable Hfr bacteria; recombinants will also grow on this medium (for the HfrC cross, this medium supplemented with methionine should be used). Label plates (3) and (4).

(3) Spread 0·1 ml of $10^{-5}$ dil. on NA + SM for viable count of F⁻ bacteria. Label plate (5).

(4) Incubate MA plates (1–4) at 37° and NA plate (5) at 30°. The smaller colonies arising overnight at the latter temperature facilitate the counting of large numbers.

## (B) To demonstrate zygotic induction

(1) As soon as possible after diluting the mating mixture, transfer 0·1 ml of the $10^{-4}$ and $10^{-5}$ dil's to separate, small tubes containing molten soft agar, in waterbath at 46°.

(2) To each add 2–3 drops of o/n broth culture of strain labelled 'I', and pour over surface of warm NA + SM plates.

(3) *Do not replace lids for a few minutes.* When agar layer has set incubate plates at 37° as quickly as possible after setting. Label plates (6) and (7). *Note.* Since both chromosome transfer and, particularly, the occurrence of zygotic induction are very dependent on the mating mixture and the zygotes, respectively, being kept at 37°, it is most important that this temperature be maintained, so far as is practicable, throughout the above operations. The experiments can be carried out in a 37° constant temperature room with advantage.

### Part 2. Day following Part 1 (1 h)

## (A) Hfr × F⁻ cross: genetic recombination

(1) Count the total number of colonies of:
(a) The Hfr parent (plates 3 and 4) and
(b) The F⁻ parent (plate 5)

(2) In each case, estimate the total number of viable bacteria initially present in the Hfr × F⁻ mixture, and record your result.

## (B) Zygotic induction

(1) Count the number of plaques of phage λ (infectious centres) on plates (6) and (7). The class should compare the results given by the crosses involving the different Hfr strains. Only the cross with strain EMG24 (Hfr$H$λ⁺) should show a significant number of plaques at the $10^{-4}$ dilution, since strain EMG25 (Hfr$C$λ⁺) does not transfer the λ prophage locus at high frequency. Note, however, that plaques may appear which do not arise from zygotic induction but from direct infection of the indicator bacteria.

(2) In the case of the Hfr$H$(λ)⁺ × F⁻(λ)⁻ cross, assume that all the plaques counted are due to zygotic induction and that it is 100 per cent efficient:
(a) Calculate the number of zygotes per ml of the undiluted mating mixture to which λ prophage was transferred;
(b) From this figure, and from the number of Hfr bacteria per ml mixture (plates 3 and 4), calculate the percentage Hfr bacteria which transferred the $O$ ——λ region of chromosome to zygotes.

*Part 3.    Day following Part 2 (1 h)*

## (A) Hfr cross: genetic recombination

(1)   Examine plates (1) and (2). Count, by marking the back of the plates, the number of prototrophic ($thr^+leu^+$) recombinant colonies arising from an appropriate dilution. Do not stab the colonies.

(2)   Calculate the number of recombinants emerging from 1·0 ml undiluted mating mixture. From this figure, and from the number of Hfr bacteria per ml of mixture (plates 3 and 4), calculate the percentage of Hfr bacteria which transferred the *O——thr leu* segment of chromosome to zygotes.

(3)   From the number of recombinants on plates (1) and (2) and the number of F⁻ bacteria per ml (plate 5), calculate the number of F⁻ bacteria yielding one recombinant and compare the result with that given by the F⁺ × F⁻ cross (experiment 8, part 3 (A) (3), p. 264). Remember that the ratio ♂/♀ bacteria was 1/1 in the F⁺ cross and 1/10 in the Hfr cross.

## (B) Purification of recombinants for genetic analysis

(1)   Touch 60 recombinant prototroph colonies on plates (1) and (2) lightly with a sterile wire and streak on $MA + B_1 + SM$ (for the Hfr*C* cross, this medium should be supplemented with methionine). Twenty colonies can usually be streaked comfortably on a single plate, ten on each half. The streaking technique need not aim at producing isolated colonies (though this is desirable), but merely at 'diluting out' contaminating F⁻ bacteria and enriching the prototrophs by a second cycle of growth on selective medium.

(2)   Label plates (8), (9) and (10) and incubate overnight.

*Part 4.    Day following Part 3 (1 h)*

## (A) Hfr × F⁻ cross: scoring of unselected markers among recombinants

(1)   Pipette 1 ml buffer to each of 60 small tubes.

(2)   Pick, with a loop, a small portion of growth furthest from the inoculation site or, preferably, an isolated colony, from each of the 60 streak cultures on supplemented minimal agar (plates 8, 9 and 10).

(3)   Suspend the growth from each streak in 1 ml buffer, so as to yield a very slight turbidity.

(4)   Streak a very small loopful of each suspension on each of the following media (20 per plate as before).

(a) *Hfr*H × *F⁻ crosses:*

   (i) $MA + SM$ to score for $thi^+$ ($B_1^+$) inheritance.

   (ii) EMB-lactose medium, to score for $lac^+$ inheritance.

(b) *HfrC × F⁻ cross:*
   (i) MA + Met + SM, to score for *thi⁺* inheritance.
   (ii) MA + B₁ + SM, to score for *met⁻* inheritance.
   (iii) EMB-lactose medium, to score for *lac⁺* inheritance.
(5)  In all crosses, the inheritance of sensitivity to phages T1 and T6 from the Hfr parent is then scored by spotting a loopful of suspensions of the phages on the streak inoculum of each recombinant on the EMB-lactose plates. Between each spotting, the loop should be flamed and cooled in buffer before recharging with phage, so as to prevent the transfer of bacteria from one streak to another.
(6)  Incubate plates overnight.

### Part 5.  Day following Part 4 (1 h)
(1)  From the plates put up under Part 4 score, among the 60 *thr⁺leu⁺* recombinants, the number which have inherited various unselected markers from the Hfr parent. The markers to be scored, in the sequence of their arrangement from *O* are as follows:
For the Hfr*H* crosses: *T1-s lac⁺T6-s thi⁺*.
For the Hfr*C* cross: *T6-s lac⁺T1-s thi⁺met⁻*.
Convert these numbers into the *percentage* of *thr⁺leu⁺* recombinants which inherit each unselected Hfr marker.
(2)  From these results deduce the order on the chromosome of those unselected markers located *distal* to the selected markers. (See under 'conclusions', below.)
(3)  Compare the results of the crosses involving the Hfr*H*(λ)⁺ and Hfr*H*(λ)⁻ ♂ strains, and assess the effect of zygotic induction on the inheritance of various unselected markers.

### (c) Conclusions

The diagram shows, for the strains Hfr*H* and Hfr*C*, the relationships of the unselected markers to the leading locus *O* and to the selected markers *thr⁺leu⁺*. The interrupted lines indicate, for each Hfr strain, the most distal chromosome region which should not be transferred at significant frequency.

1. In the Hfr$H$ crosses no $thi^+$ recombinants should be found. All the other unselected markers lie on the proximal part of the chromosome and distal to the selected markers. They should show a gradient of inheritance among the recombinants, due to chromosome breakage, in proportion to their distance from $thr^+leu^+$. Thus among the prototrophic recombinants, about 70 per cent should be $T1$-s, 40–45 per cent $lac^+$ and 20–30 per cent $T6$-s.

2. In the Hfr$C$ cross all zygotes generating prototrophic recombinants must have received the $O$——$thr^+leu^+$ region of chromosome, so that inheritance of unselected markers in this region depends on *recombination* and not on pre-zygotic exclusion. Since recombination occurs frequently in relation to the distances between most of these markers (Hayes 1968, p. 690) they will tend to segregate randomly (40–60 per cent) so that it is difficult (or impossible) to map them unambiguously. On the other hand, the frequency of inheritance of $thi^+$ and $met^-$, which lie distal to the selective markers, should show a gradient which permits mapping of their positions relative to *thr leu*.

3. In the cross involving Hfr$H$ $(\lambda)^+$, zygotic induction should eliminate all zygotes which receive the $O$——$\lambda$ segment of chromosome; but many zygotes will receive shorter segments, due to chromosome breakage, and these will generate recombinants. The closer an unselected marker is located to $\lambda$, the more likely it is to be transferred in association with $\lambda$ and to be eliminated by zygotic induction. Thus the gradient of unselected markers should be steeper in the Hfr$H$ $(\lambda)^+$ cross than in the Hfr$H$ $(\lambda)^-$ cross where zygotic induction does not occur.

## EXPERIMENT 10
### INTERRUPTED MATING: MAPPING AND THE ESTIMATION OF GENETIC DISTANCE AS A FUNCTION OF TIME

A log broth culture of an Hfr *str-s* and an F$^-$ *str-r* strain, are mixed and incubated at 37°. At appropriate intervals thereafter samples are removed, diluted, and violently shaken to separate the mating couples.

Appropriate dilutions are finally plated on various media selecting for recombinants inheriting different Hfr markers. The number of recombinants of each class are then counted and plotted on a graph as a function of the time at which the mating was interrupted.

### (a) Requirements

*Part 1 (For the whole class)*
5 ml log broth culture strain EMG27 (HfrB1), diluted 1/10 in fresh broth at 37° and maintained at 37°.

5 ml log broth culture of F$^-$ strain EMG28 (*pro$^-$thr$^-$leu$^-$thi$^-$str-r*) maintained at 37°.

10 × 100 ml s/c bottles of buffer.

10 × 5 ml s/c bottles, containing 1·8 ml buffer, and marked 0, 5, 10, 15, 20, 25, 30, 35, 40 and 50.

### (*For each group of students*)

Ten small test tubes marked 0, 5, 10, 15, 20, 25, 30, 35, 40 and 50.

Ten plates MA + Thr + Leu + B1 + SM (marked 'A') to select for inheritance of the Hfr marker *pro$^+$*.

Ten plates MA + Pro + B1 + SM (marked 'B') to select for inheritance of the Hfr markers *thr$^+$leu$^+$*.

### (b) Method

#### Part 1 (*1 h*)

(A single mating mixture is set up, and samples interrupted at intervals thereafter, by the supervisor. These samples are then diluted to the plating dilution, and aliquots distributed for plating to each member of the class. If practicable, all these operations are best carried out in a 37° room, to ensure constancy of temperature throughout.)

(1)  5 ml of the (1/10 dil.) Hfr broth culture are mixed with 5 ml of the (undil.) F$^-$ culture in a 25 ml s/c bottle, which is immediately clamped to the periphery of a 33 rev/min rotor. (A stopwatch is started at the time of mixing.)

(2)  Immediately after mixing (0 min) and at 5, 10, 15, 20, 25, 30, 35, 40 and 50 min thereafter, the bottle is gently removed from the turntable, and 0·2 ml of the mixture transferred to a 5 ml s/c bottle containing 1·8 ml buffer (= 1/10 dil.).

(3)  The sample to be interrupted is immediately attached to a 'Microid' Flask Shaker (or other agitating device, see Note 1b) and vibrated at maximum speed for 30 seconds.

(4)  As soon as the froth has sufficiently subsided (10–20 seconds), transfer 1 ml to 100 ml buffer (= 10$^{-3}$ final dil.). Shake and dispense 1 ml amounts into tubes, marked with the time of interruption; one for each group of students.

(5)  Each group now plates 0·1 ml of each sample, without further dilution on plates A and B and labels with the time of dilution.

(6)  Incubate plates for 42 to 48 hours.

### Notes

(1)  The method described here is that used in our laboratory, but there are many variations which give equally satisfactory results provided three important points are observed:

(a) Since chromosome transfer is very temperature-dependent and is optimal at 37°, every effort should be made to prevent fluctuations of the mating mixture from this temperature, such as may occur if the mixture is removed periodically from an incubator and cold pipettes are used to remove samples. Cooling of the samples once withdrawn, is not only permissible but advantageous since it inhibits further chromosome transfer, as well as the formation of secondary unions after interruptions.

(b) The use of a 'mixer' or 'blender' to interrupt the mating is cumbersome and unnecessary. On the other hand, shaking by hand or by holding the sample tube against an eccentrically rotating rubber stopper or disc driven by an electric motor (e.g. a 'Whirlimixer'), is usually inadequate. An efficient apparatus is described by Low and Wood (1965).

An alternative to mechanical separation of mating pairs is to treat the samples for 10–15 min with a high multiplicity of phage T6 to which only the Hfr bacteria are sensitive (Hayes 1957). Adsorption of the phage immediately kills the Hfr bacteria and prevents further chromosome transfer. Subsequent dilution and plating of the samples so reduces the concentration of phage on the plates that the segregation and growth of *T6-s* recombinants is not significantly affected.

(c) To prevent the secondary formation of mating pairs, it is desirable to dilute the samples of mating mixture before mechanical agitation, and to make the plating dilution as soon as possible afterwards.

(2) Any other combination of Hfr and F⁻ strains may be used, provided that suitably spaced markers are available which can be cleanly selected.

### Part 2.    Two days after Part 1 ($\frac{3}{4}$ h)

Examine the plates spread (in part 1) with samples from the Hfr × F⁻ cross interrupted at various times after mating. Count the number of *pro*⁺ (plates 'A') and *thr⁺leu⁺* (plates 'B') recombinants arising from each time sample. Construct a graph relating the number of recombinants of each class to the time at which the mating mixture was interrupted. Assess the time after mating at which the markers begin to enter the zygotes, by extending the curves back to intercept the time co-ordinates.

### (c) Conclusions

The curves obtained should approximate to those shown in figure 6·4. Strain HfrB1 begins to transfer the *pro* locus at about 4–5 min and the *thr leu* loci at about 10–12 min after mixing the parental cultures. The plateaux indicate that mating and transfer of a particular Hfr locus are complete in the population. Since HfrB1 transfers the *thr leu* loci after the *met* locus, the lower plateau level of *thr⁺leu⁺* recombinants is due to chromosome breakage between the loci during transfer. The slopes of the two curves are determined partly by the rate of pair formation and partly by heterogeneity among the Hfr bacteria in the time required to initiate chromosome transfer after mating (De Haan & Gross 1962), and actually show the same kinetics.

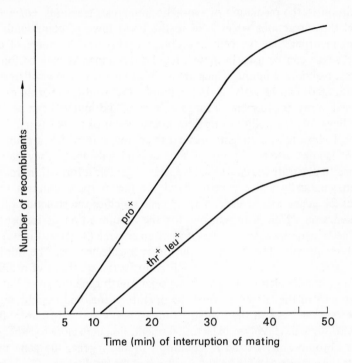

**Fig. 6.4.** Kinetics of chromosome transfer in *Escherichia coli*.

It is important to note that the curves may fail to show a plateau if equivalent numbers of parental bacteria are mated: before pair formation is complete in the population, unmated bacteria of both parental types will have multiplied, so that the curves continue to rise over a prolonged period.

For a discussion of the kinetics of the mating process, see Hayes 1968, pp. 679–685.

### References

De Haan P.G. & Gross J.D. (1962) *Genet. Res. Camb.*, **3**, 251.
Hayes W. (1968) *Genetics of Bacteria and their Viruses*. Blackwell, Oxford.
Low B. & Wood T.H. (1965) *Genet. Res. Camb.*, **6**, 300.

## EXPERIMENTS 11 AND 12
## THE CONTROL AND EXPRESSION
## OF RELATED GENES IN THE 'LAC' OPERON

In addition to the mediation of synthetic (anabolic) reactions, enzymes also control those reactions which lead to the breakdown of compounds (catabolism) in order that the cell may obtain energy and a source of carbon groups which can be used in these anabolic reactions. As may be found for synthetic reactions, mutants which are unable to perform one or other of these catabolic steps can be isolated. For example, the 'wild-type' strain of *E. coli* can break down (or ferment) a range of sugars, and mutants can be selected which have lost the ability to ferment one of more of these sugars.

The following experiments make use of mutants which are unable to ferment lactose, a disaccharide of glucose and galactose. They have been designed to illustrate the controlled transfer of genetic information from genes into enzymatically active proteins, and for this purpose mutants in three distinct *lac* genes will be used. The '*y*' gene specifies the structure of a permease-enzyme which is responsible for the accumulation of lactose in the cell. The '*z*' gene specifies the structure of an enzyme (β–galactosidase) which hydrolyses (breaks down) lactose and other β–galactosides. The third gene, the '*i*' gene, is a regulator gene, which controls the synthesis of a diffusible protein product which is thought to act on a fourth gene ('*o*' gene, but which may be part of the '*z*' gene) called the operator gene, and thereby regulates the enzyme synthesis of the two structural genes, *y* and *z*. Mutants affecting these various genes involved in lactose fermentation map close together on the *E. coli* chromosome. In the $i^+$ condition of the *i* gene, the gene product permits the expression of the *z* and *y* genes (i.e. the production of their enzyme products) only in the presence of an inducer such as lactose or another β–galactoside. Thus, mutation to the $i^-$ condition gives rise to cells which are said to be 'constitutive', in that the expression of the *z* and *y* genes is permitted even in the absence of an added inducer. The genes *z*, *y* together with the operator gene '*o*', form a co-ordinated unit of function which is termed an 'operon' (see Jacob & Monod 1961).

The essential requirement for enzymological work is a convenient and reliable assay procedure. Relatively few enzymes have the stability and convenience of assay which make β–galactosidase the system of choice for many kinds of experiment. The assay procedure of this enzyme which is used in the two following experiments is based on that described by Pardee, Jacob & Monod (1959) in which a colourless substrate (o–nitrophenyl–β–galactoside, ONPG) is hydolysed by the enzyme to give the coloured product ONP.

$$\text{ONPG} \xrightarrow{\text{β–galactosidase}} \text{Galactose} + \text{ONP (yellow)}$$

The amount of yellow colour, and therefore the amount of hydrolysis, is conveniently assayed spectrophotometrically.

### References

AMES B.N. & MARTIN R.G. (1964) *Ann. Rev. Biochem.*, **33**, 237.
BECKWITH J. (1964) *Biochim. biophys. Acta*, **76**, 162.
BENZER S. & CHAMPE S.P. (1961) *Proc. natn. Acad. Sci. Wash.*, **47**, 1025.
GILBERT W. & MULLER-HILL B. (1967) *Proc. natn. Acad. Sci. Wash.*, **56**, 1891.
JACOB F. & MONOD J. (1961) *J. molec. Biol.*, **3**, 318.
JACOB F. & MONOD J. (1965) *Biochem. biophys. Res. Commun.*, **18**, 693.
NAKADA D. & MAGASANIK B. (1964) *J. molec. Biol.*, **8**, 105.
PARDEE A.B., JACOB F. & MONOD J. (1959) *J. molec.. Biol.*, **1**, 165.

### Procedure for assay of β-galactosidase

(1) *Reagents: ONPG.* Dissolve 0·2 g ONPG in 50 ml of 0·25M sodium phosphate buffer at pH 7·0 containing 0·001M $MgSO_4$, 0·0002M $MnSO_4$ and 0·1M mercaptoethanol. Heat as little as possible to effect the solution of the ONPG and cool immediately. Store the reagent in a refrigerator.
1M solution of $Na_2CO_3$
Toluene

(2) *Method of assay.* Shortly before taking samples, take a series of small tubes each containing one drop of toluene (use same pipette throughout the experiment) and place them in an ice bath where they are maintained until the entire series has been taken. At the time of sampling, add 1 ml of bacterial culture to each tube, shake vigorously and return to the ice bath. When all the samples have been collected, the tubes are placed in a water bath at 37° for 30 min and shaken at intervals throughout this period. They are then transferred to a 28° waterbath and 0·2 ml of ONPG reagent is added to each of the tubes. The addition of ONPG is timed at intervals of fifteen seconds and hydrolysis of ONPG which then occurs is allowed to proceed until a visible yellow colour is produced (usually 15–30 min is adequate). The reaction is stopped by the addition of 0·5 ml of 1M $Na_2CO_3$ solution to the tubes, again at 15 second intervals. The exact time interval between starting and stopping the reaction is noted. (The function of the $Na_2CO_3$ is to raise the pH sufficiently so as to stop the reaction and enhance the colour due to the o-nitrophenyl ion.) The optical density of each of the tubes is then compared in a spectrophotometer using a blank containing no enzyme, at 420 nm and 550 nm. (The 550 nm reading is used to correct for absorption due to bacterial turbidity.) Enzyme activity (*e*) in units per ml can then be calculated from the expression

$$e = \frac{OD\ 420 - (1·65 \times OD\ 550)}{\text{time incub. (min)} \times 0·0075}$$

## EXPERIMENT 11
## PHENOTYPIC BEHAVIOUR OF 'LAC' MUTANTS

This experiment demonstrates the phenotypic behaviour of a number of the mutants of the *lac* region following growth in media with or without either glucose, thiomethylgalactoside (TMG) or isopropylthiogalactoside (IPTG) (either TMG or IPTG act as inducers for the lactose genes but glucose does not induce). Genetic studies show that these *lac* mutants have sites which map closely together in the lactose region of the *E. coli* chromosome.

The four mutants required are EMG7 ($i^+z^+y^+$), EMG8 ($i^+z^-y^+$), EMG9 ($i^-z^+y^+$) and EMG10 ($i^+z^+y^-$). These strains are grown on NA plates (Stock plates) and then each is inoculated in 5 ml M9-glycerol medium (M9 medium in which 0·2 per cent w/v glycerol replaces glucose), supplemented with the appropriate growth factors*, and incubated overnight.

### (a) Requirements

#### Part 1

1 ml o/n cultures of strains EMG 7, 8, 9 and 10 in M9-glycerol
6 ml ONPG solution ⎱ Reagents for assay
15 ml 1 M Na$_2$CO$_3$ ⎰
1 ml TMG solution (2 × 10$^{-2}$M) ⎫
1 ml IPTG solution (5 × 10$^{-3}$M) ⎬ Supplements
1 ml Glucose solution 2 per cent w/v ⎭
Ice bath
Two waterbaths set at 37° and 28°
Spectrophotometer
20 large tubes each containing 5 ml M9-glycerol medium plus leucine threonine and thiamin (each at 20 mg/ml)
20 Pasteur pipettes (for bubbling cultures)
20 large tubes for assay of β-galactosidase

### (b) Method

#### Part 1 (6–7 h)

(1) Five replicate sub-cultures of each strain are prepared by inoculating 5 ml sterile M9-glycerol medium with 0·05 ml of each overnight culture. Using Pasteur pipettes, and a small aeration pump, aerate each culture at 37°† until a concentration of *c.* 2 × 10$^8$/ml is achieved (3–4 h).

(2) Add, as specified by the protocol below, and where required, supplements of 0·55 ml TMG (to *c.* 2 × 10$^{-3}$M), 0·6 ml IPTG (to *c.* 5 × 10$^{-4}$M) and

---

* The supplements are: EMG7, EMG8 and EMG9–thiamin; EMG10–thiamin, threonine and leucine.

† Alternatively, aerate in Erlenmeyer flasks on rotary shaker.

| Supplement | Mutant | | | |
|---|---|---|---|---|
| | $i^+ z^+ y^+$ (EMG7) | $i^+ z^- y^+$ (EMG8) | $i^- z^+ y^+$ (EMG9) | $i^+ z^+ y^-$ (EMG10) |
| Nil | | | | |
| TMG | | | | |
| TMG + glucose | | | | |
| IPTG | | | | |
| IPTG + glucose | | | | |

0·6 ml glucose (*c.* 0·2 per cent w/v). Continue the incubation and aeration for a further 45 min.

(3) At the conclusion of the 45 min incubation, remove 1 ml samples from each culture and assay for β-galactosidase activity as previously detailed (p. 275).

Explain your results in terms of the known genotype of the organisms. Note the inhibitory effect of glucose on the synthesis of β-galactosidase (see Ames, B.N. and Martin, R.G. (1964) *Ann. Rev. Biochem.*, **33**, 237).

## EXPERIMENT 12
## GENE EXPRESSION IN THE '*LAC*' REGION
## FOLLOWING GENETIC TRANSFER

This experiment demonstrates the appearance and accumulation of the products of two genes of the lactose region of *E. coli*, following their transfer from a male strain of K12 into a female. The experiment utilizes a male 'intermediate' strain harbouring a substituted F sex factor (F prime factor) (see p. 258) carrying the lactose region (F'–*lac*), which can be infectiously transferred at high efficiencies to females on contact. The infected female cells can thus be made heterozygous for the *lac* region and complementation in this region can thereby be studied. In this experiment the male transfers an F-prime factor carrying the lactose genes $i^+z^+$ into a female which is genetically $i^-z^-$. The appearance in this female of the dominant $z^+$ and $i^+$ activities can thus be detected. The $z^+$ gene expresses itself in the structure of the enzyme β-galactosidase, whereas the $i^+$ gene product, after an initial delay, represses the synthesis of this enzyme, except when an inducer is present.

As a prelude to the experiments the two strains EMG11 (F'–*lac*$^+$ ($i^+z^+$)) and EMG12 (*lac*$^-$ ($i^-z^-$) F$^-$)* should be cultured overnight, each in 5 ml M9-glycerol medium.

* In fact EMG12 is a strain with the genotype $i^-$ $o^°$ $z^+$ $y^+$. However, since the effect of the $o^°$ mutation is to repress only the $z^+$ and $y^+$ genes which are linked to it, it is a suitable recipient in which to observe the activity of the newly-introduced $i^+$ and $z^+$ genes (which are, of course, transferred linked to an $o^+$ gene and thereby normally active).

K

## (a) Requirements

1 ml o/n cultures of EMG 11 and 12 in M9 + glycerol
60 ml M9–glycerol medium
5 × 250 ml sterile Erlenmeyer flasks
43 large tubes
0·2 ml chloramphenicol solution (CM)—2·5 mg/ml
0·2 ml 5–fluorouracil solution (FU)—2 mg/ml
0·2 ml thymidine solution (TDN)—4 mg/ml
10 ml ONPG reagent
25 ml 1M $Na_2CO_3$ solution
4 ml TMG solution ($2 \times 10^{-2}$M)
Ice bath
Waterbaths set at 37° and 28°
Spectrophotometer

## (b) Method

### (6 h)

(1) Sub-culture from the o/n cultures of the two strains provided into M9–glycerol medium by transferring 0·4 ml of EMG 12 ($lac^-F^-$) to 40 ml medium (label this flask 'A') and 0·2 ml of EMG 11 ($F'-lac^+$) to 20 ml medium (label this flask 'B').

Time (min)

| | |
|---|---|
| −90 | (2) Incubate sub-cultures at 37° in waterbath with aeration by shaking until a concentration of about $2 \times 10^8$ cells/ml is reached (about $1\frac{1}{2}$ h). |
| 0 | (3) Mix 36 ml from 'A' and 18 ml from 'B' into a sterile flask (label this flask 'C').<br>Continue incubation but withdraw aeration. |
| 0 | Take 0·5 ml from A to assay tube* (label A0).<br>Take 0·5 ml from B to assay tube* (label B0).<br>Take 1·0 ml from C to assay tube* (label C0). |
| 5 | Take 1·0 ml from C to assay tube* (label C5). |
| 10 | Take 1·0 ml from C to assay tube* (label C10). |
| 15 | Take 1·0 ml from C to assay tube* (label C15). |
| 20 | Take 1·0 ml from C to assay tube* (label C20).<br>*Gently* remove 16 ml from C into a flask. Label 'D'.<br>Add 0·16 ml CM. Incubate D at 37°.<br>*Gently* remove 16 ml from C into a flask. Label 'E'.<br>Add 0·16 ml FU + 0·16 ml TDN. Incubate E at 37°. |

---

* Large tube containing one drop of toluene in ice bath.

| | |
|---|---|
| 30 | Take 1 ml from C, D and E each into an assay tube. Label C30, D30, E30. |
| | Take 0·5 ml from A and B each to assay tube. Label A30, B30. |
| 40 | Take 1 ml from C, D and E each to assay tube. Label C40, D40, E40. |
| 60 | Take 1 ml from C, D and E to assay tube. Label C60, D60, E60. |
| | Take 0·5 ml from A and B to assay tube. Label A60, B60. |
| 80 | Take 1 ml from C, D and E to assay tube. Label C80, D80, E80. |
| 100 | Take 1 ml from C, D and E to assay tube. Label C100, D100, E100. |
| 120 | Take 1 ml from C, D and E to assay tube. Label C120, D120, E120. |
| | Take 0·5 ml from A and B to assay tube. Label A120, B120. Add TMG; 1·2 ml to C, 1·1 ml to D, 1·1 ml to E. |
| 140 | Take 1 ml from C, D and E to assay tube. Label C140, D140, E140. |
| 160 | Take 1 ml from C, D and E to assay tube. Label C160, D160, E160. |
| 180 | Take 1 ml from C, D and E to assay tube. Label C180, D180, E180. |

(4) Now assay all samples for β–galactosidase (see page 275).

(5) Plot your results against time after mixing.

### (c) Conclusions

Note that enzyme synthesis in the first part of the experiment is dependent on gene transfer (compare parental controls A and B with mixed–culture curves, C, D and E). Note also that the inhibition of synthesis by chloramphenicol and fluorouracil is not relieved by addition of the inducer TMG (curves D and E). Note also that enzyme synthesis stops (curve C) after 80–90 min due to the accumulation of an intracellular inhibitor. The activity of this intracellular inhibitor is removed by the addition of inducer TMG (curve C) (the inhibitor could be demonstrated to be a product of the $i$ gene by using an F′–$lac^-$ donor strain carrying the genes $i^-z^+$, in which case enzyme synthesis would continue linearly beyond 90 min).

If facilities exist to deal with, and count, radioactive isotopes, the experiment could be extended by following the incorporation of labelled amino acids, e.g. L–leucine labelled either with $^3H$ or $^{14}C$ during the experiment. In this case it will be found that only CM prevents net incorporation of label. FU inhibits enzyme synthesis by virtue of its incorporation into messenger–

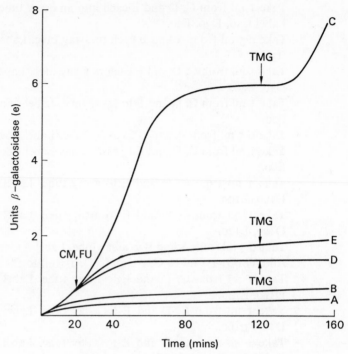

**Fig. 6.5.** Kinetics of transfer of *lac* genes.

RNA, in place of uracil, and thus giving an incorrect message to the protein-synthesizing machinery of the cell (Benzer S. & Champe S.P. (1961) *Proc. Nat. Acad. Sci. (Wash.)*, **47**, 1025).

## APPENDIX 6.1

## MEDIA

### (1) Nutrient broth (broth)

Oxoid No. 2 nutrient broth powder, 25 g.
Water, to 1 litre.
Dissolve broth powder in water.
Dispense in appropriate screw-capped bottles (e.g. 10 × 100 ml).
Autoclave at 121° for 15 min.
pH 7·4–7·6.

## (2) Nutrient Agar (NA)

Oxoid No. 2 nutrient broth powder, 25 g.
Davis New Zealand agar powder, 12·5 g.
Water, to 1 litre.
Suspend agar and broth powder in water.
Steam at 100° until dissolved ($1\frac{1}{2}$ h for 12 litres).*
Dispense into s/c bottles (500 ml).
Autoclave at 121° for 20 min.
pH 7·4 approx.

## (3) Water Agar (4/3 concentrate)

(Used as solidifying agent for minimal agar with Minimal Salts ($\times$ 4) (see No. 4), and with E.M.B. nutrient base (see No. 8) for E.M.B. agar (see No. 10).
Davis New Zealand agar powder, 20 g.
Water, to 1 litre.
Suspend and steam at 100° until dissolved (2 h for 12 litres).
Adjust pH to 7·2.†
Dispense 75 ml or 300 ml volumes into 100 ml or 400 ml bottles respectively.
Autoclave at 121° for 20 min.

## (4) Minimal salts (4 times concentrate)

$NH_4Cl$, 20 g.
$NH_4NO_3$, 4 g.
$Na_2SO_4$ anhydrous, 8 g.
$K_2HPO_4$ anhydrous, 12 g.
$KH_2PO_4$, 4 g.
$MgSO_4.7H_2O$, 0·4 g.
Water, to 1 litre.
Dissolve each salt in cold water in the order indicated, waiting until previous salt is dissolved before adding next (a light precipitate will be formed).
Filter into 25 ml or 100 ml bottles.
Autoclave at 121° for 15 min (no further precipitate should be formed).
pH 7·2.

## (5) 20 per cent glucose ($\times$ 100 concentrate)

D-glucose, 200 g.
Water, to 1 litre.

* Molten agar does not easily mix. When dissolved be particularly careful to ensure the agar is well distributed throughout the medium.
† If this step is omitted, subsequent autoclaving may lead to hydrolysis of the agar resulting in its inability to "gel" properly.

Dissolve in warm water.
Dispense into bottles (100 ml).
Autoclave at 109° for 10 min.
(Also applies for: maltose, lactose, D-galactose, D-xylose, D-mannitol, L-arabinose.)

### (6) *Liquid Minimal Medium (MM)

Minimal salts (× 4) see No. 4, 25 ml or 100 ml.
20 per cent glucose, 1 ml or 4 ml.
Sterile water, to 100 ml or 400 ml.
Mix three components under aseptic conditions just before use.

### (7) *Minimal Agar (MA)

Water agar (see No. 3), 300 ml or 75 ml.
Minimal salts (× 4) (see No. 4), 100 ml or 25 ml.
20 per cent glucose (see No. 5), 4 ml or 1 ml.
Melt agar at 100° or by autoclaving at 121° for 15 min. Add warmed sterile salts and glucose. Media is now ready to dispense in plates.

### (8) E.M.B. nutrient base

Difco Bacto casamino acids, 42·5 g.
Difco Bacto yeast extract, 5·25 g.
NaCl, 27 g.
$K_2HPO_4$, 10·5 g.
Water, to 1 litre.
Dissolve in the order indicated.
Dispense into 75 ml bottles.
Autoclave at 121° for 15 min.

### (9) E.M.B. dyes (× 100 concentrate)

Eosin yellow, 4 g.   Methylene blue, 0·65 g.
Water, to 100 ml.   Water, to 100 ml.
In each case, weigh dye out directly into final container and add water.
Autoclave at 121° for 15 min.
(Dyes obtainable from George Gurr Ltd, London S.W.6.)

### (10) *E.M.B. Agar

Water agar (see No. 3), 300 ml.
E.M.B. nutrient base (see No. 8), 75 ml.

* These media are made up immediately before use, either as liquid culture media or in order to pour plates.

20 per cent sugar (see No. 5), 20 ml.
Eosin yellow ($\times$ 100) (see No. 9), 4 ml.
Methylene blue ($\times$ 100) (see No. 9), 4 ml.
Melt agar at 100°, or by autoclaving at 121° for 15 min.
Add warmed, sterile base, sugar and dyes.
Agar is now ready to dispense in plates.

### (11) Soft agar

Difco 'Bacto' agar powder, 6 g.
Water, to 1 litre.
Steam at 100° for 1 h.
Dispense into bottles (50–100 ml).
Autoclave at 121° for 15 min.
pH 7·0.

### (12) M9 salts ($\times$ 10 concentrate)

$Na_2HPO_4$ anhydrous, 60 g.
$KH_2PO_4$ anhydrous, 30 g.
NaCl, 5 g.
$NH_4Cl$, 10 g.
Water, to 1 litre.
Dissolve in order indicated.
Dispense into 100 ml bottles.
Autoclave at 121° for 15 min.

### (13) *M9 medium

M9 Salts ($\times$ 10) (see No. 12), 100 ml.
20 per cent glucose, 20 ml.†
0·1M $MgSO_4$, 10 ml.
0·01M $CaCl_2$ 10 ml.
Sterile water, to 1 litre.
Autoclave each solution separately (121° for 15 min).
Mix all components aseptically just before use.

### (14) Buffer

$Na_2HPO_4$ anhydrous, 7 g.
$KH_2PO_4$, 3 g.

* This medium is made up immediately before use, either as liquid culture medium or in order to pour plates.
† When M9–glycerol medium is required, replace 20 ml glucose by 10 ml of 20% glycerol.

NaCl, 4 g.
$MgSO_4.7H_2O$, 0·2 g.
Water, to 1 litre.
Dissolve each salt in the order given, before adding next.
Dispense in 100 ml s/c bottles.
Autoclave at 121° for 15 min.


### (15) Phage broth

Difco Bacto peptone, 15 g.
Oxoid tryptone broth powder, 8 g.
NaCl, 8 g.
Glucose, 1 g.
Water, to 1 litre.
Dissolve in order shown.
Dispense into appropriate s/c bottles (100 ml).
Adjust pH to 7·2.
Autoclave at 121° for 15 min.


### (16) T-phage Nutrient Agar (TNA)*†

Oxoid tryptone broth powder, 10 g.
Difco 'Bacto' agar powder, 10 g.
NaCl, 8 g.
Glucose, 1 g.
Water, to 1 litre.
Suspend agar and nutrients in water.
Steam at 100° ($1\frac{1}{2}$ h for 12 litres).
Dispense into s/c bottles (500 ml).
Autoclave at 121° for 15 min.
pH 7·0.


### (17) Phage buffer

$Na_2HPO_4$ anhydrous, 7 g.
$KH_2PO_4$ anhydrous, 3 g.
NaCl, 5 g.
0·1M $MgSO_4$, 10 ml.
0·01M $CaCl_2$, 10 ml.

---

* When 'salt-free' TNA is specified (e.g. experiment 13) the medium is made up omitting the NaCl.

† When streptomycin is added to TNA to prevent phage-infected bacteria developing as infected centres, a concentration of 2000 µg ml is used, and is abbreviated to 'Strep TNA'.

Water, to 1 litre.
Dissolve first three compounds in order indicated then add prescribed volumes of next two sterile solutions.
Autoclave at 131° for 15 min.
Dispense in appropriate s/c bottles.

### (18) Subtilis Minimal Salts (SMS) (four times concentrate)

Ammonium sulphate, 8 g.
$K_2HPO_4$, 56 g.
$KH_2PO_4$, 24 g.
Sodium citrate $2H_2O$, 4 g.
$MgSO_4.7H_2O$, 0·8 g.
Water, to 1 litre.
Dissolve in order indicated.
Autoclave at 121° for 15 min.
Dispense into appropriate s/c bottles.

### (19) *Subtilis Minimal Medium (SMM) (for competent cultures)

SMS ($\times 4$) (see No. 18), 100 ml.
20 per cent glucose (see No. 5), 10 ml.
0·8 per cent sterile casein hydrolysate, 10 ml.
Sterile water, to 400 ml.
Mix aseptically.

### (20) *Subtilis Minimal Agar (SMA)

Water agar (see No. 3), 300 ml.
SMS ($\times 4$) (see No. 18), 100 ml.
20 per cent glucose (see No. 5), 20 ml.
Melt agar at 100° or by autoclaving at 121° for 15 min.
Add warmed sterile salts and glucose.
Agar is now ready to dispense in plates.

### (21) Nutrient agar slopes

Difco Bacto nutrient broth powder, 15 g.
NaCl, 5 g.
Difco 'Bacto' agar powder, 12·5 g.
Water, to 1 litre.
Dissolve agar in water by steaming (1 h).

* This medium is made up immediately before use, either as liquid culture medium or in order to pour plates.

Add and dissolve other ingredients.
Dispense 2·5 ml into 5 ml bijoux bottles.
Autoclave at 121° for 15 min.
Allow to set in a sloping position.

### (22) Stab medium

Difco nutrient broth powder, 0·9 g.
NaCl, 0·5 g.
Difco 'Bacto' agar powder, 0·75 g.
Water, to 100 ml.
Dissolve agar in water by steaming for $\frac{1}{2}$ h.
Add and dissolve other ingredients.
Autoclave at 121° for 15 min.
Dispense aseptically into sterile tubes* under ultraviolet hood (fill tube about half full).
Cap aseptically under ultraviolet hood.

### (23) Supplements

Supplements should be added to media to give the following concentrations:
Amino acids, 20 μg/ml (of L– form).
Vitamins, 1 μg/ml.
Streptomycin † (SM), 200 μg/ml.
Streptomycin ‡ (Strep), 2000 μg/ml.

* Tubes are presterilized by heat overnight, wrapped in foil. Caps are sterilized by rinsing in absolute alcohol, removing to sterile paper under ultraviolet hood to dry.
† This concentration of streptomycin is adequate for all purposes where the intention of the experiment is to prevent growth of streptomycin-sensitive bacteria (it is abbreviated to SM).
‡ In experiments with virulent phage (pp. 247–256) where streptomycin is added to kill phage-infected bacteria and prevent formation of infected centres, a concentration of 2000 μg/ml is usually added and this is abbreviated to 'Strep'.

## APPENDIX 6.2

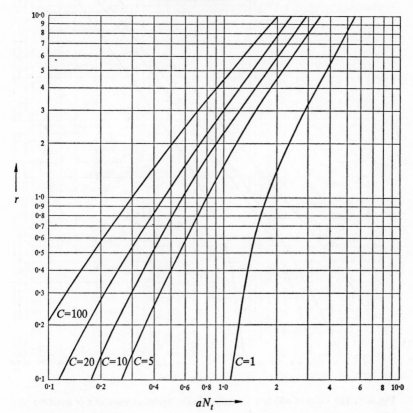

**Fig. 6.6.** The value of $aN_t$ as a function of $r$ for various values of $C$ ($r$ from 0·1 to 10).

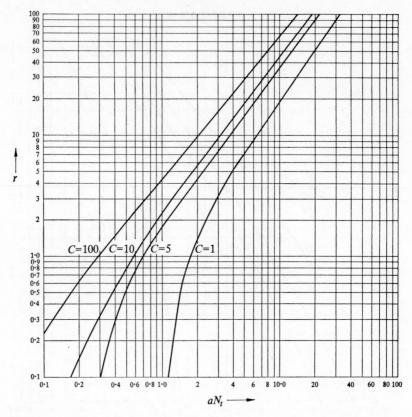

**Fig. 6.7.** The value of $aN_t$ as a function of $r$ for various values of $C(r$ from 0·1 to 100).

# APPENDIX 6.3

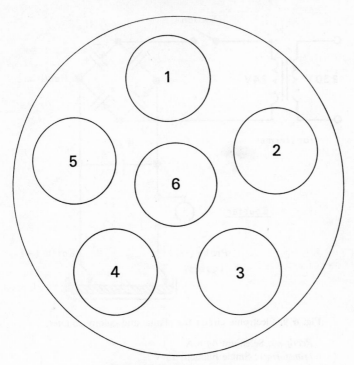

**Fig. 6.8** Template for Expt. 8.

## APPENDIX 6.4

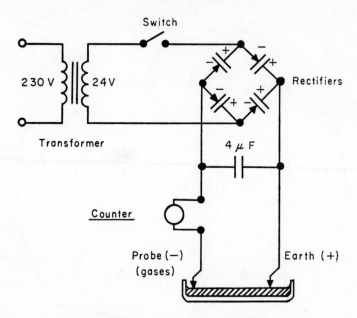

**Fig. 6.9.** Electronic circuit for plaque and colony counter.

*Rectifiers:* Selenium 65 mA
*Transformer:* Single filament 24 V
*Counter:* 'Sodeco' Impulse Counter 24 V d.c. (TCeZ4E)
A commercial apparatus is available (see p. 294).

# APPENDIX 6.5

## LIST OF COMMERCIAL SUPPLY HOUSES

**Supply Houses (U.K.)**

(1) *Rotor* (Matburn blood cell suspension mixers, 240 V, 50 cycles)
Originally made by
Matburn Ltd,
25 Red Lion Street,
London W.C.1,
now taken over by
Magnavox Company Ltd,
Alfred's Way, By-Pass Road,
Barking, Essex.

(2) *Petroff-Hauser counting chambers*
Arnold R. Horwell Ltd,
2 Grange Way, Kilburn High Road,
London N.W.6.

(3) *Whirlimix* (Whirlimixers WM/250)
Fisons Scientific Apparatus Ltd,
Loughborough, Leicestershire.

(4) *Microid shaker*
Baird and Tatlock Ltd,
Freshwater Road,
Chadwell Heath, Essex.

(5) *Hand-tally counter*
Griffen and George Ltd,
Ealing Road,
Alperton,
Wembley, Middlesex.

(6) *Electronic counter* ('*Sodeco*'—*Geneva*)
Stonebridge Electrical Co,
6 Queen Anne's Gate,
London S.W.1.

(7) *Aeration pump* (piston pump type)
Medcalf Brothers Ltd,
Cranborne Industrial Estate,
Cranborne Road,
Potters Bar, Middlesex.

(8) *Valves and manifolds* for above (airline regulator valves)
    Medcalf Brothers, Ltd,
    (see 7 above).

(9) *U.V. lamps* (Hanovia model 12 and model 6A UV lamps)
    Engelhard-Hanovia Industries Ltd,
    Slough, Bucks.

(10) *Screw-capped bottles* (various sizes)
    United Glass Ltd,
    79 Kingston Road,
    Kingston, Surrey.
    or
    Johnsen & Jorgensen Ltd,
    Herringham Road, S.E.7.

(11) *Pasteur pipettes*
    Arnold R. Horwell Ltd,
    2 Grange Way, Kilburn High Road,
    London N.W.6.
    (Distributors)
    or direct from
    Harshaw Chemicals Ltd,
    Daventry, Northants.

(12) *Aluminium caps* (for test tubes)
    Oxoid Ltd,
    Southwark Bridge Road,
    London E.C.4.

(13) *Plastic plates*
    Sterilin Ltd,
    9 The Quadrant,
    Richmond, Surrey.

(14) *Stab vials* (Trident containers series SNB)
    Camlab (Glass) Ltd,
    Milton Road,
    Cambridge.

(15) *Platinum wire* (10 per cent iridium-platinum wire, 24 s.w.g.)
    Johnson, Matthey & Company Ltd,
    78 Hatton Garden,
    London E.C.1.

(16) *Media chemicals*
Oxoid Ltd,
Southwark Bridge Road,
London E.C.4.
Baird & Tatlock Ltd (Difco products),
Freshwater Road,
Chadwell Heath, Essex.
Thomas Kerfoot Ltd (sugars),
Vale of Bardsley,
Ashton-u-Lyne, Lancs.
George T. Gurr Ltd (dyes),
136–138 New Kings Road,
London S.W.6.
British Drug Houses Ltd (various),
Poole, Dorset.
Chemicals to be imported can usually be obtained through:
V.A.Howe Ltd,
46 Pembridge Road,
London W.11,
or
Kodak Research Chemicals Ltd,
Kirkby, Liverpool.

## Supply Houses (U.S.)

(1) *Petroff-Hauser counting chambers*
E.H.Sargent & Co,
4647 West Foster Avenue,
Chicago, Illinois 60630.

(2) *Vortex Genie Mixer (Whirlimix)*
Aloe Scientific,
1831 Olive Street,
St Louis, Mo. 63103.

(3) *Burrell Wrist-Action Shaker* (Microid shaker)
New Brunswick Scientific Co,
1130 Somerset Street,
New Brunswick, N.J. 08903.

(4) *Hand-tally counter*
E.H.Sargent & Co,
(see 1 above).

(5) *Electronic counter*—Luminescent Bacterial Colony
Counter Model C-101 (complete apparatus)
New Brunswick Scientific Co,
(see 3 above).
Sodeco 12 V d.c. counter
Landis & Gyr, Inc,
34 W. 45th Street,
New York, New York 10036.

(6) *U.V. lamp*—G.E. 15 watt (G15T) 'Germicidal' low-pressure
mercury resonance lamp
General Electric,
1 River Road,
Schenectady, New York 12305.

(7) *Screw-capped bottles* (various sizes)
Aloe Scientific
(see 2 above).

(8) *Pasteur pipettes*
Bellco Glass Inc,
P.O. Box B,
Vineland, New Jersey, 08360.

(9) *Aluminum caps*
Colab Laboratories, Inc,
1526 Hallstead,
Chicago Heights, Illinois 60411.

(10) *Plastic plates*
Aloe Scientific
(see 2 above).

(11) *Stab vials*

E.H.Sargent & Co
(see 1 above).
Screwcapped
W.Hillen,
8 Habichtweg,
Widdersdorf, 5021,
West Germany.

(12) *Platinum wire*
Aloe Scientific
(see 2 above).

(13) *Media chemicals*

Agar     Davis Gelatin (NZ) Ltd,
         P.O. Box 9542,
         Woolston,
         Christchurch, New Zealand.

Media and     Colab Laboratories, Inc
sugars        (see 9 above).

              Difco Laboratories,
              920 Henry Street,
              Detroit, Michigan 48201.

Dyes     Allied Chemical Co,
         National Aniline Division,
         41 Rector Street,
         New York, New York, 10006.

Amino acids,     Nutritional Biochemicals,
vitamins, etc.   21010 Miles Avenue,
                 Cleveland, Ohio 44128.

                 Cal Biochem,
                 3625 Medford Street,
                 Los Angeles, California 90063.

Nitroso-     Aldrich Chemical Co,
guanidine,   2371 N. 30th,
aminopterin  Milwaukee, Wisconsin 53210.
Penicillin,  Difco Laboratories

streptomycin,   (see 13—Media and sugars—above).
sulphonamides

(14) *'Touchomatic' burners*

         Hanau Engineering Co,
         1235 Main TR,
         Buffalo, New York 14209.

# CHAPTER 7

# POPULATION AND ECOLOGICAL GENETICS

## P. M. SHEPPARD

## I. INTRODUCTION

Although any practical class concerned with population genetics and evolution should contain experiments to demonstrate the distribution of genes in populations and the action of natural as well as artificial selection, this is by no means always easy to arrange. Consequently, with an elementary class it is sometimes necessary to introduce these subjects by constructing a model using coloured beads or seeds to represent different allelomorphs in a population (p. 25). Care should be taken to see that the beads (or seeds) of different colour are the same size, weight and texture, unless it is desired to introduce a selective bias into the experiments.

## II. TO DEMONSTRATE THE HARDY-WEINBERG EQUILIBRIUM IN A LARGE POPULATION MATING AT RANDOM

In this account it will be assumed that black and white beads are being used to represent a pair of allelomorphs. A known number of black beads and of white beads, preferably more than 400, are put into a large bag or other container. Mix the beads thoroughly so that they are randomly distributed and without looking remove one bead at a time, note its colour and replace it in the container; mix again and withdraw the next one. Each bead represents the allelomorph contributed by a single parent to its offspring in the next generation. The contribution of the other parent to its offspring is obtained from the second bead. If two blacks are picked, the offspring is homozygous for black; if two white, homozygous for white; and if one bead of each sort is obtained the individual is a heterozygote. By repeating this process a large number of times the frequencies of the three genotypes in the next generation can be obtained and compared with the expected frequency,

296

calculated from the gene frequency in the parental population. If the proportion of black and white beads in the container is $p$ and $q$ respectively, where $p + q = 1$, then the expected proportions of the three genotypes will be $p^2 : 2pq : q^2$. $p^2$ is the proportion of black homozygotes, $q^2$ that of white homozygotes, and $2pq$ the heterozygotes. Different gene frequencies can be tried out and the results plotted on a triangular diagram, fig. 7.1. In the diagram each apex of the equilateral triangle represents 100 per cent of one of the genotypes, and the opposite base 0 per cent. Intermediate frequencies are represented by parallel lines at right angles to the perpendicular from the

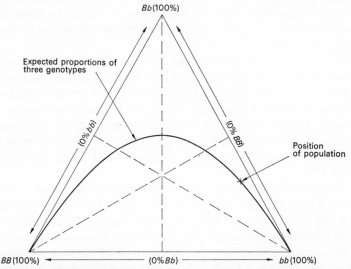

**Fig. 7.1.** Triangular diagram for plotting the frequencies of the three genotypes in a population polymorphic for a pair of autosomal allelomorphs. The curve gives the Hardy-Weinberg equilibrium for all possible gene frequencies.

apex to the base. Thus if the proportion of black homozygotes to heterozygotes to white homozygotes were 16 : 128 : 256, then the population could be plotted by measuring from the base 128/400 (0·32) of the way up the perpendicular dropped from $Bb$ and drawing a line at right angles to the perpendicular. One then measures from the base 256/400 (0·64) of the way up the perpendicular from $bb$ and draws a line at right angles to this perpendicular. Where these lines intersect is the position of the population on the diagram. If a similar line is calculated and drawn on the perpendicular from $BB$ it will intersect the other two at the same point, showing that the frequencies of the genotypes can be represented as a single point on the graph. Not only should each population be plotted, but also the curve showing the expected proportions for all frequencies of $q$ (see fig. 7.1). This is done by estimating the genotype frequencies from a number of values of $q$ and plotting the appropriate line through these points.* By plotting the results from large and small populations (number of pairs of beads withdrawn) it will be discovered that the departure of the observed genotype frequencies from expectation (the scatter of populations about the theoretical line) will be larger the smaller the population size.

The procedure can also be reversed and the gene frequency of the parental population established from the frequencies of the genotypes. If the container had an unknown frequency of black beads it can be estimated by the number of black beads divided by the total number of beads picked out. It will be apparent to the class that this is the same as twice the number of black homozygotes plus the number of heterozygotes divided by twice the total number of individuals, since the homozygotes are represented by two black beads and the heterozygotes by one.

The experiment can also be used to show that the square root of the frequency of one of the homozygotes is an estimate of the frequency of the allelomorph concerned. Thus, if one assumes that white is dominant in the model population (heterozygotes and white homozygotes cannot be distinguished) the frequency of the allelomorph for black will be estimated by the square root of the frequency of black homozygotes. These estimates can be directly compared with those obtained when heterozygotes can be recognized.

The model can be extended to situations in which two or more allelomorphs are present (p. 303) by using more than two colours of beads and the appropriate equilibrium frequencies for the genotypes determined (e.g. in the three allele case $(p + q + r)^2$).

If the locus is sex linked then the genotype of the heterogametic sex is found by extracting one bead and the homogametic sex by extracting two. It will be seen with a recessive sex linked condition that its frequency in the heterogametic sex at equilibrium is an estimate of the frequency of the allelomorph concerned in the population. However, equilibrium is not established in one generation of random mating as it is with an autosomal locus. To demonstrate this, use two bags, one for males the other for females, and start with different gene frequencies in the two bags. Sample *one* bead at a time from the female bag to represent the male genotype in the next generation until sufficient males have been produced. Sample one bead from the female bag and one from the male bag to obtain the genotype of each female in the next generation. Finally, adjust the male and female bags to their new gene frequencies and repeat until an approximation to equilibrium is reached. The proportions of the male genotypes can be plotted against generation. The frequencies of the three female genotypes in each generation can be plotted on a triangular diagram, numbering each point with the appropriate generation.

---

* Note that the particular population of beads falls exactly on the theoretical line.

## III. GENETIC DRIFT

To illustrate the effect of genetic drift in populations of different size a procedure similar to that used to investigate the Hardy–Weinberg equilibrium can be utilised. Having decided on the size of the population to be investigated a sample of that number of genotypes is obtained as previously, and the gene frequency in the sample calculated by counting the number of each genotype as explained above. The frequencies of beads in the container are then adjusted to the gene frequencies of the new population by taking out and adding the appropriate numbers of beads. Thus, if the frequency of black beads in the sample was 0·54 and the original container had 200 black beads and 200 white beads (a gene frequency of 0·5) 16 black beads would have to be added and 16 white beads removed from the container. The frequencies of genotypes and of the allelomorphs in each subsequent generation can be determined by repeating the process until one or other of the allelomorphs becomes lost from the population, that is to say when the other allelomorph becomes fixed.

It will often be found convenient to start the population at a gene frequency of 0·5 and then run the experiment until one allelomorph becomes fixed, or until 15 generations have passed, whichever happens first. These final results in terms of gene frequencies of one allelomorph can then be plotted in the form of a histogram and compared with that obtained from a population of different size. Using this procedure populations of five and 20 produce a considerable contrast.

To demonstrate the interaction between genetic drift and selection, one can modify the model by removing a proportion of the individuals before determining the gene frequency of the next generation. Thus with a population of 20 individuals consisting of six white homozygotes, ten heterozygotes, and four black homozygotes, 25 per cent selection against the black homozygotes would require the removal of one such homozygote giving a gene frequency for black of 16/38, consequently a suitable number of beads to represent the gene frequency in the next generation would be 176 black and 242 white. It will be noted that this is the smallest number of beads that can be used to give the correct gene frequencies and keep the number of beads greater than 400. The required number of beads can be determined quickly by dividing the number of beads in the sample into 400 and taking the nearest integer equal to or greater than the result. To obtain the correct number of beads of a particular colour this number is multiplied by the number of that colour of bead in the sample. Thus, in the present example, the number of black beads will be 11 × 16 and the number of white beads 11 × 22, 400/38 being greater than ten and less than 11. The procedure can be repeated from generation to generation and changes in gene frequency plotted.

To obtain a suitable comparison with the situation in which there is no selection the same size populations as before (five and 20) can be used, but with 50 per cent selection against both homozygotes. This is achieved by removing half the number of each type of homozygote before estimating the new gene frequency. If there are an odd number of homozygotes in one class one can determine whether one removes the odd one by using a random number table or by tossing a coin. However, it is more convenient, but not strictly legitimate, to remove one of the two beads, thus removing half the contribution of the particular homozygote to the next generation.

In the procedure outlined above (model 1) the breeding population varies somewhat in size from one generation to the next when selection is acting. It is perhaps better, therefore, to use a different model and remove every other homozygote as it is obtained (regardless of which type it may be) and to continue sampling beads until the correct number of surviving individuals is obtained (model 2). In the present example this would be either five pairs of beads or 20 pairs of beads representing five or 20 individuals.

With model 2 the distribution of gene frequencies after 15 generations in populations of five individuals will be very similar to that in populations of 20 without selection, whereas populations of 20 with selection will show much less variability in gene frequency and will tend to be grouped around a frequency of 0·5. Incidentally, a comparison of the two models (with or without replacement of individuals which died due to selection) can lead to a discussion of both genetic load and density-dependent factors in the control of population size.

By using populations much larger than 20 with selection against both homozygotes it is possible to deduce that such selection tends to produce stability with both allelomorphs maintained in the population. Furthermore, by using different initial gene frequencies and different selective values the relationship between the relative selective values of the two homozygotes and the gene frequency at stable equilibrium can be investigated.

In order to get a sufficient number of results to make meaningful comparisons between populations of different sizes and different magnitudes of selection it may be necessary to accumulate results over several years. Alternatively, the procedures can be simulated by a computer and a very large number of results accumulated in a short time. A program (approximating to model 2), written in KDF9 Algol, is given in appendix 7.1 together with some results. The program is written in a very simple form and is therefore rather slow. However, it may serve as a basis for constructing more efficient programs either in Algol or some other language. If such a program is used the results are worth comparing with the class histograms, since it may well be found that the proportion of populations in which one or other allelomorph has become fixed by generation 15 is smaller using the computer than in the class experiment. The possible reasons for this can be discussed. The most likely reason is that the students failed to mix the beads properly, so that a non-random sample is obtained. This is equivalent to a small amoun of inbreeding and illustrates that real populations may well have effective population sizes smaller than the number of actual breeding individuals. The more advanced student may also suggest that the program is not exactly simulating the class experiment and this may lead him on to learning programming in order to check his hypothesis.

The same basic procedures using either beads or a computer can be extended to examine the changes in gene frequency in some isolated populations where there is both immigration and emigration and also differences in selective pressures from place to place. When one wishes to investigate the effects of different breeding systems or the equilibrium resulting from different systems of incompatibility it is better to use a computer than beads, since the latter procedure is too tedious and time consuming, except perhaps when investigating equilibria involving incompatibility alleles in plants.

## IV. ARTIFICIAL AND NATURAL SELECTION IN ANIMALS

The fruit fly *Drosophila melanogaster* (see chapters 1 and 3) is a convenient organism for demonstrating the action of natural selection. The disadvantages of almost all major mutants can be shown by starting a population of homozygotes in a bottle and then introducing a few wild type males. The percentage of wild type flies in each generation can be counted and plotted to show the rate of elimination of the mutant. It is convenient to transfer the flies to a new bottle at the beginning of each generation and to remove the parents before the next generation appears, so that the phenotype frequency in each generation can be determined without ambiguity. It is possible to investigate the relevance of the environment by setting up populations in which a particular aspect of the environment, such as density of flies, the temperature, or the composition of the food medium, is varied and to see whether such differences affect the rate of elimination of the mutant (e.g. try *vg* in wet medium where it tends to do well since the wings do not stick to the sides of the container). Convenient mutants for such experiments are those given in chapter 1. Some mutants, notably ebony and sepia, may not be entirely eliminated but establish a stable equilibrium with the homozygote at low frequency. For advanced class work it is worth taking individuals from such colonies and preparing new populations homozygous for the mutant in some of which a single chromosome carrying the mutant is used as the original founder of the mutant stock, and others in which two or more such chromosomes have been included. The experiment can then be repeated. If all the populations give the same results it suggests that the effect of the mutant is determining the selective pressures, but if they do not then it may be concluded that the rest of the genotype is also important.

Populations using homozygous lethals such as stubble *Sb* (p. 8) can also

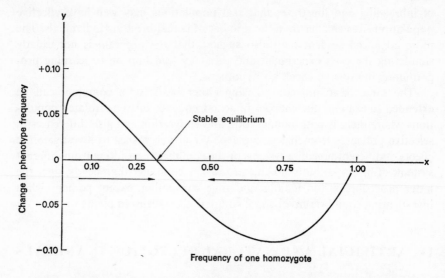

**Fig. 7.2.** Diagram showing the expected change in the frequency of one homo-zygote when its selective advantage (probability of survival) is 0·569, and that of the other homozygote is 0·264. The arrows indicate the direction of change of pheno-type frequency and the stable equilibrium point.

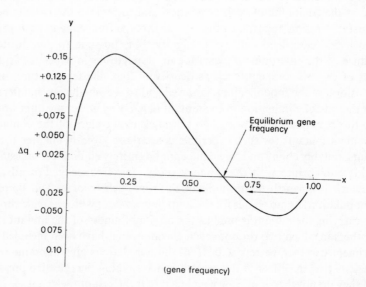

**Fig. 7.3.** The expected change in gene frequency of the same population whose change in phenotype frequency is plotted in fig. 7.2.

be very imformative since one can easily calculate the expected change in gene frequency each generation and compare it with the actual one. Thus:

$$q_1 = \frac{pq + q^2 \times 0}{p^2 + 2pq + q^2 \times 0}$$

since the homozygote $q^2$ is lethal. This simplifies to $q_1 = q/(1+q)$ (remembering that $p + q = 1$) and the change in gene frequency, $\Delta q$, is $q_1 - q$. If stubble is eliminated more rapidly than expected one can deduce that the wild type homozygote $Sb^+Sb^+$ is at an advantage to the heterozygote $Sb^+Sb$. If, on the other hand, it disappears more slowly then the wild type homozygote is at a disadvantage and a stable equilibrium should eventually be obtained. The stable gene frequency, $p$, of the wild type allelomorph will be

$$\frac{1 - c}{(1 - a) + (1 - c)}$$

where $c$ is the probability of survival of $q^2$ (in this case 0) and $a$ is the probability of survival of the other homozygote ($p^2$) and is less than 1. With one of the homozygotes lethal, the selective value of the other ($a$) can be estimated since the new gene frequency $q_1$ in each generation is

$$\frac{pq + cq^2}{ap^2 + 2pq + cq^2}$$

However, if $c$ is not equal to 0 and one of the phenotypes is dominant then it requires observations over more than one generation to estimate both $a$ and $c$. Furthermore, the calculations require iterative procedures and are best done on a computer. Figs. 7.2 and 7.3 give the results of such calculations from a class experiment using different initial frequencies of ebony $e$ and wild type $e^+$. The relative selective values of the homozygote wild type ($a$) is estimated as 0·264 and that of ebony ($c$) as 0·569.

# V. GENE FREQUENCIES IN NATURAL POPULATIONS

It is possible to relate the bead experiments to natural populations if there is a polymorphic species available where the genetic control of the morphs is known. Samples can be taken and the gene frequencies estimated. It will usually be found that dominance is complete for such characters, so that the gene frequencies will have to be estimated by taking the square root of the frequency of the recessive class. There are appropriate modifications for situations in which there are multiple allelomorphs. Thus, if the ABO blood groups of the class are determined (p. 69), the frequencies of the three allelomorphs, $I^A$, $I^B$ and $i$, will be given by $1 - \sqrt{\bar{O} + \bar{B}}$, $1 - \sqrt{\bar{O} + \bar{A}}$, and $\sqrt{\bar{O}}$, where $\bar{A}$, $\bar{B}$ and $\bar{O}$ are the frequencies of the A, B and O phenotypes respectively.*

White clover, *Trifolium repens*, can be found in most temperate climates, and the frequencies of the genes *Ac* $(p)$, *ac* $(q)$ and *Li* $(r)$ and *li* $(s)$ (p.42) can be determined. It is then possible to estimate the expected frequencies of the four phenotypes, *AcLi*, *Acli*, *acLi* and *acli*, since their ratio is $p^2r^2 + 2p^2rs + 4pqrs + 2pqr^2$, $p^2s^2 + 2pqs^2$, $q^2r^2 + 2q^2rs$ and $q^2s^2$ respectively (table 7.1).

Table 7.1

| Geno-types | Fre-quencies | $Ac$ $p^2$ | $Ac$ $2pq$ | $ac$ $q^2$ | Genotypes Frequencies |
|---|---|---|---|---|---|
| *Li* | $r^2$ | $p^2r^2$ $AcLi$ | $2pqr^2$ $AcLi$ | $q^2r^2$ $acLi$ | |
| *Li* | $2rs$ | $2p^2rs$ $AcLi$ | $4pqrs$ $AcLi$ | $2q^2rs$ $acLi$ | |
| *li* | $s^2$ | $p^2s^2$ $Acli$ | $2pqs^2$ $Acli$ | $q^2s^2$ $acli$ | |

Any significant departure from the joint frequencies of the phenotypes may be due to their incorrect determination, to local inbreeding and thus heterogeneity in the data, or to selection. In this example selection will be an unlikely explanation since the loci are unlinked. However, in other polymorphic species where the genes are linked, considerable linkage disequilibrium can be found. Returning to the example of clover, it is useful in hilly regions to take samples at various altitudes to see if there is a change in gene frequencies. The results can then be discussed in the light of the work of Daday (1965) and Bishop & Korn (1969) on this species.

## VI. QUANTITATIVE CHARACTERS

The methods of showing the effect of artificial selection on quantitative characters, such as bristle number in *Drosophila*, are given in chapter 3. Similar experiments to determine the action of 'natural' selection can be designed by exposing the larvae to noxious substances such as salt (NaCl), salts of heavy metals such as copper, zinc and lead, or some other substance such as an insecticide. The amount that should be added to the medium should be just enough to kill about 60 per cent of the larvae. This concentration can

*The frequencies may not quite equal 1, since the phenotype AB is not used. An improved estimate for the frequencies of $I^A$, $I^B$ and $i$ is respectively $(t-s)/v$; $(u-s)/v$ and $s/v$ $(s = \sqrt{O}, t = \sqrt{O + A}, u = \sqrt{O + B}, v = t + u - s)$ using the *numbers* in each phenotype.

be determined by adding various amounts to the food medium and then placing eggs or very young larvae on it. The proportion surviving can then be estimated as the proportion emerging compared with a control culture without the poison. The results can be plotted on probability paper and the concentration which kills 60 per cent of the insects (LC60) is read off the resulting graph. Alternatively, more accurate estimates may be obtained using a Probit analysis (for references to suitable statistical texts see chapter 3). In the selection experiment the adults should be discarded from the bottles, kept at 25°C, after seven days from the beginning of the culture and the flies, which will be parents of the subsequent generation, collected on day 14. The LC60 of the stock can be determined from time to time, perhaps at five generation intervals, and plotted to determine whether selection is effective, that is to say whether there is a significant positive regression of LC60 on generation. The experiment is useful since it can be related to the adaptation of natural populations to environments with high concentrations of such substances (see p. 308, also Ford 1971; Bradshaw, McNeilly & Gregory 1965; Allen & Sheppard 1971).

## VII. NATURAL SELECTION, PROTECTIVE COLOURATION, AND FREQUENCY DEPENDENT SELECTION

Natural selection in the wild is very difficult to demonstrate to a large class. However, selected students can undertake relatively simple but time consuming experiments to illustrate some of the principles involved.

### 1. ARTIFICIAL PREY

Perhaps the simplest method is to make artificial caterpillars out of a paste of $\frac{1}{3}$ lard and $\frac{2}{3}$ flour. This can be given various colours by adding vegetable dyes normally used for culinary purposes. The caterpillars can be made up by squeezing out 30 mm lengths of the paste from an icing bag used for decorating confectionery. By placing caterpillars out on the lawn and observing which are taken by birds and which are left, the relative degrees of camouflage of the various colours can be determined (Turner 1961).

This experiment can be further refined by not using equal numbers of each type of caterpillar but by varying the proportions. If frequency dependent selection is operating then the proportion of each type taken will change with its frequency. Since birds apparently form hunting images similar to those found in man, it is frequently found that the commoner form is taken more often than would be expected from its relative abundance and degree of camouflage.

The principle of warning colouration and mimicry can also be demonstrated (see O'Donald & Pilecki 1970). Thus, if an effective bird repellant (e.g. anthraquinone) is added to one of the types of caterpillar, the frequency with which it is taken will drop as the birds begin to learn to avoid that colour. The effectiveness of Batesian mimicry can also be studied if various proportions of caterpillars of the same colour as the distasteful ones but containing no repellant are included in the experiment. For the really advanced student, mimics resembling but not identical in colour to their distasteful models can also be used.

## 2. NATURAL PREY

If a moth trap is available to catch large numbers of moths, or if suitable species can be bred from caterpillars, experiments similar to those using artificial caterpillars can be undertaken using natural prey. The moth or insect to be used is killed by putting it in an atmosphere of carbon dioxide or by freezing. The dead insects are kept, until required, in air-tight containers in a deep freeze at $-20°C$. In industrial areas where there are industrial melanics (e.g. *Biston betularia*) these prove very useful material. The melanic and typical forms should be taken out of the deep freeze and placed in natural attitudes on tree trunks or other suitable sites. They can be secured there by a small spot of transparent waterproof glue placed on the substrate. The underside of the thorax of the moth is placed on this adhesive. The moths should be spread over a considerable area (e.g. not more than two moths per 200 m$^2$). Twenty-four hours after they are put out the sites can be revisited and a record made of which moths have disappeared. Thus the 24 hour survival rate of the forms can be estimated (see Bishop 1972). In order to get a significant result habitats should be used where one of the morphs is very much more conspicuous than the other. Care should be taken to use an area where predatory insects such as wasps or ants are not common since otherwise there will be a considerable non-selective elimination due to them.

To show the effect of warning colouration a number of species of insect can be put out in an area where birds are used to feeding, such as a bird tray or lawn, and the order in which the species are taken noted. Cryptic moths can be used for the palatable prey and warningly coloured moths, butterflies or Hymenoptera for the inedible ones. The Arctiidae and the Zygaenidae among moths, as well as the Danaidae amongst butterflies, and the bees and wasps among Hymenoptera are usually highly distasteful and suitable for this experiment. If mimics of any of these models are available they can also be included in the experiment.

# VIII. THE RESULTS OF ARTIFICIAL AND
## NATURAL SELECTION

## 1. DOMESTICATION

Mice can be used to illustrate the effect of artificial selection on behaviour. It is only necessary to compare the behaviour of very young wild mice to similar tame mice, and if possible the $F_1$ between them, to illustrate this point. The very young wild mice and the $F_1$ explode in all directions if picked up in the hand, whereas domestic mice remain quiescent. Care should be taken to do the experiment in a mouse-proof room or the wild mice may well escape. Wild mice may also be found more difficult to breed than their tame counterpart, but the female can sometimes be persuaded to do so if a mouse wheel is available in the cage.

The effect of domestication can also be demonstrated by comparing cultivated varieties of plants with their wild progenitors. Particularly demonstrative is the difference between the wild and cultivated sunflower, in which the former has many small heads and is much branched, whereas the latter has one or a few very large heads and a single stem. Alternatively the various domestic races of cabbage and carrot can be compared with their wild ancestors. Where the $F_1$ and $F_2$ generations can be produced the heritability of these characters can be estimated (see chapter 3).

Perhaps one of the most interesting and easily demonstrated effects of domestication in plants is an alteration in the breeding system. The Cruciferae are particularly useful for demonstrating the change from outbreeding to inbreeding under domestication. It is only necessary to protect the flowers from cross pollination (p. 64) and see which set seed and which do not. The matter can be taken further by studying the chromosome numbers (chapter 4) to determine whether polyploidy is involved, since this sometimes promotes inbreeding.

## 2. WILD POPULATIONS

Higher plants are particularly convenient organisms for demonstrating adaptation to specific environments as the result of natural selection. It is only necessary to take plants or seeds of one species from two contrasting environments and grow them under constant conditions. Any genetic differences (see chapter 3, p. 124) can be determined and their adaptive significance (if any) discussed. For example a transect from a very exposed to a sheltered site often shows that plants from the exposed site are genetically more dwarf. This would be expected since it reduces wind damage, whereas greater height in a sheltered location aids competitiveness with other species (Aston & Bradshaw 1966).

Adaptation to toxic substances such as heavy metals in the soil has been experimentally studied (see Bradshaw, McNeilly & Gregory 1965; Allen & Sheppard 1971) and can be investigated by advanced students if tolerant and non-tolerant stocks are available from the local mine tips or areas of industrial waste. If such stocks are not available they can sometimes be produced by selection. Thus Bradshaw has shown that commercial seed of the grass *Agrostis tenuis* will produce about three plants in a thousand which are heavy metal tolerant. These can be selected for by sowing seed on contaminated soil, or soil to which salts of the appropriate metal have been added. It should be borne in mind that if metal is to be added to the soil a considerable amount may be necessary since it tends to be complexed in the soil and therefore unavailable to the plant. The amount to be added can be judged by using a series of concentrations and determining the LC99·9. Alternatively it is quite simple to select for copper tolerance in yeast by adding appropriate quantities of copper sulphate to a complete yeast medium (p. 180). The tolerant and non-tolerant strains can be studied by the class by observing their growth and estimating their survival in various concentrations of copper.

One of the best methods of estimating metal tolerance in higher plants is to choose a species which roots well from cuttings, and examine root growth in aqueous solutions containing the heavy metal and not containing it. A method of measuring tolerance in grasses is given in appendix 7.2 and it can easily be adapted for other species such as the American plant *Mimulus guttatus* (Allen & Sheppard 1971). If adaptation is being studied in plants it is important to measure it in adults found *in situ* and in the progeny of seed samples from the same population. The seed sample may give a different mean value suggesting directional selection, or a greater variance suggesting stabilizing selection. Furthermore, the spatial distribution of ill adapted plants with respect to the habitat to which they are adapted and the prevailing wind (or other pollen or seed dispersing factor) may reveal the effect of gene flow and its interaction with selection (see Bradshaw 1971).

## IX. CONCLUSION

The study of gene frequencies in populations, the effects of genetic drift and the operation of natural selection is not easy for a student since among other things it is very time consuming. However, if he can manage it, it is very rewarding for it will illustrate the great difficulties encountered in interpreting field data. Furthermore, it will allow him to apply many of the techniques learnt in other parts of the genetics course. This chapter is only meant to be a brief sketch of some of the experiments and observations that can be made using artificial and natural populations, and no attempt has been made to cover such important phenomena as genotype/environment interactions (see p. 120), gene flow, breeding systems, the evolution of sex, of isolating mechanisms, or the phenomenon of introgression. However, ideas of how to study these can be obtained from the literature if a little enterprise is used. Furthermore, many of the ways in which adaptation can be studied experimentally will be found in Ford (1971) and other standard works on population and ecological genetics.

## X. REFERENCES

ALLEN W.R. & SHEPPARD P.M. (1971) Copper tolerance in some Californian populations of the Monkey Flower, *Mimulus guttatus. Proc. R. Soc. Lond. B.*, **177**, 177-196.

ASTON J.L. & BRADSHAW A.D. (1966) Evolution in closely adjacent plant populations. II. *Agrostis stolonifera* in maritime habitats. *Heredity*, **21**, 649-664.

BISHOP J.A. (1972) An experimental study of the cline of industrial melanism in *Biston betularia* (L.) (Lepidoptera) between urban Liverpool and rural North Wales. *J. Anim. Ecol.*, **41**, 209-243.

BISHOP J.A. & KORN M.E. (1969) Natural selection and cyanogenesis in white clover *Trifolium repens. Heredity*, **24**, 423-430.

BRADSHAW A.D. (1971) Plant evolution in extreme environments. In Creed R. (ed) *Ecological Genetics and Evolution*, pp. 20-50. Blackwell Scientific Publications, Oxford & Edinburgh.

BRADSHAW A.D., McNEILLY T.S. & GREGORY R.P.G. (1965) Industrialization, evolution and the development of heavy metal tolerance in plants. *V Symp. Brit. Ecol. Soc.*, 327-343. Blackwell, Oxford.

DADAY H. (1954) Gene frequencies in wild populations of *Trifolium repens. Heredity*, **8**, 61-78.

DADAY H. (1965) Gene frequencies in wild populations of *Trifolium repens* L. IV. Mechanisms of natural selection. *Heredity*, **20**, 355-365.

FORD E.B. (1971) *Ecological Genetics*. Chapman and Hall, London.

O'DONALD P. & PILECKI C. (1970) Polymorphic mimicry and natural selection. *Evolution*, **24**, 395-401.

TURNER E.R.A. (1961) Survival values of different methods of camouflage, as shown in a model population. *Proc. zoo. Soc. Lond.*, **136**, 273-284.

L

## APPENDIX 7.1

```
begin   integer array da,pa1,pa2 [1:100],of[1:2];
        real q1,q2,c,B;
        integer i,j,n,x,aa,bb,r1,r2,z,G,A,k,L,m,o,p,q,s,f1,f2,f3;
        open(20); open(30); writetext(30,[population*simulation[c]]);
        f1:=format([nd.ddddd]); f2:=format([nddd.d]); f3:=format([ndddddddddc]);
        x:=read(20);m:=read(20);
        for L:=1 step 1 until m do begin copytext(20,30,[:;:]);
        n:=read(20); aa:=1;bb:=n;
        for i:=1 step 1 until n do da [i]:=read(20);
        q1:=read(20);q2:=read(20);q:=read(20);p:=read(20);
        c:=(bb+1−aa)/67101323;
        r1:=entier(q1×67101323);
        r2:=entier(q2×67101323);
        for o:=1 step 1 until p do begin s:=A:=G:=0; writetext(30,[[2c]]);
        for i:=1 step 1 until n do begin pa1[i]:=da[i];A:=A+pa1[i] end;
        write(30,f1,A/n);writetext(30,[[s]q]);write(30,f2,G);writetext(30,[[s]gen[c]]);
        k:=j:=1;
LI:     x:=8192×x; x:=x−(x÷67101323)×67101323;
        z:=entier(aa+c×x);
        of[j]:=pa1[z];
        if j=1 then begin j:=2; goto LI end else j:=1;
        x:=8192×x; x:=x−(x÷67101323)×67101323;
        if of[1]+of[2]=0 then begin
        if x<r1 then goto MI
        else goto LI end;
        ifof[1]+of[2]=2 then begin
        if x>r2 then goto LI end;
MI:     pa2[k]:=of[1];k:=k+1; pa2[k]:=of[2];k:=k+1;
        if k<n then goto LI;
        G:=G+1;A:=0;
        for i:=1 step 1 until n do begin A:=A+pa2[i];pa1[i]:=pa2[i]
        end;B:=A/n;
        write(30,f1,B);writetext(30,[[s]q]);write(30,f2,G);writetext(30,[[s]
        gen[2s]]);s:=s+1;
        if s=5 then begin writetext(30,[[c]]);s:=0 end;
        if B>0 andB<1 andG<q then begin k:=1; goto LI end
        end end;
        writetext(30,[[3c]end**=x]);write(30,f3,x);
        close(20); close(30)
        end→
```

```
513575;4;
POP*OF*5*50*PERCENT*SELECTION;
10;
1;1;1;1;1;
0;0;0;0;0;
0·5;0·5;15;200;
POP*OF*20*50*PERCENT*SELECTION;
40;
1;1;1;1;1;1;1;1;1;1;1;1;1;1;1;1;1;1;1;1;
0;0;0;0;0;0;0;0;0;0;0;0;0;0;0;0;0;0;0;0;
0·5;0·5;15;200;
POP*OF*5*NO*SELECTION;
10;
1;1;1;1;1;
0;0;0;0;0;
1;1;15;200;
POP*OF*20*NO*SELECTION;
40;
1;1;1;1;1;1;1;1;1;1;1;1;1;1;1;1;1;1;1;1;
0;0;0;0;0;0;0;0;0;0;0;0;0;0;0;0;0;0;0;0;
1;1;15;200;→
```

The data consist of an odd positive integer less than 67101323 to start the pseudo-random number generator. This should be different every time the program is run and the number for the next run is printed on the output as x=... The next integer is the number of different population sizes and selective values to be simulated. Then comes the title of the first group of populations, followed by the number of genes (twice the population size). The next row (or rows) gives the genes present represented as 0 or 1. The genotypes of the individuals are read off in pairs, that is the first population has two homozygotes of one type, a heterozygote, and two homozygotes of the other type. The next two values are the probability of survival of the homozygotes 00 and 11 respectively. Then comes the integer giving the generation at which simulation is to stop if prior fixation has not occurred. Finally, comes the number of populations under the particular heading to be simulated. In the example given it is 200. The title for the next set of conditions then follows, and so on. If the homozygote 00 and the heterozygote 10 are to have relative survival values of less than 1, and the homozygote 11 a value of 1, this can be achieved by replacing in the program = 2 by = 1 in the line

$$\text{if of } [1] + \text{ of } [2] = 2 \text{ then begin}$$

This will convert the second of the two probabilities of survival in the data from that of the homozygote 11 to that of the heterozygote. If the population is to be greater than 50 then the number 100 in line one of the program must be increased to a value equal to or greater than the number of genes in the population.

Some results obtained by using the program are given below. Table 7.2

**Table 7.2** Computer simulation results

| Gene frequency | Populations of 5 | | Populations of 20 | |
|---|---|---|---|---|
| | No selection | Selection | No selection | Selection |
| 0·000 | 360 | 74 | 26 | 0 |
| 0·025 | | | 15 | 0 |
| 0·050 | | | 18 | 0 |
| 0·075 | | | 12 | 0 |
| 0·100 | 27 | 20 | 20 | 0 |
| 0·125 | | | 17 | 0 |
| 0·150 | | | 21 | 1 |
| 0·175 | | | 15 | 1 |
| 0·200 | 28 | 52 | 17 | 0 |
| 0·225 | | | 17 | 1 |
| 0·250 | | | 27 | 1 |
| 0·275 | | | 24 | 2 |
| 0·300 | 26 | 105 | 24 | 5 |
| 0·325 | | | 37 | 16 |
| 0·350 | | | 20 | 33 |
| 0·375 | | | 32 | 35 |
| 0·400 | 35 | 157 | 32 | 52 |
| 0·425 | | | 30 | 83 |
| 0·450 | | | 31 | 99 |
| 0·475 | | | 37 | 111 |
| 0·500 | 38 | 164 | 21 | 118 |
| 0·525 | | | 35 | 102 |
| 0·550 | | | 22 | 85 |
| 0·575 | | | 29 | 74 |
| 0·600 | 40 | 155 | 42 | 56 |
| 0·625 | | | 26 | 44 |
| 0·650 | | | 29 | 38 |
| 0·675 | | | 26 | 23 |
| 0·700 | 24 | 111 | 32 | 12 |
| 0·725 | | | 22 | 5 |
| 0·750 | | | 27 | 1 |
| 0·775 | | | 24 | 0 |
| 0·800 | 19 | 59 | 25 | 2 |
| 0·825 | | | 17 | 0 |
| 0·850 | | | 16 | 0 |
| 0·875 | | | 24 | 0 |
| 0·900 | 32 | 26 | 24 | 0 |
| 0·925 | | | 15 | 0 |
| 0·950 | | | 15 | 0 |
| 0·975 | | | 12 | 0 |
| 1·000 | 371 | 77 | 45 | 0 |

shows the number of populations at each gene frequency after 15 generations using the computer model. The breeding population in every generation was either five or 20 individuals. Note that in the populations of five the gene frequency has become fixed (0 or 1) in more than half of them, but in those of 20 the distribution of gene frequencies is almost flat. In the presence of 50 per cent selection against the two homozygotes none of the populations of 20 individuals have become fixed and most are grouped around the starting point of 50 per cent gene frequency. In the populations of five with the same selection a number of populations have become fixed but many fewer than in the same sized ones without selection.

## APPENDIX 7.2

## THE MEASUREMENT OF
## HEAVY METAL TOLERANCE

The effect of heavy metals in general expresses itself in a suppression of root growth at moderately low (e.g. 0·5 ppm Cu) concentrations, and total inhibition at slightly higher concentrations (e.g. 2·0 ppm Cu).

## THE INDEX OF TOLERANCE

To quantify the effect of the heavy metal upon root growth, root growth made in the presence of the metal ion in solution is compared with root growth (ramets of the same plant) in solution not containing the metal. Generally the solution without the metal contains calcium nitrate at 0·5 g per litre. For copper and lead this addition gives better rooting. For zinc, however, tap but preferably deionized or distilled water will do just as well. The metal containing solution must, of course, be the same as the control, differing only in the addition of the metal at suitable concentrations.

Grasses are good material for testing as adults. Single plants can be split up into tillers, and provided each tiller has a node at the base it will produce roots readily in solution. Root growth in the absence of, and in the presence of, the metal can thus be tested many times over for a single individual.

## ADULTS

### Method

Material performs best under test in spring, autumn and winter. (In summer the flowering process interferes with rooting which becomes very erratic.)

Plants should preferably be allowed to grow in ordinary soil for a time prior to testing. However, tillers may be taken directly from the field and put straight into test.

Tillers are supported in plastic tubing, approximately 10 mm in diameter, the tubes themselves being supported in thin polystyrene film which can be drilled suitably to take a varying number of tubes. A slight lip on the top of the tubes will prevent them from falling out of the polystyrene support. The tubes should be about 20–30 mm long.

The polystyrene support holding the tubes is then placed across the top of a plastic beaker containing the solution and the tillers placed in position.

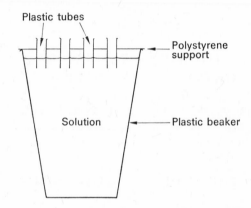

**Fig. 7.4.** Apparatus for testing heavy metal tolerance in adult plants.

Points of importance
1. Tiller must have a node at the base.
2. Tiller base must be under surface of solution.
3. Container for solution must be plastic, since glass will absorb ions from solution.
4. The solutions in the containers should be changed twice per week to allow some aeration and to keep ion levels up.
5. All roots should be removed from tillers beforehand.
6. The beakers and tubes into which the tillers from any one plant are placed should be random (remember to make a note which is which!).

Root growth as represented by the length of the longest root, should be measured after about two weeks from the start of the experiment. (The greater the number of tillers measured the greater the accuracy with which the index of tolerance may be measured.) In practice 15 tillers in each treatment (metal present/metal absent) has been found to be sufficient.

The index of tolerance for each plant is then found as follows.

$$\frac{\text{Index}}{\text{of}}_{\text{tolerance}} = \frac{\text{mean length of longest root in metal containing solution}}{\text{mean length of longest root in solution without metal}} \times 100$$

The procedure may be summarized as follows:

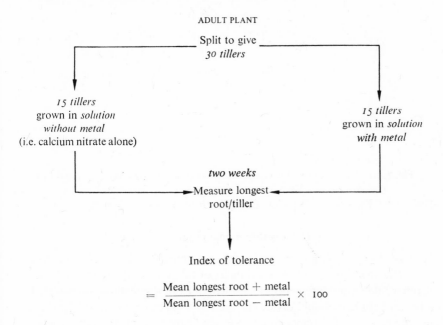

ADULT PLANT

Split to give
*30 tillers*

*15 tillers*
grown in *solution*
*without metal*
(i.e. calcium nitrate alone)

*15 tillers*
grown in *solution*
*with metal*

*two weeks*

Measure longest
root/tiller

Index of tolerance

$$= \frac{\text{Mean longest root} + \text{metal}}{\text{Mean longest root} - \text{metal}} \times 100$$

## SEED MATERIAL

A large number of seeds of any genus may be tested fairly quickly for tolerance, or alternatively individual seeds may be examined. In the first place information is collected only in the form of mean values for populations, in the second, indices of tolerance for individuals within populations are obtained and may be presented as histograms.

### To test whether a population (a seed sample) is tolerant

Solutions used are as for adults. Seed should be pregerminated on moist blotting paper in Petri dishes. When growth of the radicle commences the seeds should be transferred carefully to the 'apparatus' used for seedling testing. This consists of nylon stocking material supported at the surface of the solution in a plastic beaker. The stocking material may then be placed over the top of the beaker and held in position with a rubber ring. A more refined method is to cut the top off several beakers, and stick the stocking material over the lower surface using polystyrene cement. This is then placed inside an entire beaker (see below) and the solutions added to reach the surface of the stocking. It is important that the stocking material be kept tight.

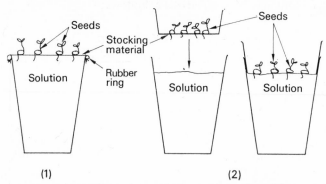

(1)                                               (2)

Fig. 7.5. Apparatus for testing heavy metal tolerance in newly germinated seedlings.

Seeds may then be grown and root growth measured as before.

### To test whether a population is tolerant

Seed need simply be grown in a metal containing solution, with a control in which the metal is absent. A mean index of tolerance may be obtained in this manner taking measurements of root length as in the method for adults. The advanced student should divide the seed into several samples so that he can use the analysis of variance to test the significance of any differences between test and control solutions.

Alternatively root growth of several populations may be compared with the root growth of 'control' population samples from known non-contaminated sources. The amount of rooting reflects tolerance.

### To obtain the index of tolerance for individual seedlings

This gives more work, and it is necessary to use the apparatus with the removable top (2). Longest root length is measured after ten days' growth, and again at 15 days from the beginning. This gives a value for root elongation per day. The metal solution is then added at the appropriate concentration, and after a further five days root length is again measured, and a second value, for root elongation with the metal present, is obtained. The index of tolerance is calculated as in the method for adults.

$$\text{Index} = \frac{\text{daily increase in length} + \text{metal}}{\text{daily increase in length} - \text{metal}} \times 100$$

If seed and adult samples are to be compared (p. 308) the index of tolerance for the adults should be measured in the same way as that for seedlings (by root elongation) and not by the index previously described for adults.

## SOLUTIONS

| Metal | Salt used | Solution: stock | For use |
|---|---|---|---|
| Zinc | $ZnSO_4 . 7H_2O$ | 4·977 g in 250 ml deionized or distilled water | 1 ml in 300 ml deionized or distilled water |
| Nickel | $NiSO_4 . 7H_2O$ | 0·172 g in 200 ml deionized or distilled water | 1 ml in 300 ml solution of calcium nitrate at 0·5 g/litre |
| Lead | $Pb(NO_3)_2$ | 1·150 g in 200 ml deionized or distilled water | 1 ml in 300 ml solution of calcium nitrate at 0·5 g/litre |
| Copper | $CuSO_4 . 5H_2O$ | 0·300 g in 500 ml deionized or distilled water | 0·5 ml in 300 ml solution of calcium nitrate as for lead |
| Calcium nitrate solution | | 0·5 g in 1 litre deionized or distilled water | |

# NAME INDEX

Numbers in **bold** face denote pages on which full references appear

320 NAME INDEX

Gall J.G.  162, **171**
Gibson J.B.  **30**
Gilbert W.  **275**
Gill J.J.B.  159, 170
Glass B.  **254**
Gluecksohn-Waelsch S.  **84**
Grant V.  **84**
Gregory R.P.G.  305, 308, **309**
Gregson N.M.  *fp.* 148
Grell E.H.  8, **23**
Griffith F.  242, **243**
Gross J.D.  **273**

Hagen C.W.  **67**
Hardy G.H.  299
Haskins K.P.  **36**
Hayes W.  **236, 248,** 258, 259, 260, **261,**
  272, **273**
Head J.J.  **84**
Hershey A.D.  **251**
Hoste R.  **36**
Hughes W.L.  134, **172**
Humphrey L.M.  134, **172**
Hungerford D.A.  153, **171**

Jackson R.W.  220, **223**
Jacob F.  274, **275**
Jinks J.L.  87, 108, 113, 115, 119, **129**
Johansson C.  166, **170**
John B.  170
Jones W.H.  152

Keer W.E.  **84**
Kelly P.J.  **30**
Komai T.  **84**
Kopriwa B.M.  135, **171**
Korn M.E.  304, **309**
Krug C.A.  **84**
Kudynowski J.  166, **170**
Kuwada Y.  134, **171**

La Cour L.F.  130, 134, 136, 137, 141,
  149, **171**
Laidlaw H.H.  **84**
Lambert R.J.  52, 55, 84
Lawrence M.J.  84
Lawrence W.J.C.  53, 54, **55, 82**
Leblond C.P.  135, **171**
Levan A.  132, **171**
Levine P.  **84**
Lilly L.J.  136, **171**
Lindley D.V.  12, **23, 129**

Lindsley D.L.  8, **23**
Lovis J.D.  170
Low B.  272, **273**
Luria S.E.  239, **240,** 242
Lush J.L.  **84**

McCarty M.  242, **243**
Macleod C.M.  242, **243**
McElroy W.D.  **254**
McKelvie A.D.  **33,** 84
McLeish J.  136, **171**
McNeilly T.S.  305, 308, **309**
McQuown F.R.  82, **84**
Magasanik B.  **275**
Mahoney R.  60, 61, **84**
Martin R.G.  **275, 277**
Mather K.  12, 17, **23,** 87, 102, 108, 113,
  **129**
Matthews S.  83
Mellman W.J.  153, **171**
Mendel G.  24, 83, 86
Meyer J.R.  133, **171**
Miles U.J.  137, **171**
Miller J.C.P.  12, **23, 129**
Mitchell H.K.  **64**
Modest E.J.  166, **170**
Monod J.  274, **275**
Moore K.L.  146, **171**
Moorhead P.S.  153, **171**
Morgan T.H.  1
Moses L.E.  113, **129**
Müller A.J.  31, 32, 33, **34**
Müller H.J.  20, 135, **172**
Muller-Hill B.  **275**

Nagao S.  **84**
Nakada D.  **275**
Nakamura T.  134, **171**
Nowell P.W.  153, **171**
Nowinski W.W.  130, **171**

Oakford R.V.  113, **129**
O'Donald P.  306, **309**
Oldroyd H.  **84**
O'Mara J.G.  133, **172**
Owen R.D.  8, **23**

Pardee A.B.  274, **275**
Pelc S.R.  135, **172**
Pilecki C.  306, **309**
Postlethwait S.N.  82, **84**
Prescott D.M.  134, 135, **170, 172**
Purdum C.E.  21, **23**
Pusey J.G.  **43,** 83

# SUBJECT INDEX

Page numbers in italics refer to diagrams or tables.

322

6656666

Fungi (*cont.*)
  handling 174–83
    growth tubes 176–7
    mass transfers 174
    media 177–82
    replica plating 176
    spreading spore suspensions 174–5
    stock maintenance 182–3
  heterokaryosis 184–5, 207, 211
  heterothallic, dipolar 186–8
    tetrapolar 188–92, Figs. 5.1 and 5.2
     *fp. 189*
  homothallic 183, 184, 185–6
  host-parasite relationship 222–3
  imperfect 183, 211
  induction of cytoplasmic mutant
    'petite' 196–8
  infections of *Drosophila* 6, 7
    of maize 51
  interallelic complementation 215–6,
    221–2
  life cycles and breeding systems
    183–92
  mutations induced by ultraviolet
    radiation 193–4
  parasexual cycle 210–5
  ordered tetrad analysis 204–6
  quantitative genetics, demonstration of
    194–6
    effect of medium and temperature on
     growth 124–6
    estimate number of genes involved
     128–9
    mean size of gene effects 129
    random mating, common parents
     114–9
  random spore analysis 199–202
  segregation 183, 188, 190–1, 200
  sexual compatibility 114, 184, 186–92,
    Plate 5.1 *fp. 189*, Plate 5.2,
    *fp. 189*
  uniparental inheritance of a cytoplasmic
    character 209–10
  unordered tetrad analysis 202–4
  variation and mutation 192–8

$\beta$-Galactosidase, assay 274–5
  induction of 276–7
  synthesis after genetic transfer 277–80
$\gamma$-rays, mutagenesis 135
Gelatin, medium 149
Gene flow 308, 309
Gene frequencies, changes in 300
  equal, random matings 110–14
  estimate of 119, 298, 303–4
  genetic drift 299, 300
  Hardy-Weinberg equilibrium 297, 298

  in natural populations 303–4
  selection and 301–3
  unequal 117
Genes, additive effects of 107, 108, 110,
    117, 123
  complementary 42–3, 78
  interactions, illustrated by beads 27
    in maize 45
    non-allelic (epistasis) 35, 64, 79, 118,
     119
  major 81
  modifying 29, 64, 81
  operator 274
  regulator 274
  structural 274
Genetic background 81
  mean effect of 100–1, 105
Genetic drift 299–301, 312–3
  selection and 299, 300, 308, 312–3
Genetic mapping, bacteria 258–9, 260–1,
    269–70, 270–3
  bacteriophages 250
  *Drosophila* 1, 8, 12, 15, 21–2, 23
  fungi, fine structure analysis 207–9
  meiotic 199–209
  mitotic, parasexual cycle 211–5
Genetics, basic, aims of teaching 24–5,
    37, 56–7, 77–8, 82–3
  course design 56–7, 77–8
    schools 78–9
    university 79–83
  crosses *qv* 25–30
  segregation *qv* 37–78
  statistics *qv* 75–8
Genetics, human, blood groups 69–71,
    303–4
  individuality charts 74–5
  pedigrees 68, 78, 81
  phenylthiocarbamide tasting 68–9
  secretor testing 71–2
  stature 86–7
  use of other characters 72–4
Genetics, population, gene frequencies in
    a natural population 303–4
  genetic drift 299–301, 312–3
  Hardy-Weinberg equilibrium 296–8
  selection *see* Selection
Genetics, quantitative, components of
    family means 100–10
  demonstration and cause of quantitative
    variation 88–97
  genotype-environmental interaction
    81, 87, 109, 120–6
  mean size of gene effects 129
  number of genes involved 127–9
  random matings, common parents
    114–20

Plants (*cont.*)
pollen grains   56, 61, 79
pressed   57
seeds   57
DNA synthesis   134–5, 165–6
fixation   138
formation of hybrids   158–9
maceration of tissue   137, 139
meiosis   133, 134, 154–6, 158–9,
Plate 4.5 *fp. 148*, Plate 4.6 *fp. 148*
Plate 4.8 *fp. 148*
mitosis   133, 147–50, Plate 4.2 *fp. 148*
sex chromosomes   162
tolerance and adaptation   308, 313–17
X-ray effects on   166
*See also Arabidopsis*; Maize; *Vicia faba etc.*
Pleiotropy   24, 28, 32
*see also* epistasis
Pneumococcus (*Streptococcus pneumoniae*) 242–3
Ploidy, changes in   133, 159, 183, 211–15
Pollen grains, mitosis   149–50, 166
starchy-waxy, of maize   56, 61, 79
Pollen mother cells   134, 154
Pollen tubes, mitosis   149–50, 166
Polygenes   81
*See also* Genetics, quantitative,
Polymorphism   38–43, 58, 71, 81, 135, 303
Polyploidy   133, 159, 307
Polytene *see* Chromosomes
Population, gene frequency in natural 303–4
size, density dependant factors   300
effective   301
genetic load and   300
*See also* Genetics, population
Potassium chloride   137, 152, 154
Potassium cyanide   136
Potato blight *see Phytophthora infestans*
Potence   107–8
Presporulation medium (PSP)   181, 203
Pretreatment methods *see* Cytogenetics
Prey, artificial   305–6
natural   306
*Primula*, possible research on   82
Probability paper   305
Probit analysis   305
Propionic acid, fixative   138, 142
Propionic-orcein   142, 170
Propiono-carmine   142
PSP *see* Presporulation medium
PTC (Phenylthiocarbamide), tasting 68–9
Pyridoxine   199–202, 215

Quantitative genetics *see* Genetics, quantitative; Variation, quantitative
Quinacrine hydrochloride   141, 166
Quinacrine mustard   141, 166

R *see* Rye extract medium
Radish (*Raphanus sativus*), segregation 38, *53*
Radioactive chemicals   134–5, 165, 169
*Rana* (Frog), mitosis   150
*Ranunculus acris*, accessory chromosomes 163
mitosis, Plate 4.2 *fp. 148*
*Ranunculus ficaria*, accessory chromosomes 163
polyploidy   159
*Ranunculus marginatus*, hybrid meiosis 158
*Ranunculus muricatus*, hybrid meiosis   158
*Ranunculus* spp., mitosis   147–8
*Raphanus sativus* (Radish), segregation 38, *53*
Rats, preserved skins   58
Recessivity, *Arabidopsis*   32
double   80
*Drosophila*   10, 13, 18, 22
humans   73
notation   8
teaching   80
*Tribolium*   35
Reciprocal cross   9, 18, 97, 100
Recombination, bacteria   256–61, 263, 264, 266–70
crossing over frequency and   17
double   20, 202
*Drosophila*   16–7
fraction   17, 18–20, 26, 192
trequency   17, 19–20, 202, 204, 208–9, 264
fungi   191–2, 199, 207–9, 210–15
inversions and   20
mitotic   213, 214–5
phage   250, 252–3
*Tribolium*   35
Regression analysis   119, 126
coefficient   126
Respiratory differential medium (YPD) 180, 183, 197
*Rhoeo discolor*, chromosome mutations 159
*Rhododendron*, possible research on   82
RNA, staining   142
Root tips, in cytogenetics   136, 139, 147–9, 165, 166
*Rosa*, possible research on   82
*Rubus idaeus* (Blackberry), segregation *54*